Gabriele Broll
Beate Keplin
Mountain Ecosystems
Studies in Treeline Ecology

Gabriele Broll
Beate Keplin (Eds.)

Mountain Ecosystems

Studies in Treeline Ecology

With 96 Figures

Springer

Prof. Dr. Gabriele Broll
Department of Geo- and Agroecology, ISPA
University of Vechta
P.O. Box 1553
49364 Vechta
Germany

Dr. Beate Keplin
Institute of Landscape Ecology
University of Muenster
Robert-Koch-Str. 26-28
48149 Muenster
Germany

E-mail: gbroll@ispa.uni-vechta.de
keplin@uni-muenster.de

ISBN 3-540-24325-9 **Springer Berlin Heidelberg New York**

Library of Congress Control Number: 2004117857

Springer is a part of Springer Science+Business Media
springeronline.com

Printed in The Netherlands

Cover design: E. Kirchner, Heidelberg
Production: A. Oelschläger
Typesetting: Camera-ready by the Authors
Printing: Krips, Meppel
Binding: Litges+Dopf, Heppenheim
Printed on acid-free paper 30/2132/AO 5 4 3 2 1 0

Preface

Mountain ecosystems belong to the most endangered ecosystems in the world. Especially, the treeline ecotone acts as an indicator for environmental change. However, ecological processes in the treeline ecotone are not yet completely understood. The studies provided in this book may contribute to a better understanding of the interactions between vegetation, climate, fauna, and soils in the treeline ecotone. An introductory chapter is given on plants living under extreme conditions, climate change aspects, and methods for characterization of alpine soils. The following articles focus on mountainous areas in America, Europe and Asia.

The Working Group on Mountain and Northern Ecosystems at the Institute of Landscape Ecology, University of Münster (Germany), has been working on topics related to the treeline ecotone for several decades. This period under the chairmanship of Friedrich-Karl Holtmeier has come to an end now when he retired in 2004. He initiated numerous studies in high mountains and in the North. Many of his students, who became infected by the 'mountain virus', will continue these investigations on ecological processes in the altitudinal and northern treeline ecotones. With this compilation of studies in mountain ecosystems we want to thank Friedrich-Karl Holtmeier for his excellent guidance in these cold and fascinating environments.

This book could not have been edited without much valuable help of many people. We gratefully acknowledge the interesting contributions of the authors and also the constructive comments from those colleagues who reviewed earlier versions of the manuscripts. We are grateful to Dr. Hans-Jörg Brauckmann, Maja Masanneck and Marta Jacuniak (University of Vechta, Germany) for the careful preparation of the final version of the papers. Not last our thanks go to Dr. Christian Witschel and his staff (Geosciences, Springer Publishers) for the very good cooperation.

Gabriele Broll and Beate Keplin

Contents

Regional Treeline Studies in Europe

Regional Treeline Studies in Asia

Editors

Gabriele Broll
Department of Geo- and Agroecology (ISPA)
University of Vechta
P.O. Box 1553
D-49364 Vechta
Germany
gbroll@ispa.uni-vechta.de

Beate Keplin
Institute of Landscape Ecology
University of Münster
Robert-Koch-Str. 26
D-48149 Münster
Germany
keplin@uni-muenster.de

Authors

Frank Bednorz	Institute of Landscape Ecology University of Münster Robert-Koch-Str. 26 D-48149 Münster Germany bednorz@web.de
Sascha Beißner	Institute of Geobotany University of Hannover Nienburger Str. 17 D-30167 Hannover Germany
Gabriele Broll	Department of Geo- and Agroecology (ISPA) University of Vechta P.O. Box 1553 D-49364 Vechta Germany gbroll@ispa.uni-vechta.de
Jan Cermak	Faculty of Geography University of Marburg Deutschhausstraße 10 D-35032 Marburg Germany cermak@staff.uni-marburg.de
Robert M. M. Crawford	Plant Science Laboratory Sir Harold Mitchell Building University of St Andrews St Andrews KY16 9A1 U.K. rmmc@st-andrews.ac.uk

Renate Hildebrand	Zum Hüggel 3 D-49205 Hasbergen Germany hildebrandrenate@compuserve.de
Bettina Hiller	Centre for Environmental Research University of Münster Mendelstr. 11 D-48149 Münster Germany hillerb@uni-muenster.de
Andreas J. Kirchhefer	Department of Biology University of Tromsø N-9037 Tromsø Norway andreas.kirchhefer@ib.uit.no
Sabine Mellmann-Brown	P.O. Box 1044 Cooke City, MT 59020 At time of research affiliated with: U.S.D.A. Forest Service Montana State University Bozeman, MT 59717-0278 U.S.A. smelbrown@earthlink.net
Georg Miehe	Faculty of Geography University of Marburg Deutschhausstraße 10 D-35032 Marburg Germany miehe@staff.uni-marburg.de
Gerald Müller	Haus Vogelsang GmbH Umweltconsulting Vogelsangweg 21-23 D-45711 Datteln-Ahsen Germany gerald.mueller@hvg.mbh.de

Andreas Müterthies	EFTAS Remote Sensing Ostmarkstr. 92 D-48145 Münster Germany andreas.mueterthies@eftas.com
Lars Opgenoorth	Faculty of Geography University of Marburg Deutschhausstraße 10 D-35032 Marburg Germany opgenoor@mailer.uni-marburg.de
Pietro Piussi	Dipartimento di Scienze e Tecnologie Ambientali Forestali Via S. Bonaventura, 13 50145 Firenze Italy Pietro.piussi@unifi.it
William Pollmann	Department of Geography University of Colorado Boulder, CO 80309-0260 U.S.A. william.pollmann@t-online.de
Richard Pott	Institute of Geobotany University of Hannover Nienburger Str. 17 D-30167 Hannover Germany pott@geobotanik.uni-hannover.de
Thomas Reineke	Institute of Geography RWTH-Aachen Templergraben 55 D-52056 Aachen Germany Thomas.Reineke@geo.RWTH-Aachen.de

Udo Schickhoff

Institute of Geography
University of Bonn
Meckenheimer Allee 166
D-53115 Bonn
Germany
schickhoff@giub.uni-bonn.de

Hans-Uwe Schütz

Biologische Station Rieselfelder Münster
Coermühle 181
D-48157 Münster
Germany
Uwe.gopher@t-online.de

Diana F. Tomback

Department of Biology
CB 171 University of Colorado at Denver
P.O. Box 173364
Denver, CO 80217 – 3364
U.S.A.
Diana.Tomback@cudenver.edu

Gian-Reto Walther

Institute of Geobotany
University of Hannover
Nienburger Str. 17
D-30167 Hannover
Germany
walther@geobotanik.uni-hannover.de

Mountain Ecosystems

Studies in Treeline Ecology

General Aspects of Vegetation and Soils in Cold Environments

Guideline for Describing Soil Profiles in Mountain Ecosystems

Gabriele Broll, Bettina Hiller, Frank Bednorz, Gerald Müller and Thomas Reineke

1 Introduction

This guideline for describing soil profiles in mountainous ecosystems is intended to provide scientists around the world with other than soil science expertise to collect useful soil data such as soil profile descriptions and soil sampling. From the collected soil profile descriptions the scientists should be able to calculate important parameters such as field capacities. In addition, the main objective of this guideline is to streamline methods for soil data collection in mountainous terrain throughout the world, which would result in comparable soil data.

The field book Schoeneberger et al. (2002) is recommended as basis for the guideline. This manual was used providing a minimum data set for descripting soil profiles in mountainous areas. Before describing a soil profile a representative site should be selected. A representative site is defined by the objective of the study. This could consist of parameters such as vegetation communities, microtopography etc. Only those parameters have to be considered, which are essential for a minimum data set. We tried to focus on the specific site conditions in mountainous areas with great heterogeneity in many ways and added special recommendations for their description and sampling. We focussed on those soil parameters, which are necessary to investigate ecological processes, like interactions between plants and soil. In this manual we do not consider genetic purposes. Interpretation of pedogenesis as well as soil mapping should be done in cooperation with soil scientists only. In connection with soil profile description some data, e. g. texture and slope gradient, are collected, which are necessary for erosion risk assessment. Examples of soil profile descriptions in alpine areas of Europe and Asia are given in order to improve the clarity of the guideline.

Basic Manual:

Schoeneberger PJ, Wysocki DA, Benham EC and Broderson WD (2002): Field book for describing and sampling soils. National Resources Conservation Service version 2.0, USDA, National Soil Survey Center, Lincoln, NE. ftp://ftp-fc.sc.egov.usda.gov/NSSC/Field_Book/FieldBookVer2.pdf (03.09.04)

2 Field Work

2.1 Site Description

2.1.1 Name

2.1.2 Date

2.1.3 Profile Number

- •Photo and sketch of the profile are recommended.

2.1.4 Location

- •Location: Country, latitude / longitude (GPS coordinates)
- •Physiographic location
- •Elevation [m a.s.l.]

2.1.5 Topography / Relief

- •Landform. For detailed definitions and further landforms see also Schoeneberger et al. (1998a).

Depressional landforms

basin floor	saddle
col	trough
depression	valley
intermontane basin	valley floor
mountain valley	

Eolian landforms

blowout	loess hill
deflation basin	sand sheet
dune	

Erosional landforms – Water erosion (overland flow) related and excluding fluvial, glaciofluvial, and eolian erosion

- arete (sharp ridge)
- col
- meander scar
- peak
- pediment
- saddle
- scarp slope

Fluvial landforms – dominantly related to concentrated water (channel flow), both erosional and depositional processes, and exclusing glaciofluvial landforms

- bar
- delta
- fan
- flood plain
- levee
- meander
- pediment
- stream terrace

Glacial landforms (including glaciofluvial forms)

- arete
- cirque
- col
- drumlin
- esker
- glacial drainage channel
- glacial lake (relict)
- hanging valley
- kame
- moraine
- end moraine
- ground moraine
- lateral moraine
- medial moraine
- outwash
- till plain
- U-shaped valley

Mass movement landforms (including creep forms)

- block glide
- fall
- flow
- debris flow
- earth flow
- mud flow
- sand flow
- landslide
- slide

Slope landforms

- dome (rounded summit)
- escarpment (steep slope)
- gap
- headwall
- hill
- horn
- mountain
- peak
- plain
- plateau
- ridge
- rim

horst
knob (round-shaped mass)
knoll (small hill)
meander scar
mesa
scarp
spur
U-shaped valley
V-shaped valley

Tectonic, structural and volcanic landforms

anticline
caldera
dome
graben
horst
lava plain
shield volcano
stratovolcano
syncline

Wetland terms and landforms

bog
fen
peat plateau

•Slope aspect
•Slope gradient
•Slope shape (figure 1)
•Site position on slope

Crest
Upper slope
Middle slope
Lower Slope
Toe slope
Depression

•Microfeature

Gilgai
Gully
Mound
Rib
Solifluction lobe
Solifluction sheet
Solifluction terrace
Terracettes

Periglacial patterned ground microfeatures

circle
non-sorted circles
sorted circles
earth hummocks
peat hummocks
palsa, palsen
polygons
high-center polygons
ice wedge polygons
low-center polygons
stripes
trough (hollow)

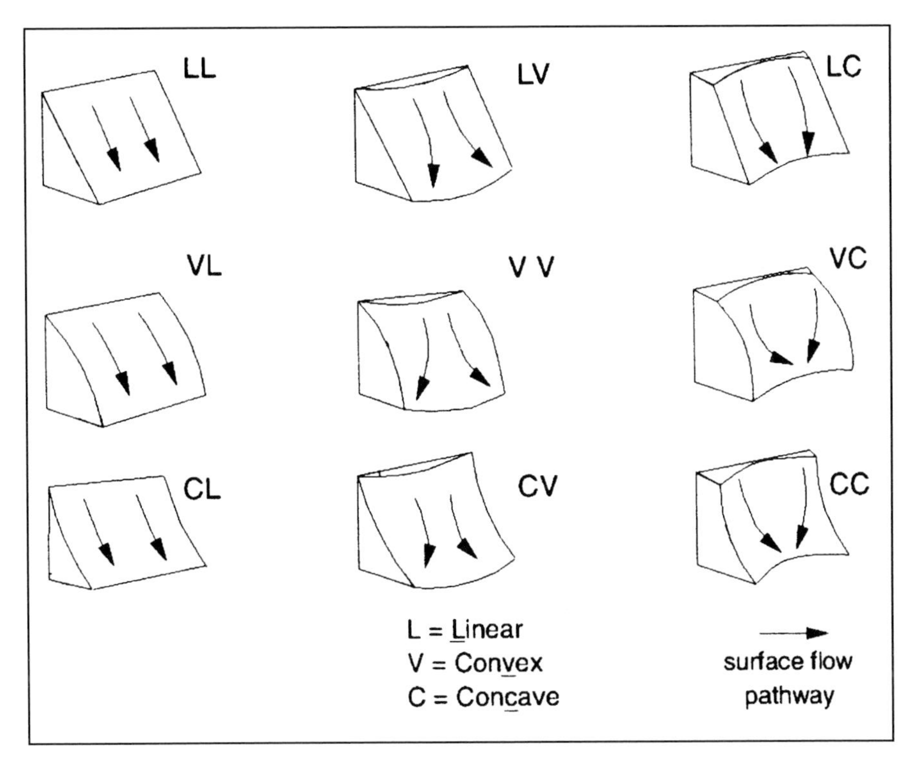

Figure 1 Slope shape is described in two directions: up-and-down slope (perpendicular to the contour), and across slope (along the horizontal contour); e.g. *Linear, Convex* or LV (Schoeneberger and Wysocki 1996; cited in Schoeneberger et al. 2002, adapted)

2.1.6 Water Status

•Drainage classes (See figure 2 next page)
•Depth to water table [cm]
•Depth to impermeable layer [cm]
•In permafrost regions: Depth of thaw [cm]

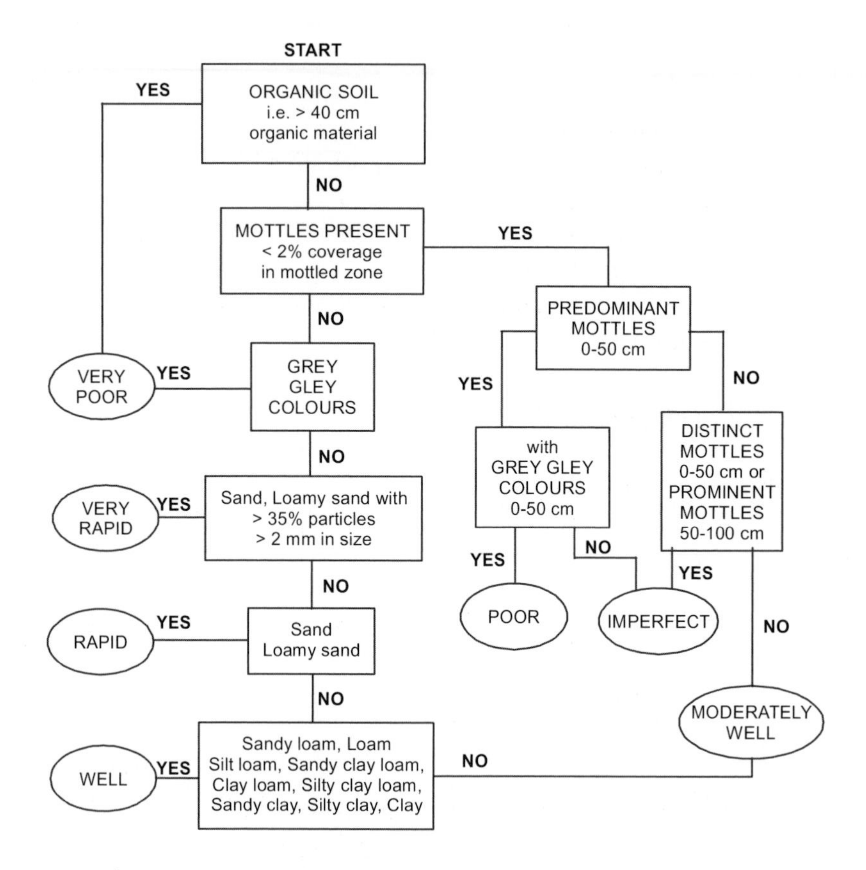

Figure 2 Drainage classes (Denholm and Schut 1993, modified)

2.1.7 Vegetation - Land Use

- Percentage of ground cover

  Total trees [%]
  Total shrubs [%]
  Total vascular plants [%]
  Total lichen [%]
  Total moss [%]
  ___
  Total vegetation [%]
  Bare ground [%]

  For further informations see also Miehe and Miehe (2000).

•Land use (FAO 1990, modified)

Crop agriculture (e.g. annual field cropping, shifting cultivation)
Animal husbandry (e.g. extensive grazing, intensive grazing)
Forestry (e.g. natural forest and woodland)
Mixed farming (e.g. agro-forestry)
Extraction and collection (e.g. hunting and fishing, exploitation of natural vegetation)
Nature protection (e.g. parks, wildlife management)
Not used and not managed

2.1.8 Parent Material and / or Bedrock

•Kind of parent material

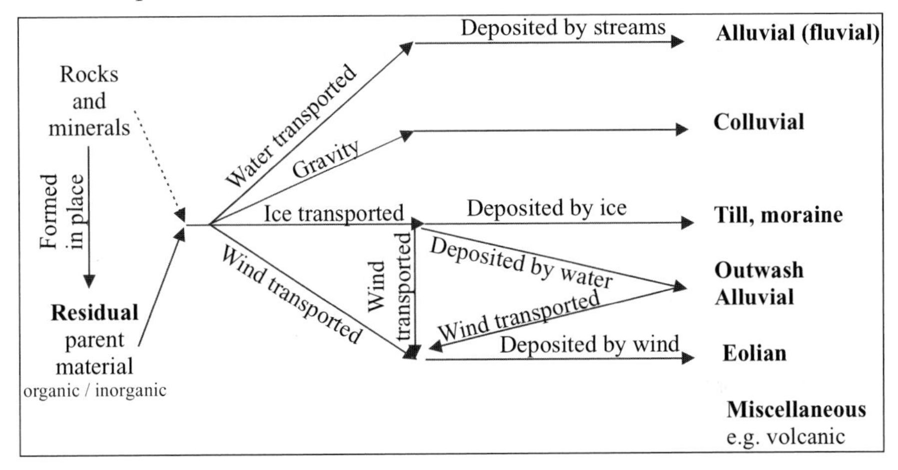

Figure 3 How various kinds of parent material are formed, transported, and deposited (Brady and Weil 1998, modified)

Following is recommended:

Percentage of saprolite (cf. glacial deposits mixed with saprolite)
For more detailed informations see also Catt (1986).

•Kind of bedrock material

Igneous
Sedimentary
Metamorphic
Pyroclastic

Stratigraphic and petrographical classification is recommended.

2.1.9 Surface fragments

•Classes of percentage of surface cover (for surface coarse fragments and rock outcrops) (FAO 1990)

None	0 %	Many	15 - 40 %
Very Few	0 - 2 %	Abundant	40 - 80 %
Few	2 - 5 %	Dominant	> 80 %

•Size classes (FAO 1990)

Fine gravel	0.2 - 0.6 cm	Boulders	20 - 60 cm
Medium gravel	0.6 - 2.0 cm	Large	
Coarse gravel	2.0 - 6.0 cm	Boulders	60 -200 cm
Stones	6.0 - 20 cm		

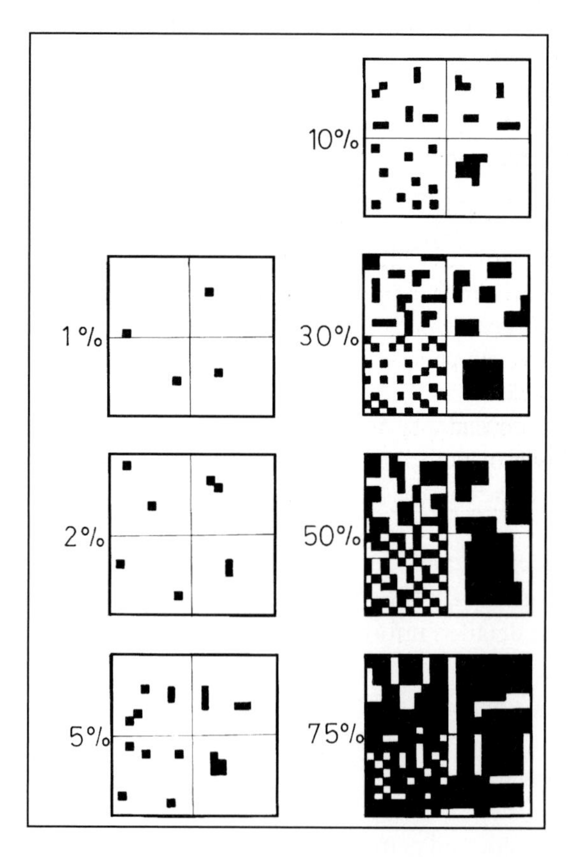

Figure 4 Estimation of percentage of area covered (AK Standortskartierung 1996)

2.2 Profile Description

2.2.1 Horizon Nomenclature

•Master, transitional and common horizon combinations. Only the most important master horizons and horizon suffixes have been considered.

Horizon	Criteria
O	Predominantly organic matter (litter and humus)
A	Mineral, organic matter (humus) accumulation
A/B (or E/B)	Discrete, intermingled bodies of A (or E) and B material; majority of horizon is A (or E)
E	Mineral, loss of Si, Fe, Al, clay, or organic matter
B/A (or B/E)	Discrete, intermingled bodies of B and A (E) material; majority of horizon is B material
B	Subsurface accumulation of clay, Fe, Al, Si, humus, $CaCO_3$, $CaSO_4$; or loss of $CaCO_3$ or accumulation of sesquioxides (e.g. Fe_2O_3)
BC	Dominantly B horizon characteristics but also contains characteristics of the C horizon
B/C	Discrete, intermingled bodies of B and C material; majority of horizon is B material
CB	Dominantly C horizon characteristics but also contains characteristics of the B horizon
C/B	Discrete, intermingled bodies of C and B material; majority of horizon is C material
C	Little or no pedogenic alteration, unconsolidated material, soft bedrock
R	Hard, continuous bedrock
W	A layer of liquid water (W) or permanently frozen water (Wf) within the soil (excludes water/ice above soil)

•Horizon suffixes

Horizon suffix	Criteria
b	Buried genetic horizon (not used with C horizon)
d	Densic layer (physically root restrictive)
f	Permanently frozen soil or ice (permafrost); continuous ice; not seasonal
ff	Permanently frozen soil ('Dry' permafrost); no continuous ice; not seasonal
g	Strong gley
h	Illuvial organic matter accumulation
jj	Evidence of cryoturbation
k	Pedogenic carbonate accumulation
m	Strong cementation (pedogenic, massive)
o	Residual sesquioxide accumulation (pedogenic)
p	Plow layer or other artificial disturbance
r	Weathered or soft bedrock
s	Illuvial sesquioxide accumulation
t	Illuvial accumulation of silicate clay
w	Weak color or structure within B (used only with B)

For further information see Soil Survey Staff (2003) and Soil Survey Staff (1999).

2.2.2 Horizon Thickness [cm]

•Horizon thickness is recommended instead of horizon depth because of complications with cryoturbated soils.
•Horizon thickness of the organic layer is also recommended.

2.2.3 Horizon Boundary

•Distinctness of horizon boundaries (Schoeneberger et al. 2002)

Distinctness class	Abruptness of vertical changes [cm]
Very abrupt	< 0.5
Abrupt	0.5 to < 2.0
Clear	2.0 to < 5.0
Gradual	5.0 to < 15.0
Diffuse	≥ 15.0

•Topography (Schoeneberger et al. 2002)

Topography	Variations of boundary plane
Smooth	Planar with few or no irregularities
Wavy	Width of undulation is > than depth
Irregular	Depth of undulation is > than width
Broken	Discontinuous horizons; discrete but intermingled, or irregular pockets

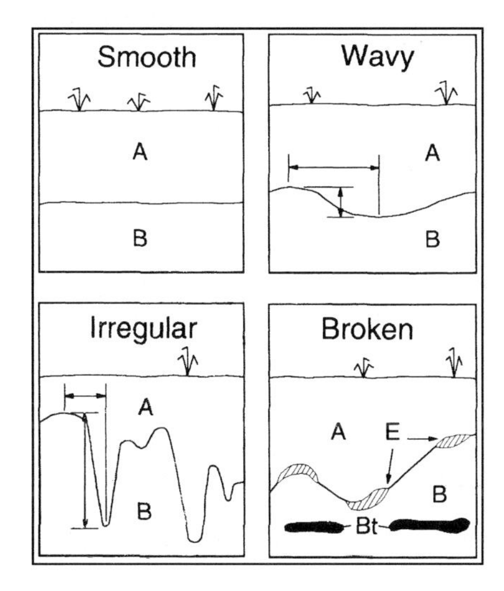

Figure 5 Topography of horizon boundaries (Schoeneberger et al. 2002)

2.2.4 Soil Color

•Munsell Color Charts (Hue, Value, Chroma), moist soil
Soil matrix color

Mottles
Color of mottles (Use Munsell Color Charts)

Quantity classes of mottles

Quantity class	Criteria: range in percent
Few	< 2 % of surface area
Common	2 to < 20 % of surface area
Many	≥ 20 % of surface area

2.2.5 Soil Texture

The particle sizes for silt and sand are different in Europe and North America. In case a particle size analysis should be done, sieves with different mesh diameters are necessary depending on the taxonomy which is used (cf. Table: Particle size classes).

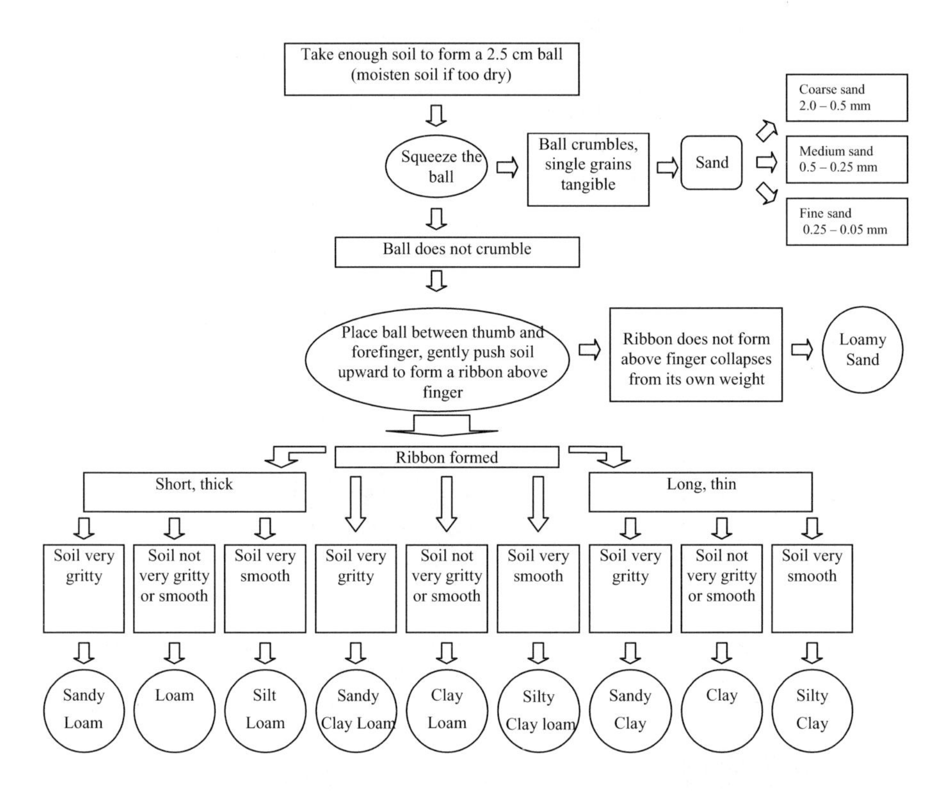

Figure 6 Soil texturing by feel (Thien 1979; cited in Tiner 1999, modified)

•Particle size classes (Schoeneberger et al. 2002, modified)

Fine Earth (USDA)							
Clay	Silt		Sand				
	fine	coarse	very fine	fine	medium	coarse	very coarse

0.002 0.02 0.05 0.1 0.25 0.5 1 2
[mm]

•Coarse and other fragments / Texture modifiers

Content: Estimate the quantity of gravel, cobbles, stones and/or boulders on a volume percent basis (Schoeneberger et al. 2002)

Roundness (simplified): 3 classes: 1. angular, 2. subangular, subrounded, 3. rounded

Sieving in the field is recommended in case a better quantification is necessary (Mosimann 1985)

Size	Noun
> 2 – 75 mm diameter	gravel
> 75 – 250 mm diameter	cobbles
> 250 – 600 mm diameter	stones
> 600 mm diameter	boulders

2.2.6 Soil Structure

Single grain

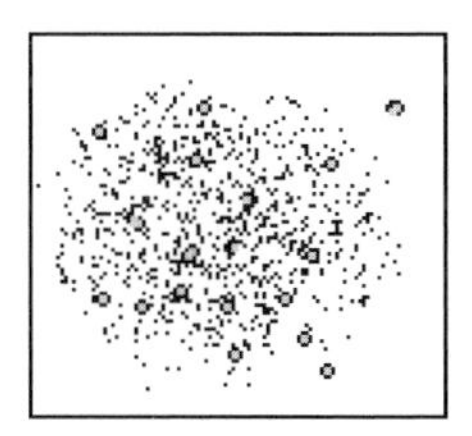

Massive, common in cemented horizonts, e.g. Ortstein

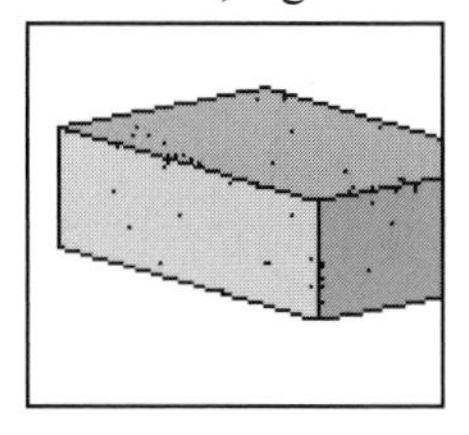

Granular, characteristic of surface (A) horizons, showing high biological activity

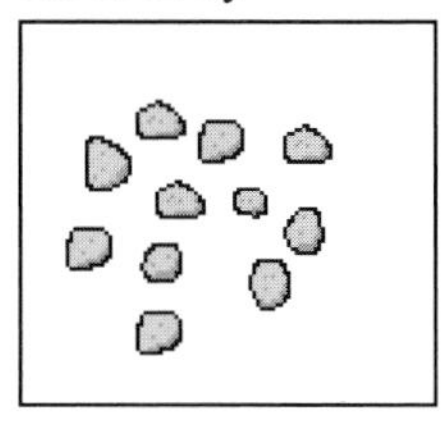

Subangular blocky, common in B horizons particularly in humid regions

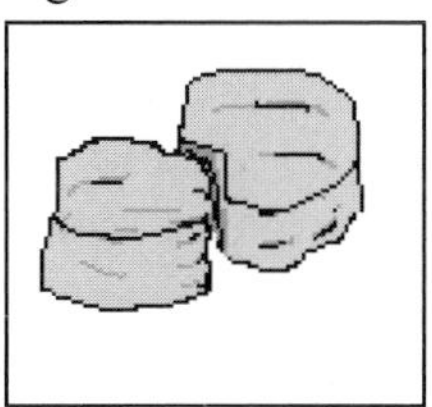

Angular blocky, common in B horizons, particulary in humid regions

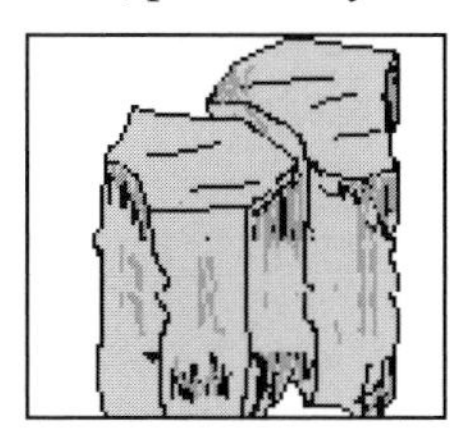

Prismatic, usually found in B horizons. Most common in soils of arid and semiarid regions

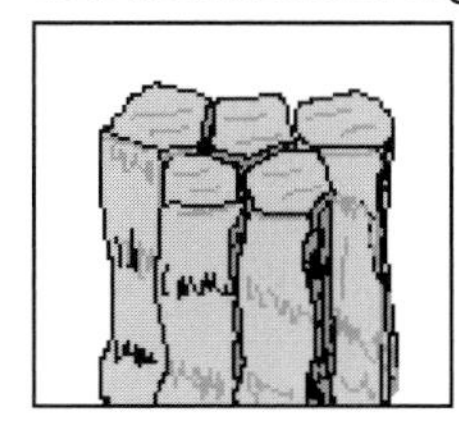

Platy, common in E horizons, may be in any part of the profile. Often inherited from parent material of soil, or caused by compaction

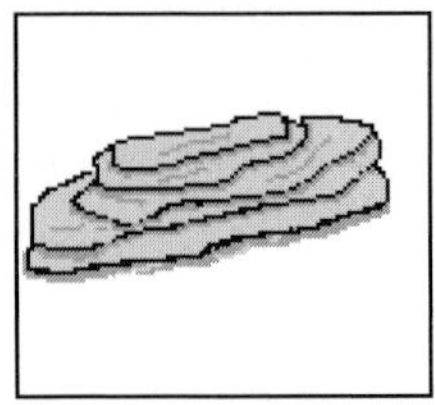

Figure 7 Soil structure types (Brady and Weil 1998; Schoeneberger et al. 2002, modified)

2.2.7 Calcarousness

•Effervescence and $CaCO_3$ content (Day 1983)

	$CaCO_3$ equivalent [%]	Effervescence (using 10% HCl)
No carbonate	0	No bubbles
Weakly calcareous	< 5	Few bubbles
Moderately calcareous	5 - 15	Numerous bubbles
Strongly calcareous	15 - 25	Bubbles form low foam
Very strongly calcareous	25 - 40	Bubbles form thick foam

2.2.8 Penetration Resistance / Bulk Density

•Penetration resistance tested in the field with a pencil, knife or penetrometer (Brady and Weil 1998, modified)

Soil at in situ moisture	Penetration resistance	Field penetration test
Soft	1	Blunt end of pencil penetrates deeply with ease
Medium firm	2	Blunt end of pencil can penetrate about 1.25 cm with moderate effort
Firm	3	Blunt end of pencil can penetrate about 0.5 cm
Very firm	4	Blunt end of pencil makes slight indentation
Hard	5	Blunt end of pencil makes no indentation

- Bulk density (ρt) of mineral soils (bulk density = the ratio of the mass of dry solids to the bulk volume of the soil after drying at 105 °C in g cm^{-3} (Blake and Hartge 1986, AG Boden 1994))

Penetration resistance	**ρt [g cm^{-3}]**	**Interpretation**
1	< 1.25	very low
1 – 2	1.25 – 1.45	low
2 – 3	1.45 – 1.65	middle
3 – 4	1.65 – 1.85	high
4 – 5	> 1.85	very high

2.2.9 Roots

Quantity *	Size *	Location
Few	Fine	Throughout
Common	Medium	Matted on top of horizon
Many	Coarse	In cracks

* in detail: Schoeneberger et al. 2002: 2-56, 2-57

Example: common fine roots matted on top of horizon

2.2.10 Root Restricting Depth

- Definition: Depth of the soil at which root growth is strongly inhibited.
- Classes of root-restricting depth

Extremly shallow	0 - 5 cm
Very shallow	5 - 15 cm
Shallow	15 - 30 cm
Moderately deep	30 - 50 cm
Deep	50 - 100 cm
Very deep	> 100 cm

2.2.11 Remarks

- For example:
  - Cracks
  - Roots in cracks of bedrock
  - Crusts
  - Biological features, like earthworm casts

Cryoturbation
Salt
Redoximorphic features
(Test: α-α-dipyridyl, cf. Schoeneberger et al. 2002: 2-66)
Charcoal

2.3 Soil Classification

It is recommended to use the US Soil Taxonomy (Soil Survey Staff 2003) because it is worldwide distributed. Moreover, the suitability of the US Soil Taxonomy has been proved at many high mountain sites.
Notice: Soil temperature data at a depth of 50 cm are necessary.

Humus Forms

The description of humus forms requires:

- •Separation between organic layers ($\geq$ 30 % organic matter mass, AG Boden (1994); > 17 % organic carbon mass, Green et al. (1993)) and the A horizon.
- •Identification of the different organic horizons (see key).
- •Determination of the thickness of the organic horizons as well as the A horizon.
- •Determination of the structure of the A horizon (Chapter 2.2.6).

Key

Organic horizons			Description
Green et al. (1993)	AG Boden (1994)	Soil Surv. Staff (2003)	
L	L	Oi	Relatively fresh plant residues, not fragmented, usually discolored.
F	Of	Oe	Fragmented plant residues predominate over fine substances (< 70 Vol.-% organic fine substances, AG Boden 1994).
H	Oh	Oa	Organic fine substance predominate. Fragmented plant residues are generally not recognizable. The color is typically black.

L= Litter; **H**, **h**= humified; **F**, **f**= fermented; **i**= fibric; **e**= hemic; **a**= sapric

- The small scale variability of the site conditions in high mountain ecosystems is responsible for a high spatial heterogeneity of humus forms. Thus, it is necessary to get an overview of this variability in order to create an adequate sampling design (cf. grid point sampling within a 20 x 20 m grid).
- The sedimentation of mineral material transported by wind or water (e. g. alpine loess, 'Flugsand') may modify the properties of organic horizons. Thus, the identification of organic horizons and the differentiation of organic layer from the mineral soil may be aggravated. In case of sedimentation of mineral material the term 'mineric' can be used in the classification of Green et al. (1993).
- In high mountain ecosystems humus forms which are influenced by erosion and/or human impact are very common. Especially at exposed sites or steep slopes erosion is very effective. Erosion may destroy only the upper horizons or the whole humus profile. Within an eroded area often residues of former humus profiles are common. Some of these humus forms are called 'Hagerhumusformen' according to AG Boden (1994).
- In the European Alps some terms of Kubiëna (1953) are still used: 'Tangelhumus' and 'Pechmoder'. Both humus forms are characterized by an organic layer which overlies solid limestone. They might be interpretated as special raw humus / mor or moder humus forms. 'Tangelhumus' occurs typically in subalpine coniferous forest and dwarf shrub ecosystems. Commonly 'Pechmoder' is found under alpine plant communities.

2.4 Soil Sampling

- Especially in mountain ecosystems soil sampling should be done very carefully because of the high spatial heterogeneity.
- Figure 8 shows different sampling strategies. The kind of sampling depends on the aim of the study.
    - Catena: A sequence of soils of about the same age, derived from similar parent material, and occurring under similar climatic conditions but having different characteristics due to variation in relief and drainage (SSSA 1997).
    - Transect
    - Composite depth sampling

2.4.1 Sampling of Soil Horizons

- Composite mixed samples
  - Each horizon has to be sampled separately.
  - Composite samples
  - The rock fragment content can be determined by sieving (2 mm) and weighing in the field (Mosimann 1985).
- Undisturbed sampling

  Core samples with steel cylinders (size usually 100 cm^3)

  Sampling horizontally or vertically possible

2.4.2 Stratified Sampling for Composite Samples

- Composite depth sampling
  - The sample site should be subdivided into parts which are as homogeneous as possible. The dominant vegetation type and/or microtopography can be used for subdivisions (figure 8).
  - number of samples of each component: suggestion 20 (randomly distributed)
  - replicates in the field: suggestion 3
  - Do not mix major horizons. If an organic layer exists, sample it separately.

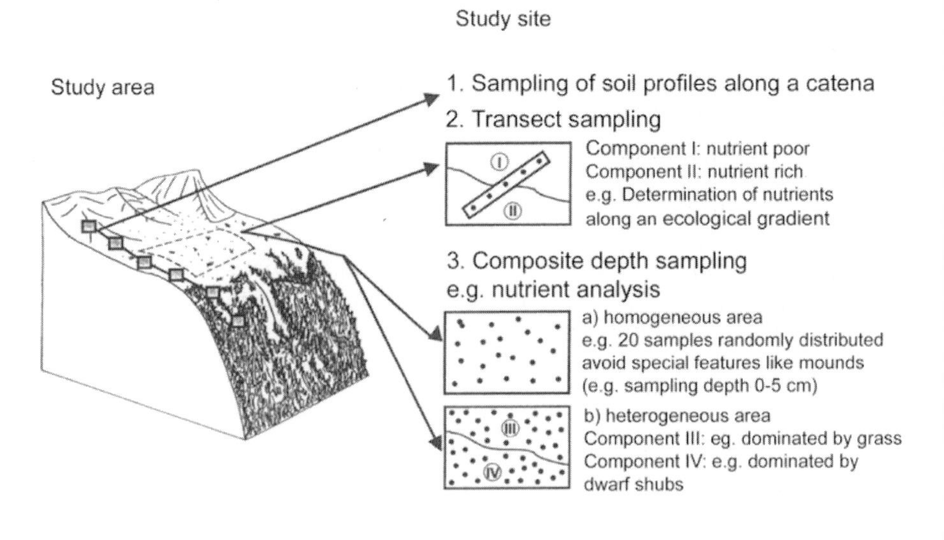

Figure 8 Sampling strategies

- •Undisturbed soil samples
  - •Core samples with steel cylinders (size usually 100 cm^3), e.g. for bulk density
  - •Sampling horizontally or vertically possible.
  - •Soils with a high percentage of rock fragments: After taking disturbed samples the volume of the hole is determined by tapemeasure. A more exact method would be the refilling with sand or PU-foam (Deutschmann et al. 1994; Schlichting et al. 1995).

2.5 Field Measurements

2.5.1 pH-Value

- •The pH-value can be determined by pH-papers or by electrometric measurement with a glass electrode. Soil pH is measured in water or in a 0.01 mol L^{-1} $CaCl_2$-solution (soil-solution ratio = 1:2.5) (Hendershot et al. 1993; Schlichting et al. 1995; Thomas 1996).

2.5.2 Electrical Conductivity

- •The determination of the electrical conductivity is especially important for the investigation of soils in arid regions. The electrical conductivity is measured in the field in a soil-water solution (soil-solution ratio = 1:2.5) (DVWK 1995; Schlichting et al. 1995; Rhoades 1996).
- •Salinity class (Schoeneberger et al. 2002)

Salinity class	Electrical conductivity $dS\ m^{-1}$
Non-saline	< 2
Very slightly saline	2 to < 4
Slightly saline	4 to < 8
Moderate saline	8 to < 16
Strongly saline	> 16

2.5.3 Soil Temperature

- •Electrical sensors with data loggers are used for continuous soil temperature measurements.
- •The depths for continuous measurements are depending on the aim of the investigation. To classify soils after the US Soil Taxonomy soil temperature regime at a depth of 50 cm from the soil surface is essential.

- If there is an organic layer, soil temperature should be measured in the organic layer and in the mineral topsoil.
- Surface temperatures (organic layer or mineral soil) should be measured at a depth of 2.5 cm.

2.5.4 Soil Water

Gravimetric measurement

- Water content measurements by gravimetric methods involve weighting the wet sample, removing the water by oven drying (105°C), and reweighting the sample to determine the amount of water removed. The water content (mass basis) can be calculated as a percentage of the mass of dry soil. To determine the volumetric water content undisturbed soil core samples are needed (z. B. Gardner 1986, Schlichting et al. 1995).

Time Domain Reflectrometry (TDR)

- The TDR technique measures the velocity of propagation of a high-frequency signal along metal waveguides. With the TDR technique water content (volume basis) measurements in the field are possible (Topp 1993, Schlichting et al. 1995).
- Complications with high contents of rock fragments are possible.

3 Interpretation

3.1 Soil Water and Soil Aeration

- Air porosity (Brady and Weil 1998)
  The proportion of the bulk volume of soil that is filled with air (pores > 50 μm)
- Field capacity (Brady and Weil 1998)
  The percentage of water remaining in a soil two or three days after its having been saturated and after free drainage has practically ceased (pores ≤ 50 μm)
- Plant available water (Brady and Weil 1998)
  The water retained in soil between the states of field capacity and wilting point (pores 0.2 – 50 μm)

Necessary Data

- For mineral soils in general: texture, penetration resistance, soil organic matter.
- For mineral soils with high content of rock fragments (recommendation ≥ 20 %) the following factor

$$\frac{100\text{-(Vol.-\% rock fragments)}}{100}$$

has to be multiplicated with volume percent of air porosity, field capacity and plant available water.
- The calculation has to be carried out for every horizon, e.g. for root restricting depth.

Example:

Ah: loamy sand with low penetration resistance (2) and 40 Vol.-% of rock fragments
Soil organic matter content of about 4 %
Factor for rock fragments: 0.6
Air porosity: 16.5 Vol.-% * 0.6 = 9.9 Vol.-%
Field capacity: 33.5 Vol.-% * 0.6 = 20.1 Vol.-%
Plant available water: 22 Vol.-% * 0.6 = 13.2 Vol.-%

Plant Available Water for the Root Restricting Depth Calculation Example for an Inceptisol with Loamy Sand (AG Boden 1994):

a) Profile description

Ah: 20 cm loamy sand with low penetration resistance (2) and a soil organic matter content of about 4 %,
Bv: 60 cm loamy sand with middle penetration resistance (3)
Root restricting depth = 80 cm

b) Calculation

Ah: 20 cm with 24 Vol.-% [1)]	= 480 cm * (mm * 10 cm^{-1})
Bv: 60 cm with 16 Vol.-% [1)]	= 960 cm * (mm * 10 cm^{-1})
Root restricting depth : 80 cm	= 1440 cm * (mm * 10 cm^{-1})
Plant available water for the root restricting depth	= 144 mm

[1)] Vol.-% = mm * 10 cm^{-1}

•Air porosity, plant available water, field capacity in Vol.-% depending on texture and penetration resistance (AG Boden 1994, modified).

Texture (USDA)	Air porosity			Plant available water			Field capacity		
	1-2	3	4-5	1-2	3	4-5	1-2	3	4-5
Coarse sand*	22.5	19.5	–	12.0	10.5	–	18.5	16.5	–
Medium sand*	28.0	25.0	–	9.0	7.5	–	13.5	11.0	–
Fine sand*	17.5	10.0	–	21.5	20.0	–	25.5	24.0	–
Loamy sand	17.5	14.0	10.0	19.0	16.0	15.0	26.0	23.5	21.0
Sandy loam	11.5	9.5	5.5	22.5	19.5	17.0	32.0	28.0	25.5
Loam	9.0	6.5	5.5	23.0	17.0	14.5	40.0	33.0	29.0
Sandy clay loam	17.0	11.5	9.0	17.5	16.0	12.5	24.0	25.5	23.0
Clay loam	6.5	5.0	3.5	17.5	14.5	11.0	48.0	40.5	33.0
Silt loam	8.5	5.5	25.0	25.0	22.5	19.0	39.0	35.0	32.5
Silt	6.5	3.0	–	28.5	26.0	–	39.5	36.5	–
Clay	–	2.5	1.5	–	16.0	11.0	–	56.0	47.5
Sandy clay	5.0	3.5	2.0	18.5	14.5	11.0	56.5	50.5	43.5
Silty clay	4.0	3.0	2.0	21.5	14.5	11.0	58.0	49.0	42.5
Silty clay loam	6.5	5.0	4.0	20.0	16.0	12.0	49.0	39.0	34.0

* coarse sand: Ø 0.5 – 2.0 mm
medium sand: Ø 0.25 – 0.5 mm
fine sand: Ø 0.05 – 0.25 mm
– no data available

•Modifications of plant available water, air porosity, field capacity and total porosity in Vol.-% depending on texture and soil organic matter (AG Boden 1994, modified)

Texture	Soil organic matter [%]	Plant available water	Air porosity	Field capacity	Total porosity
Sand	< 1.0	0	0	0	0
Loamy sand	≥ 1.0 – 2.0	+ 0.5	- 1.5	+ 1.5	0
Silt	≥ 2.0 – 4.0	+ 1.0	- 1.0	+ 3.5	+ 2.5
	≥ 4.0 – 8.0	+ 3.0	- 1.0	+ 7.5	+ 6.5
	≥ 8.0 – 15.0	+ 3.5	0	+ 10.0	+ 10.0
Sandy loam	≥ 1.0 – 2.0	+ 0.5	0	+ 1.5	+ 1.5
Loam	≥ 2.0 – 4.0	+ 1.0	+ 1.0	+ 3.5	+ 4.5
Silt loam	≥ 4.0 – 8.0	+ 3.0	+ 2.0	+ 8.0	+ 10.0
	≥ 8.0 – 15.0	+ 4.0	+ 2.5	+ 11.5	+ 14.0
Sandy clay	≥ 1.0 – 2.0	+ 0.5	+ 0.5	+ 2.5	+ 3.0
Clay loam	≥ 2 .0– 4.0	+ 1.5	+ 1.5	+ 4.0	+ 5.5
Sandy clay loam	≥ 4.0 – 8.0	+ 4.0	+ 3.0	+ 10.0	+ 13.0
	≥ 8.0 – 15.0	+ 7.0	+ 5.0	+ 13.5	+ 18.5

Texture	Soil organic matter [%]	Plant available water	Air porosity	Field capacity	Total porosity
Silty clay loam	≥ 1.0 – 2.0	+ 1.0	+ 0.5	+ 2.5	+ 3.0
Silty clay	≥ 2.0 – 4.0	+ 2.5	+ 1.5	+ 5.0	+ 6.5
	≥ 4.0 – 8.0	+ 5.5	+ 2.5	+ 10.5	+ 13.0
	≥ 8.0 – 15.0	+ 10.0	+ 4.5	+ 15.0	+ 19.5
Clay	≥ 1.0 – 2.0	+ 2.0	0	+ 3.5	+ 3.5
	≥ 2.0 – 4.0	+ 5.0	0	+ 7.5	+ 7.5
	≥ 4.0 – 8.0	+ 10.5	+ 1.0	+ 13.0	+ 14.0
	≥ 8.0 – 15.0	+ 16.0	+ 2.0	+ 18.0	+ 20.0

Total porosity = air porosity + field capacity

3.2 Soil Acidity

3.2.1 pH-Value

Descriptive term	Criteria: pH ($CaCl_2$) range	US Soil Taxonomy	
Very strongly alkaline	> 9.0		
Strongly alkaline	8.5 to 9.0		
Moderately alkaline	7.9 to 8.4		
Slightly alkaline	7.4 to 7.8		
Neutral	6.6 to 7.3		
Slightly acid	6.1 to 6.5		
Moderately acid	5.6 to 6.0		
Strongly acid	5.1 to 5.5	Nonacid	
Very strongly acid	4.5 to 5.0	Acid	**5.5**
Extremely acid	3.5 to 4.4		
Ultra acid	< 3.5		

•Descriptive terms for soil reaction (Schoeneberger et al. 2002)

3.2.2 Base Saturation

•The knowledge of the base saturation is necessary for the soil classification after US Soil Taxonomy.

pH ($CaCl_2$) < 5.5 → base saturation < 50 %
pH ($CaCl_2$) ≥ 5.5 → base saturation ≥ 50 %

4.1 Example Central Alps (Switzerland)

Site description

Name:	Frank Bednorz
Date:	30.09.1991
Profile no:	1
Photo no.:	1
Location:	
Location:	Switzerland; x 139.6, y 799.5 (System Switz.)
Physiogr. location:	Central Alps, Puschlav, Berninapaß
Elevation [m.a.s.l.]:	2206
Topography / Relief	
Landform:	mountain slopes
Slope aspect [°]:	2
Slope gradient [°]:	15
Slope shape:	LV
Site position on slope:	upper slope
Microfeature:	(micro-)depression
Water status:	
Drainage class:	well drained
Depth to water table [cm]:	-
Depth to imperm. layer [cm]:	-
Root restricting depth [cm]:	33
Depth of thaw [cm]:	-
Land use - Vegetation:	
Land use:	extensive grazing
Total trees [%]:	-
Total shrubs [%]:	50
Total vasculary plants [%]:	40
Total lichen [%]:	10
Total moss [%]:	-
Total vegetation [%]:	90
Bare ground [%]:	10

Parent material and / or bedrock:

Kind of parent material:	till
Kind of bedrock material:	granitemylonite (Series: Cavaglia-Balbalera, Berninadecke)

Surface fragments:

Classes of of surface cover:	common
Size classes:	cobbles

Soil profile

Example Central Alps (Switzerland)

Description of profile no 1

Soil classification: Typic Haplocryod

Humus form: Mormoder

Table 1

Horizon	Thickn. [cm]	Boundary Dist.	Boundary Topo.	Matrix Color	Mottles Color	Mottles Quant.	Texture	Rock fragments
Oi/Oe	2	clear	wavy	-	-	-	-	subr. cobbles 10 % subr. gravel 4 %
Oa	12	gradual	wavy	5YR2/1	-	-	-	subr. cobbles 10 % subr. gravel 4 %
E	8	gradual	wavy	7.5YR6/3	-	-	loamy sand	subr. cobbles 38 % subr. gravel 4 %
Bhs	11	gradual	wavy	10YR2/1	-	-	loamy sand	subr. cobbles 38 % subr. gravel 11 %
Bhs/C	26	gradual	wavy	10YR2/1 2.5YR6/4	-	-	loamy sand	subr. cobbles 18 % subr. gravel 11 %
C	-	-	-	2.5YR6/4	-	-	loamy sand	subr. cobbles 18 % subr. gravel 11 %

Table 1 continued

Horizon	Calcarous-ness	Structure	Penetration resistance	Roots Quant.	Roots Size	Roots Location
Oi/Oe	0% no	-	-	-	-	-
Oa	0% no	1	soft	many few	fine medium	through.
E	0% no	2	soft	common few	fine medium	through.
Bhs	0% no	3	firm	few few	fine medium	through.
Bhs/C	0% no	3	firm	-	-	-
C	0% no	3	firm	-	-	-

Example Central Alps (Switzerland)

Analytical data and interpretation

Soil classification: Typic Haplocryod

Humus form: Mormoder

Table 2

Horizon	Thickness [cm]	pH ($CaCl_2$)		Base saturation [%]	Soil organic matter [%]
Oi/Oa	2	-	-	-	-
Oa	12	3.4	ultra acid	<50%	42.6
E	8	3.7	extremely acid	<50%	4.3
Bhs	11	4.0	extremely acid	<50%	8.3
Bhs/C	26	4.3	extremely acid	<50%	3.9
C	-	-	-	-	-

Table 2 continued

Horizon	Air porosity	Plant available water	Field capacity	Total porosity
		[Vol.-%]		
Oi/Oa	-	-	-	-
Oa	-	-	-	-
E	19.6	12.8	19.4	51.8
Bhs	7.1	10.0	17.1	34.2
Bhs/C	9.2	12.1	19.2	40.5
C	-	-	-	-

Available water holding capacity for the rooting zone [mm]: 21.2 (organic layer not considered)
Root restricting depth: 33 cm, moderately deep

4.2 Example Tibetan Plateau (China)

Site description

Name:	Gabriele Broll (Smith et al. 1999)
Date:	05.08.1995
Profile no:	2
Photo no:	2

Location:

Location:	China, 29° 41' 39" N; 91° 6' 59" E
Physiogr. location:	Tibetan Plateau, Lhasa Region
Elevation [m.a.s.l.]:	3650

Topography / Relief

Landform:	valley
Slope aspect [°]:	-
Slope gradient [°]:	-
Slope shape:	-
Site position on slope:	-
Microfeature:	-

Water status:

Drainage class:	well drained
Depth to water table [cm]:	-
Depth to imperm. layer [cm]:	-
Root restricting depth [cm]:	99
Depth of thaw [cm]:	-

Land use - Vegetation:

Land use: irrigated cropland (Tibetan barley)

Total trees [%]:	0
Total shrubs [%]:	0
Total vasculary plants [%]:	30
Total lichen [%]:	0
Total moss [%]:	0
Total vegetation [%]:	30
Bare ground [%]:	70

Parent material and / or bedrock:

Kind of parent material: strongly weathered alluvium from igneous - granite

Kind of bedrock material: -

Surface fragments:

Classes of of surface cover: none

Size classes: -

Soil profile

Example Tibetan Plateau (China)

Description of profile no 2

Soil classification: Aridic Ustochrept

Humus form: -

Table 3

Horizon	Thickn. [cm]	Boundary Dist.	Boundary Topo.	Matrix Color	Mottles Color	Mottles Quant.	Texture	Rock fragments
Ap	23	clear	wavy	10YR 3/4	-	-	sandy loam	10% subangular gravel
Bw1	8	diffuse	smooth	10YR 4/4	-	-	sandy loam	-
Bw2	28	diffuse	smooth	10YR 4/4	-	-	sandy loam	5% subangular gravel
Bw3	40	clear	wavy	10YR 4/4	-	-	sandy loam	-
2C	35	gradual	wavy	10YR 4/6	-	-	coarse sand	20% angular cobbles 30% subangular gravel

Table 3 continued

Horizon	Calcarous-ness	Structure	Penetration resistance	Roots Quant.	Roots Size	Roots Location
Ap	0% no	granular / single grain	2	common	fine	matted on top
Bw1	0% no	subangular blocky	3	-	-	-
Bw2	0% no	subangular blocky	2	-	-	-
Bw3	0% no	subangular blocky	2	-	-	-
2C	0% no	single grain	-	-	-	-

Remarks and special notes

Horizon	Remarks
Ap	-
Bw1	slightly compacted plowpan
Bw2	-
Bw3	-
2C	-

Example Tibetan Plateau (China)

Analytical data and interpretation

Soil classification: Aridic Ustochrept

Humus form: -

Table 4

Horizon	Thickness [cm]	pH ($CaCl_2$)		Base saturation [%]	Soil organic matter [%]
Ap	23	7.4	slightly alkaline	> 50	0.98
Bw1	8	7.2	neutral	> 50	0.41
Bw2	28	7.2	neutral	> 50	0.46
Bw3	40	7.1	neutral	> 50	0.15
2C	35	7.1	neutral	> 50	0.09

Table 4 continued

Horizon	Air porosity	Plant available water	Field capacity	Total porosity
		[Vol.-%]		
Ap	11.5	23.0	33.5	45.0
Bw1	9.5	19.5	28.0	37.5
Bw2	11.5	22.5	32.0	43.5
Bw3	11.5	22.5	32.0	43.5
2C	-	-	-	

Available water holding capacity for the rooting zone [mm]: 221.50
Root restricting depth: 99 cm, deep

4.3 Example Karakorum (North-Pakistan)

Site description

Name:	Thomas Reineke
Date:	13.09.1993
Profile no:	3
Photo no:	3

Location:

Location:	south flank of Rakaposhi-Diran massive
Physiogr. location:	Karakorum, upper Bagrot-valley, tributary to Gilgit-river
Elevation [m.a.s.l.]:	3715

Topography / Relief

Landform:	mountain slope
Slope aspect [°]:	SSW
Slope gradient [°]:	34
Slope shape:	LV (linear/convex)
Site position on slope:	backslope
Microfeature:	solifuction lobes (bound to alpine mats) 'Viehgangeln' (due to extensive grazing)

Water status:

Drainage class:	moderately well drained
Depth to water table [cm]:	-
Depth to imperm. layer [cm]:	-
Root restricting depth [cm]:	100
Depth of thaw [cm]:	-

Land use - Vegetation:

Land use:	grassland
Total trees [%]:	0
Total shrubs [%]:	2
Total vasculary plants [%]:	95
Total lichen [%]:	1
Total moss [%]:	2
Total vegetation [%]:	100
Bare ground [%]:	0

Parent material and / or bedrock:

Kind of parent material: loessic colluvium overlying slope debris

Kind of bedrock material: metamorphic schist

Surface fragments:

Classes of of surface cover: very few

Size classes: large boulders

Soil profile

Example Karakorum (North-Pakistan)

Description of profile no 3

Soil classification: Typic Cryoboroll

Humus form: Cryic Rhizomull

Table 5

Horizon	Thickn. [cm]	Boundary Dist.	Boundary Topo.	Matrix Color	Mottles Color	Mottles Quant.	Texture	Rock fragments
Ah	18	gradual	wavy	10YR 2/2	-	-	silt loam	-
Bw	27	gradual	wavy	10YR 4/3	-	-	silt loam	15-25% subangular blocky
2C	55	abrupt	wavy	10YR 5/4	-	-	silt loam	35% subangular blocky

Table 5 continued

Horizon	Calcarous-ness	Structure	Penetration resistance	Roots Quant.	Roots Size	Roots Location
Ah	0% no	granular / single grain	3 high	many	fine - very fine	matted on top
Bw	0% no	subangular blocky	2-3 moderate-high	common	fine - very fine	-
2C	0% no	subangular blocky/ single grain	2-3 moderate-high	few - very few	fine	-

Remarks and special notes

Horizon	Remarks
Ah	non-sticky and non-plastic silt loam
Bw	very sticky and very plastic silt loam, medium roots in cracks and matted around rock fragments
2C	solifluction layer, partly big boulders > 20cm

Example Karakorum (North-Pakistan)

Analytical data and interpretation

Soil classification: Typic Cryoboroll

Humus form: Cryic Rhizomull

Table 6

Horizon	Thickness [cm]	pH ($CaCl_2$)		Base saturation [%]	Soil organic matter [%]
Ah	18	5.2	strongly acid	high	11.0
Bw	27	5.5	strongly acid	high	3.1
2C	55	5.9	moderately	high	1.2

Table 6 continued

Horizon	Air porosity	Plant available water	Field capacity	Total porosity
	[Vol.-%]			
Ah	11	33	49	60
Bw	16	27	39	55
2C	18	24	31	50

Available water holding capacity for the rooting zone [mm]: 180

Root restricting depth: 100 cm, deep

4.4 Example Scandes (Norway)

Site description

Name:	Jörg Löffler
Date:	15.08.1995
Profile no:	4
Photo no:	4
Location:	
Location:	Norway, x 512315, y 6863385 (UTM)
Physiogr. location:	Scandes, Vaga/Oppland, Blahö-Jetta
Elevation [m.a.s.l.]:	1540
Topography / Relief	
Landform:	Mountain slope
Slope aspect [°]:	285° W
Slope gradient [°]:	10
Slope shape:	concave/concave
Site position on slope:	lower slope
Microfeature:	micro-depression
Water status:	
Drainage class:	moderately drained, temporary: sheet flow
Depth to water table [cm]:	-
Depth to imperm. layer [cm]:	-
Root restricting depth [cm]:	38
Depth of thaw [cm]:	-
Land use - Vegetation:	
Land use:	extensive grazing (1-2 month/year)
	Polytrico communis - Caricetum bigelowii
Total trees [%]:	-
Total shrubs [%]:	-
Total vasculary plants [%]:	-
Total lichen [%]:	-
Total moss [%]:	-
Total vegetation [%]:	-
Bare ground [%]:	-

Parent material and / or bedrock:

Kind of parent material:	fine fluvial sediments overlying solifluction debris (phylite material)
Kind of bedrock material:	phylite

Surface fragments:

Classes of of surface cover:	35%
Size classes:	blocky and stony

Soil profile

Example Scandes (Norway)

Description of profile no 4

Soil classification: Lithic Cryorthent, Lithic Dystrocrept

Humus form: Mineric Mormoder

Table 7

Horizon	Thickn. [cm]	Boundary Dist.	Boundary Topo.	Matrix Color	Mottles Color	Mottles Quant.	Texture	Rock fragments
Fai-Hhi	2	diffuse		-	-	-	-	-
II Bw(jj)	26	sharp		7,5YR 4/3	-	-	sandy loam	< 63 mm 75% (50% > 200, 27% 63-200, 23% < 63)
IIR								

Table 7 continued

Horizon	Calcarous-ness	Structure	Penetration resistance	Roots Quant.	Roots Size	Roots Location
Fai-Hhi	-	single particle	2	common	fine	matted on top
II Bw(jj)	-	angular blocky / single particle	2	few	very fine	
IIR						

Remarks and special notes

Horizon (US Soil Tax)	Horizon Löffler (1998)	Remarks
Fai-Hhi	Vi (Pol) + jwMC	moss (Polytrichum commune) and fresh plant residues, fragmented plant residues mainly from Carex bigelowii, very few organic fine substance, with accumulation of mineral sheet erosion material (sandy loam)
C	IIwMC	accumulation of mineral sheet erosion material
IIBw(jj)	IIIkMBv-bxC	solifluction material, some vertical stones
IIR	IVmCn	massive phylite bedrock

Example Scandes (Norway)

Analytical data and interpretation

Soil classification: Lithic Cryorthent, Lithic Dystrocrept

Humus form: Mineric Mormoder

Table 8

Horizon	Thickness [cm]	pH ($CaCl_2$)		Base saturation [%]	Soil organic matter [%]
Fai-Hhi	2	5.1	strongly acid	< 50	-
C	10	5.0	very strongly acid	< 50	0.08
IIBw(jj)	26	4.6	very strongly acid	< 50	0.96

Table 8 continued

Horizon	Air porosity	Plant available water	Field capacity	Total porosity
		[Vol.-%]		
Fai-Hhi	-	-	-	-
C	8.0	15.8	22.4	30.4
IIBw(jj)	2.9	5.8	8.4	11.3

Available water holding capacity for the rooting zone [mm]: 30.9
Root restricting depth: 38 cm, moderately deep

Acknowledgements

We thank PD Dr. Jörg Löffler (Oldenburg, Germany) for adding one example of his field work on alpine soils in Norway, Prof. Dr. Frank Lehmkuhl (Aachen, Germany) as well as other members of the Working Group on Mountain Geoecology, (Association of German Geographers VGDH) for helpful discussions.

Also, our thanks go to Dr. Robert Ahrens (USDA, Lincoln, USA), Dr. Claudia Erber (University of Muenster, Germany), Nancy Steffen (Whitehorse, Canada), Prof. Dr. Hans Sticher (ETH Zürich, Switzerland) and Dr. Charles Tarnocai (Agriculture and Agri-Food Canada, Ottawa) for comments and reviewing earlier versions of this guideline. Not last we thank those members of the Cryosol Working Group of the International Union of Soil Science/ International Permafrost Association, who made valuable contributions during field trips.

References

AG Boden (1994) Bodenkundliche Kartieranleitung, Hannover

AK Standortskartierung in der AG Forsteinrichtung (1996) Forstliche Standortsaufnahme, 5. Aufl., Münster

Blake GR and Hartge KH (1986) Bulk density. In: Klute A (ed) Methods of soil analysis. Part 1: Physical and mineralogical methods. Soil Science Society of America. Book Series No. 5, Part 1. ASA, SSSA, 363-375. Madison, Wisconsin

Brady NC and Weil RR (1998) The nature and properties of soils. Prentice Hall International, New Jersey

Catt JA (1986) Soils and quaternary geology. A handbook for scientists. Clarendon Press, Oxford

Day JH (ed) (1983) The Canada Soil Information System (CanSIS). Manual for describing soils in the field. Expert Committee on Soil Survey. Ottawa, Ontario

Denholm KA and Schut LW (eds) (1993) Field manual for describing soils in Ontario. 4th edition. OCSRE Publication No. 93-1. Ontario Centre for Soil Resource Evaluation, Ontario

Deutschmann G, Malessa V and Rummenhohl H (1994) Bestimmung der Lagerungsdichte in stark skeletthaltigen Böden. Zeitschrift für Pflanzenernährung und Bodenkunde 157, 77-79

DVWK (Deutscher Verband für Wasserwirtschaft und Kulturbau) (ed) (1995) Bodenkundliche Untersuchungen im Felde zur Ermittlung von Kennwerten zur Standortcharakterisierung Teil 1: Ansprache der Böden. Regeln zur Wasserwirtschaft, 129/1995, Bonn

FAO (Food and Agriculture Organization of the United Nations) (1990) Guidelines for soil description. Soil Resources, Management and Conservation Service. Land and Water Development Division, Rome.

Gardner WH (1986) Water content. In: Klute A (ed) Methods of soil analysis. Part 1: Physical and mineralogical methods. Soil Science Society of America Book Series No. 5, Part 1. ASA, SSSA, 493-544. Madison, Wisconsin

Green RN, Trowbridge RL and Klinka K (1993) Towards a taxonomic classification of humus forms. Forest Science Monograph, 29

Hendershot WH, Lalande H and Duquette M (1993) Soil reaction and exchangeable acidity. In: Carter MR (ed) Soil sampling and methods of analysis, 141-145. Lewis Publishers, Boca Raton

Kubiëna WL (1953) Bestimmungsbuch und Systematik der Böden Europas. Enke, Stuttgart

Livingston NJ (1993) Soil temperature. In: Carter MR (ed) Soil sampling and methods of analysis, 673-682. Lewis Publishers, Boca Raton

Löffler J (1998) Geoökologische Untersuchungen zur Struktur mittelnorwegischer Hochgebirgsökosysteme. Oldenburger Geoökologische Studien Band 1. Biblio-theks- und Informationssystem der Carl v. Ossietzky Universität Oldenburg, Oldenburg

Miehe G and Miehe S (2000) Guidelines for the physiognomic classification of plant formations in mountain areas. In: Working group of mountain geoecology (Associaton of German Geographers VGDH) (ed) Field guide for landscape ecological studies in high mountain environments (Draft)

Mosimann T (1985) Untersuchungen zur Funktion subarktischer und alpiner Geoökosysteme (Finnmark (Norwegen) und Schweizer Alpen). Physiogeo-graphica - Basler Beiträge zur Physiogeographie 7. Basel

Rhoades JD (1996) Salinity: Electrical conductivity and total dissolved solids. In: Sparks DL (ed) Methods of soil analysis. Part 3: Chemical methods. Soil Science Society of America Book Series No. 5, Part 3. ASA, SSSA, 417-435. Madison, Wisconsin

Schlichting E, Blume H-P and Stahr K (1995) Bodenkundliches Praktikum. Eine Ein-führung in pedologisches Arbeiten für Ökologen, insbesondere Land- und Forst-wirte und für Geowissenschaftler. Pareys Studientexte 81. Blackwell Wissen-schafts-Verlag, Berlin

Schoeneberger PJ and Wysocki DA (1996) Geomorphic descriptors for landforms and geomorphic components: effective models, weakness and gaps [Abstract]. American Society of Agronomy, Annual Meetings, Indianapolis, IN

Schoeneberger PJ, Wysocki DA and Olson CG (1998) Glossary of landform and geologic terms. Part 629 (Version 2.0) of the national soil survey handbook

Schoeneberger PJ, Wysocki DA, Benham EC and Broderson WD (2002) Field book for describing and sampling soils, version 2.0. National Resources Conservation Service, USDA, National Soil Survey Center, Lincoln, NE

Smith CAS, Clark M, Broll G, Ping CL, Kimble JM and Luo G (1999) Characterization of selected soils from the Lhasa region of Qinghai-Xizang Plateau, SW China. Permafrost and Periglacial Processes 10:211-222

Soil Survey Division Staff (1993) Soil survey manual. USDA Handbook No. 18, Was-hington

Soil Survey Staff (1999) Soil taxonomy. A basic system of soil classification for making and interpreting soil surveys. USDA, Washington

Soil Survey Staff (2003) Keys to soil taxonomy. USDA, Washington

SSSA Soil Science Society of America (1997) Glossary of soil science terms. Soil Science Society of America, Madison

Thien SJ (1979) A flow diagram for teaching texture-by-feel analysis. Journal of Agricultural Education 8, 54-55

Thomas GW (1996) Soil pH and soil acidity. In: Sparks DL (ed) Methods of soil analysis. Part 3: Chemical methods. SSSA, ASA, 475-490. Madison, Wisconsin

Tiner RW (1999) Wetland indicators. A guide to wetland identification, delineation, classification, and mapping. Lewis Publishers, Boca Raton

Topp GC (1993) Soil water content. In: Carter MR (ed) Soil sampling and methods of analysis, 541-557. Lewis Publishers, Boca Raton

Peripheral Plant Population Survival in Polar Regions

Robert M.M. Crawford

Abstract

All species have limits to their distribution and individuals that demarcate margins demonstrate an end-point in adaptation to a changing environment. Limits to plant survival arise either from a failure to grow or from an inability to reproduce. Consequently, matching climatic data with geographical distribution has to consider the biological causes for the failure of plants to survive outside any given area. Physical factors, such as thermal time (day-degrees) and growing season length, differentiate between the broadly recognised vegetation zones of High, Low, and Subarctic. However, within these zones the limitations to plant distribution are less readily related to climatic factors. In peripheral areas, relating physical climatic factors to species occurrence requires a knowledge of the many interactions between physiology and genetics of populations as they approach their territorial margins. Examination of the extent of variation found in plants growing in cold habitats reveals large differences both between and within populations and at different seasons in metabolic rates, temperature responsiveness and metabolic efficiency. The High Arctic represents a peripheral zone where conditions are prone to large fluctuations. Consequently, plant distribution will be affected by aspects of the environment that are not permanent, but which occur irregularly with varying frequency and intensity. This paper considers the physiological and genetic properties of polar plant populations that may facilitate persistence in uncertain and heterogeneous adverse environments. Particular attention is given to those species that appear to have maintained a presence at high latitudes since pre-Pleistocene times and have therefore survived many past environmental fluctuations.

1 Introduction

Limits to plant distribution in the Arctic have always excited curiosity as they represent the ultimate peripheral locations for high latitude survival.

Biologically, plants are ideally suited for the study of peripheral situations as their sedentary nature facilitates mapping and the demographic recording of births and deaths. Many atlases record limits to plant distribution both past and present at high latitudes (Löve and Löve 1975; Huntley and Birks 1983; Hultén and Fries 1986; Meusel and Jäger 1992). There is also a growing corpus of information on the longevity and reproductive capacity of polar vegetation (Arft et al. 1999; Jonsdottir et al. 1999). The circumpolar distribution maps are unique for in no other area of the World is it possible to examine across adjacent continents both the north-south and east-west extensions of species ranges. The ease with which temperature records can now be collated and compared with past and present plant distribution maps can create an impression that cartographic representation combined with mathematical modeling, and some experimental studies on growth and reproduction are all that is needed to relate climatic limits with plant distribution. Unfortunately, as pointed by the pioneering Norwegian eco-physiologist Eilif Dahl 'the many indices that have been used for correlation between the northern and alpine limits of plants and meteorological factors are rarely based on any eco-physiological foundation'. As he expressed the problem 'the question is in which way does temperature affect plant performance' (Dahl 1998).

Recently, considerable attention has been given to measuring plant responses to climatic change through the International Tundra Experiment (see ITEX web reference) which established a collaborative, multi-site experiment using a common temperature manipulation to examine variability in species response across climatic and geographic gradients of tundra ecosystems. This study provided much useful information on the short-term phenotypic responses of a large number of arctic species and showed that Low Arctic sites have the strongest growth responses to climatic amelioration, while colder sites produced a greater reproductive response. This was interpreted as suggesting that greater resource investment in vegetative growth could be a successful survival strategy in the Low Arctic, where there is more competition for resources, while in the High Arctic, heavy investment in producing seed under a higher temperature scenario could provide an opportunity for species to colonize patches of bare ground during a favourable climatic window (Arft et al. 1999). Useful as these experiments are in gauging the extent of short-term phenotypic resilience of vegetation to climatic oscillations, the relatively short period over which the plots were studied provides only a limited insight into the potential evolutionary responses of plants to environmental alteration. Furthermore, advances in meta-population studies and molecular genetics now provide an improved perception of the range of adaptive responses that are available to plants in marginal situations.

Much of the physiological information that is gathered from plants growing in the Arctic depends on information that can be recorded by external sensors. Thus, there are numerous studies of photosynthetic rates, respiratory activity, transpiration and overall productivity, but relatively little research examining internal metabolic efficiency and whether or not this is affected by the limitations and fluctuations of the arctic environment. Dahl's question as to how plants are affected by temperature is therefore still a valid point which requires further investigation.

In low temperature environments plants employ different life strategies as compared with animals both in terms of physiology and genetics. The existence of many highly localised populations is the norm in plants and therefore demands attention in assessing potential genotypic as well as phenotypic responses, particularly in marginal situations. Plants also differ from animals in their reproductive responses to climatic oscillations. For animals, climatic deterioration, if it brings about a continued failure to reproduce, will inevitably lead to the abandonment of the breeding site. Similarly, climatic improvement can lead to rapid re-establishment when conditions improve. It is noticeable, particularly at high latitudes, that whenever there is climatic warming that the insects are usually the first species to respond and expand their geographical range (Atkinson et al. 1987). In contrast to animals, plants are not automatically faced with extinction when sexual reproduction fails. Many species can maintain viable and sometimes expanding populations without sexual reproduction. The Arctic Salt Grass (*Puccinellia phryganodes*) has a circumpolar distribution and is a dominant component of salt marsh vegetation, yet throughout its entire range the species is almost always completely sterile (Lid and Lid 1994).

Plants also differ from most animals in that they do not even need to grow in order to survive. Whole populations can lie dormant for decades, or even centuries, buried in the soil seed bank. A recent study in Svalbard found that 71 of the 161 species indigenous to Svalbard had the capability of persisting in the soil seed bank (Cooper et al. 2004).

Studies of climatic limits to species distribution frequently present data in the form of maps. While maps create a global perception of distribution they may be misleading in relation to determining the causes of boundaries as these may change from one location to another. This same problem can affect the distribution of entire plant communities. The southern limit of the Tundra is generally determined by the northern limits of the boreal forest. However, the northern limit of the boreal forest may be controlled by differing ecological circumstances in different regions (Callaghan et al. 2002). Temperature, as it affects the length of the growing season, is related to the position of the boreal tree line and especially in regions with marked continental climates (see also below). However, in extensive regions of western Siberia the northern limit of the boreal forest is displaced southwards by the development of vast bogs. It follows therefore that although the boreal treeline (the tundra-taiga interface) is a global phenomenon that can be mapped and even observed by remote sensing, the reasons for its position may differ from one area to another (Crawford et al. 2003). Consequently, if the different factors controlling distribution are not understood then the information that is provided by maps will be open to misinterpretation. In an attempt to improve comprehension of the nature of climatic boundaries this paper examines the variety of biological responses of plants that can now be found in populations in marginal areas and which may be potentially advantageous in uncertain climatic conditions. Particular attention is given to the interaction between physiological limits and genetic responses in marginal areas. Barriers to distribution always present an evolutionary challenge and what may be one plant's physiological limit can be another's ecological opportunity.

2 Metabolic Efficiency and Temperature Variation

Life on Earth is driven by the metabolism of carbon which requires high energy levels for the spontaneous making and breaking of covalent bonds. It is therefore one of the marvels of biochemistry that enzymatically catalysed reactions proceed at a rates that can be 10 orders of magnitude greater than non-catalysed reactions. As the distinguished 20th century animal physiologist, Sir Joseph Barcroft once wrote, 'nature has learnt so to exploit the biochemical situation as to escape from the tyranny of a single application of the Arrhenius equation' (Barcroft 1934). Given the thermodynamic problems that have had to be overcome for life to exist at all on a planet with a mean temperature of only 15°C, the colder polar regions represent only the cooler end of a thermally unpromising environment. That life on Earth is ubiquitous from the boiling thermal vents at the bottom of the oceans to the polar deserts of the High Arctic is evidence enough that that evolution can triumph over the dictates of the Arrhenius equation.

The low number of species and the diminutive size of plants in the High Arctic can lead to a natural presumption that providing energy for plant growth is the limiting factor in polar regions. Numerous investigations have therefore sought evidence for adaptations that allow plants of cold regions to metabolise efficiently at low temperatures. In particular, respiratory activity has been examined by many authors in the intuitive belief that survival in cold climates should require a specialised metabolic capacity for delivering energy at low temperatures. Here again Eilif Dahl was a pioneer in his attempts to relate respiration with growth by devising an ingenious method for relating meristem respiration with the daily growth of shoots based on the relationships between temperature, respiration and growth in Norway spruce (Dahl and Mork 1959). Integrated 'respiration sums' or respiration equivalents (Re-sums) were calculated based on the total dark respiration at 10°C for one day in Sitka Spruce meristems as representing one unit, having allowed for a basal respiration rate for maintenance and transport as measured at 2.8°C (the threshold temperature which has to be exceeded for growth in spruce to take place - for further details see Dahl (1998)). Using this formula Dahl prepared maps showing isolines for Re-values for spruce across Europe (figures 1-2). However, this method of evaluating respiration potential makes a number of assumptions that are now difficult to justify. These include assuming that the rate of change of respiration with temperature is constant across a thermal gradient and between species. The Dahl model also assumed that growth-

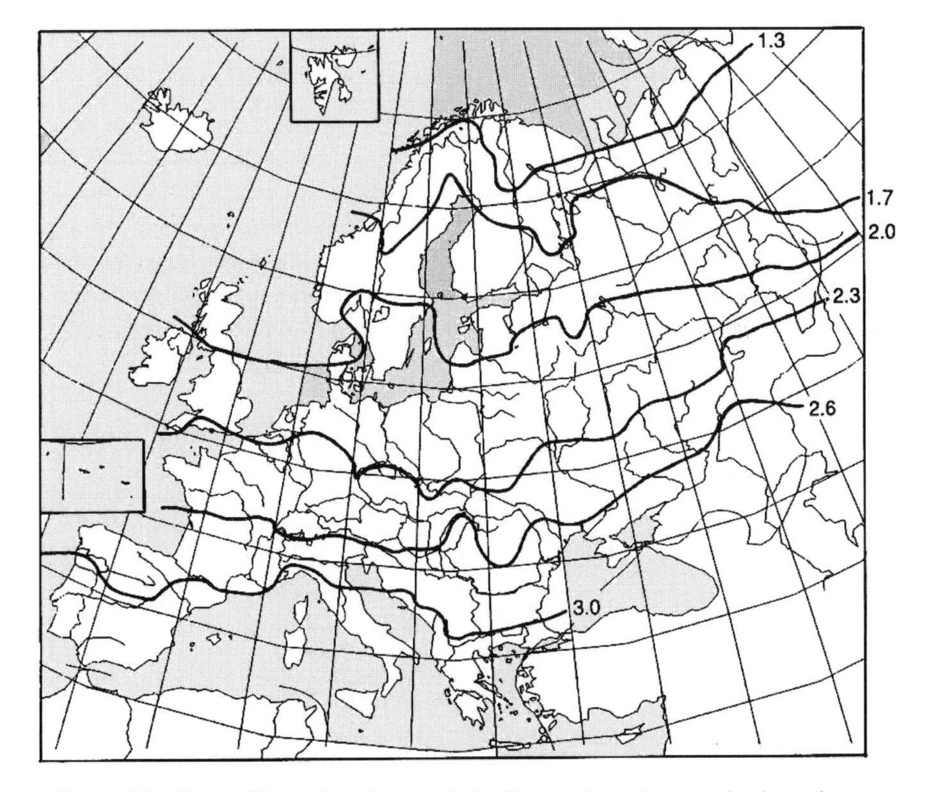

Figure 1 Isolines of **Re** values (see text) for Europe based on respiration of spruce meristems calculated for the lowest point in each of the Flora Europea squares (reproduced with permission from Dahl 1998)

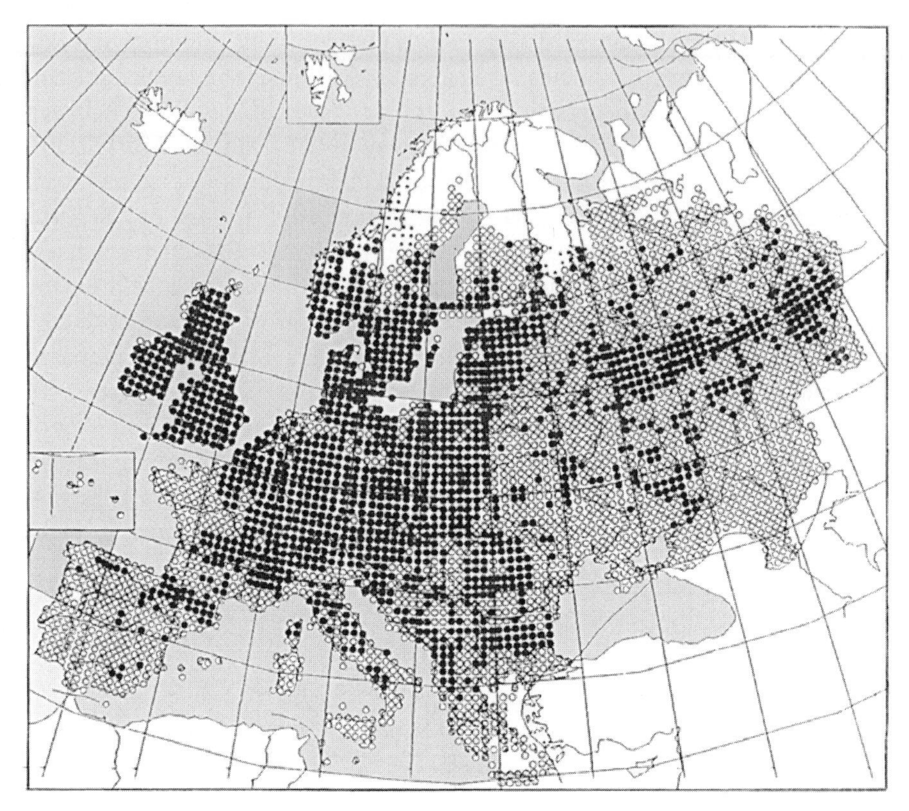

Figure 2 Distribution of *Ulmus glabra* in relation to the 1.7 **Re** value (see text). Filled circles show species presence, open circles the other Flora Europea grid squares where the Re value is 1.7 or higher (reproduced with permission from Dahl, 1998)

respiration is an entity that is distinct from other aspects of respiratory activity. This distinction between maintenance and growth respiration has been made for many years in an attempt to distinguish the fraction of carbon reserves used to maintain an existing biomass and the fraction used for producing new growth (Schulze 1982). The physiological basis for this division rested on the belief that anabolic and catabolic processes are separate and distinguishable and that by measuring respiration at temperatures too low for growth an estimation can be made of the energy requirements of resting or non-growing tissues. The constant recycling of cell constituents and the maintenance of a steady level of energy charge (see figure 3) even in the absence of growth requires a constant interaction between anabolic and catabolic activities which inevitably varies with temperature and makes the distinction between maintenance and growth respiration a doubtful reality. The variable coupling between oxidation and ATP synthesis was also not fully understood at the time when Dahl first carried out his pioneering research (Dahl and Mork

1959). When the mitochondrial electron transport chain is working at maximum efficiency, as it is when matrix-generated NADH is oxidised via complex I, the ADP/O ratio is 3.0. However, plant mitochondria possess multiple routes for NADH oxidation, and an additional NADH dehydrogenase located on the matrix face of the inner membrane by-passes complex I and the resulting ADP/O ratio falls to 2.0. Cytosolic NADH can also be oxidised, this time by a NADH dehydrogenase located on the outer surface of the inner mitochondrial membrane. In this case, the ADP/O ratio is also 2.0. If electrons then pass from ubiquinone into the alternative oxidase, thus by-passing the cytochrome pathway, the ADP/O ratio falls to 1.0 for substrates oxidised through complex I and zero for those oxidised through either of the additional NADH dehydrogenases.

The measurements that are readily made of plant respiration namely, oxygen uptake (**R_{O_2}**) and carbon dioxide production (**R_{CO_2}**) do not provide the same information and give no indication of metabolic efficiency in relation to substrate use and energy gain. The addition of a third component, the use of calorimetry to estimate respiratory heat rate (**R_q**) makes it possible to estimate metabolic efficiency, and provides essential metabolic information, especially as changes in respiration rate are often opposed by alterations in efficiency (Hansen et al. 2002). By placing plant tissue in a micro-calorimeter together with a phial of sodium hydroxide it is possible to measure the carbon dioxide

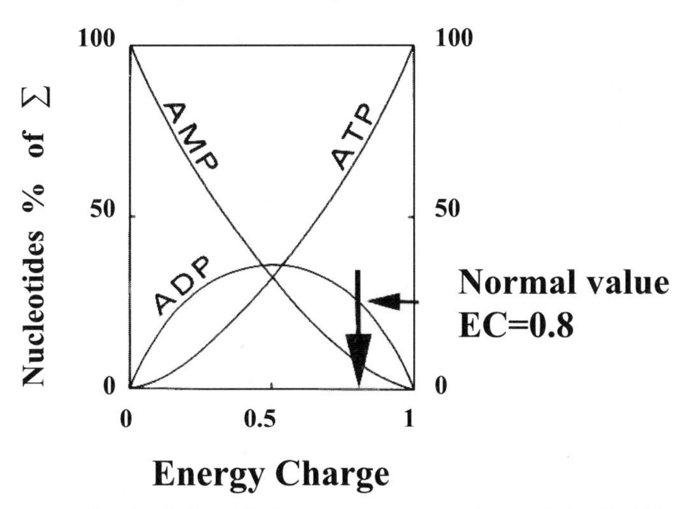

Figure 3 Generalised relationship between concentrations of AMP, ADP and ATP based on Pradet and Bomsel (1978). The arrow indicates the values calculated to give an energy charge value (EC) of 0.8 where:

$$EC = \frac{(ATP) + \frac{1}{2}(ADP)}{(ATP) + (ADP) + (AMP)}$$

produced simultaneously with the amount of heat generated by the respiring tissue (figure 4). The rate of heat production **(R_q)** is equal to the product of the reaction rate and enthalpy change and the ratio of heat production to carbon dioxide evolved R_q/R_{CO_2} estimates metabolic efficiency. The ratio of Cytochrome oxidase activity to alternative oxidase activity, contrary to common perceptions does not significantly alter the R_q/R_{CO_2} or R_q/R_{O_2} ratios (Breidenbach et al. 1997). However, increasing alternative oxidase activity does cause a loss of efficiency which is reflected in small increases in these ratios, but is generally accompanied by an increase in respiration rate which compensates to maintain an approximately constant growth rate.

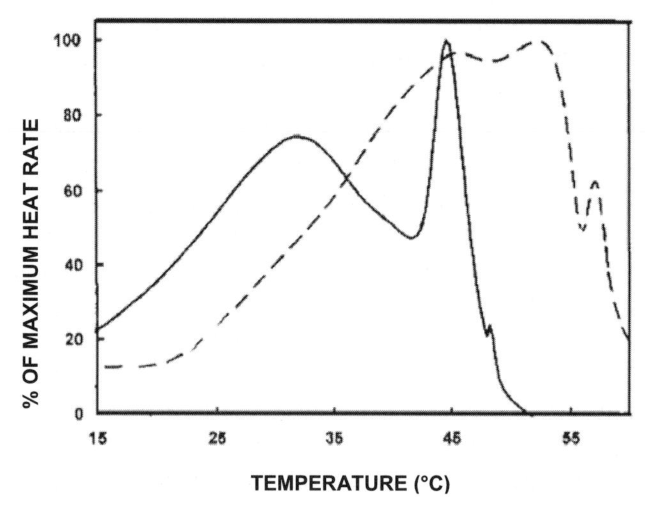

Figure 4 Continuous temperature-scanning thermograms for tissues of a tropical tree *Eremocarpus setigerus* (continuous line) and a temperate tree *Betula pendula* (dashed line) (reproduced with permission from Smith et al. 1999)

Applying these methods to the metabolic responses of several populations of the boreal cold-desert sub-shrub *Eurotia lanata* has shown (figure 5) that very great differences can be found between different populations both in the temperature responses of respiration and metabolic heat rate. It follows therefore that metabolic efficiency is equally variable (Thygerson et al. 2002).

The physiological significance of the alternative oxidase pathway is still a subject of discussion. The necessity of regulating adenylate levels (see figure 3) in order to maintain a stable energy charge is essential for metabolic regulation (Atkinson 1968). The control of futile pathways that appear to waste ATP or produce it less efficiently may contribute to the stability of cellular energy levels. It is also possible that the alternative oxidase pathway prevents the formation of toxic reactive oxygen species by maintaining mitochondrial electron transport at low temperatures which would otherwise inhibit the main phosphorylating pathways. This role is supported by the observation that alternative oxidase protein levels often increase when plants are subjected to growth at low temperatures (Gonzalez-Meler et al. 1999). In arctic plants there is also evidence of increased respiration associated with activity of the alternative oxidase pathway (McNulty et al. 1988).

Another aspect of the regulation of plant metabolism is the degree of metabolic response to changes in temperature. The exponential rise in plant respiratory rates with increasing temperature is frequently still referred to as doubling with each 10 degrees rise in temperature ($Q_{10} = 2$). Although this approximation is broadly true a more accurate representation for species comparisons of change in metabolic rate in response to change in temperature is obtained by use of the Arrhenius plot. In this plot the log of reaction rate *versus* the reciprocal of the absolute temperature in degrees Kelvin provides straight line relationships and thus makes inter-specific comparisons more clearly than exponential plots. By comparing coastal species with contrasting northern and southern distributions it is possible to compare Arrhenius plots and estimate both the magnitude of respiratory activity and its temperature dependence in species with similar growth habits and ecology but different temperature limitations. *Ligusticum scoticum* and *Mertensia maritima* are both species that have distributions that extend south from the Arctic to the Subarctic and reach their southern limits in central Scotland. By contrast, *Crithmum maritimum* and *Limonium vulgare*, are also coastal species with a distribution that extends north from the Mediterranean to central Scotland (figure 6). When the respiration of the over-wintering root systems of these species was examined between January and March it was found that the roots of the northern species had higher respiration rates than the southern species.

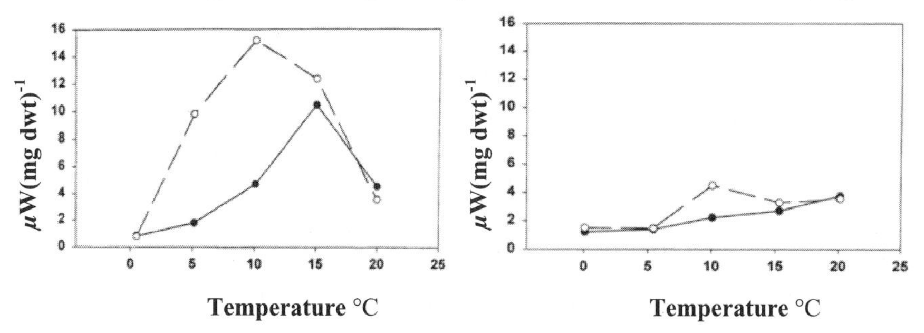

Figure 5 Metabolic heat rate (• - the loss of heat from metabolic processes) and respiration rate (o - estimated from CO_2 evolution) as measured in germinating root radicles from seeds of *Eurotia lanata* at two different montane locations in North America (adapted with permission from Thygerson et al. 2002)

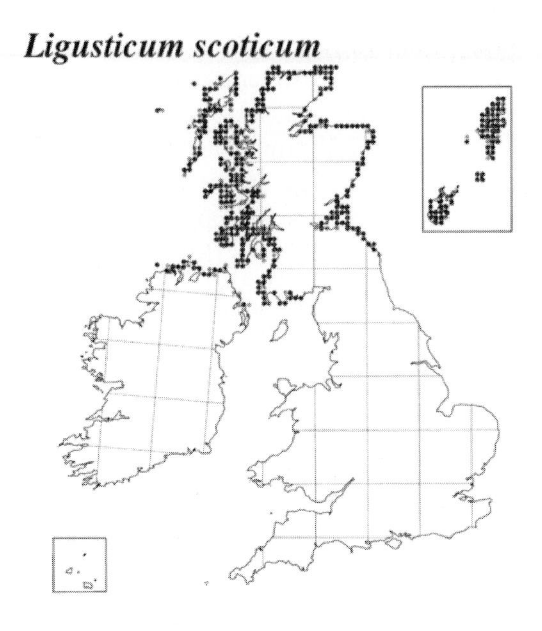

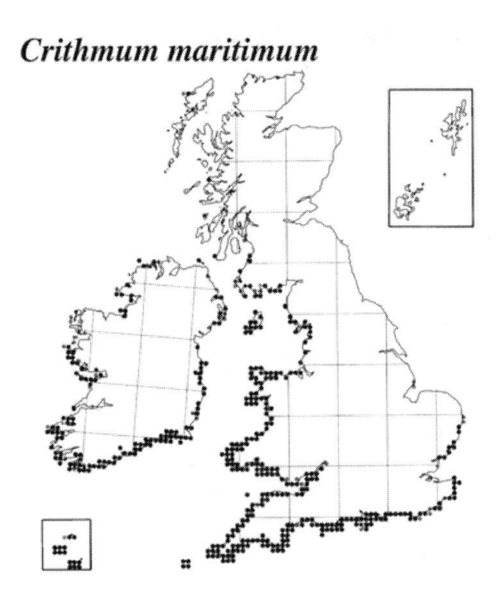

Figure 6 Distribution in the British Isles of four coastal species; two species which occur in the Arctic, *Ligisticum scoticum* and *Mertensia maritima* and reach their southern British limits in Scotland; *Crithmum maritimum* and *Limonium vulgare* extend northwards from the Mediterranean and reach the northern limits of their distribution in Scotland (reproduced with permission from the Atlas of the British Flora, Preston et al. 2002)

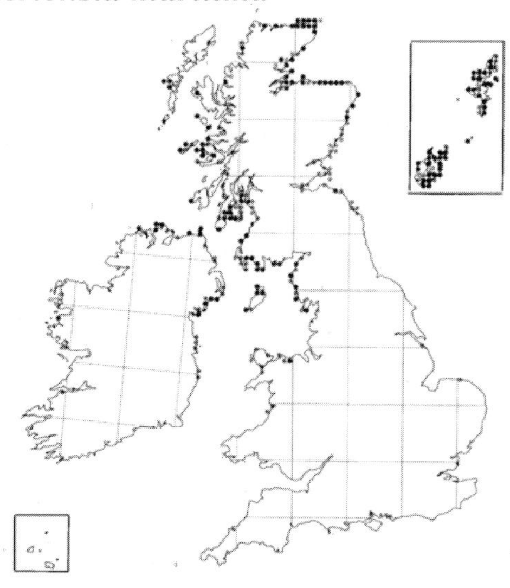

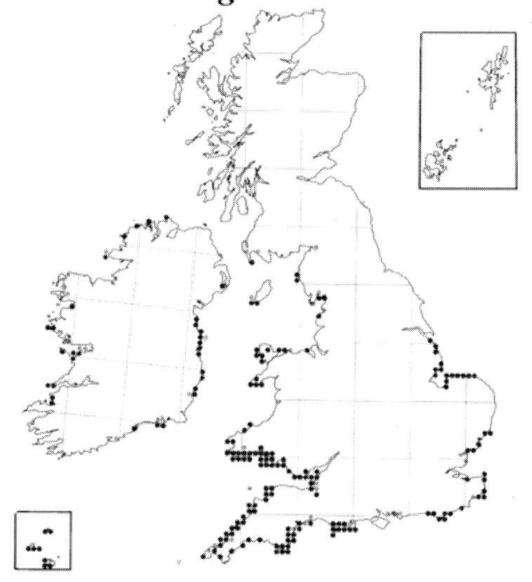

Figure 6 continued

Arrhenius plots of the log of oxygen uptake against the reciprocal of temperature in °K revealed consistent breaks so that the northern species always had lower activation energies at higher temperatures (Crawford and Palin 1981). The high respiratory activity of the northern species matched the rapid growth rate of these species in early spring. It was also shown to require a significant proportion of the carbohydrate reserves of the overwintering roots. If the Arrhenius plot had been a straight line then the draw-down of summer carbohydrate content would have been very high. Due to the change in the gradient of the Arrhenius plot the energy of activation above 20°C was always lower for the northern species. It was suggested from these studies that it was the inability of the northern species to conserve over-wintering carbohydrate supplies in mild winters that accounted for their inability to colonise more southerly coastlines (figures 7-9).

A more extensive study during the growing season in North America compared low-latitude and low-elevation woody perennial species with congeneric high-latitude and high-elevation species (Criddle et al. 1994). Distinct differences were found in the temperature coefficient of metabolism (the slope of the Arrhenius plot) of growing meristems from woody perennial plants from high latitudes and high elevations and closely related low-latitude and low-elevation plants. Low-latitude and low-elevation woody perennials had Arrhenius activation energies for metabolism that were larger than those for congeneric high-latitude and high-elevation plants.

Figure 7 *Ligusticum scoticum* growing on a beach in Orkney (59°N)

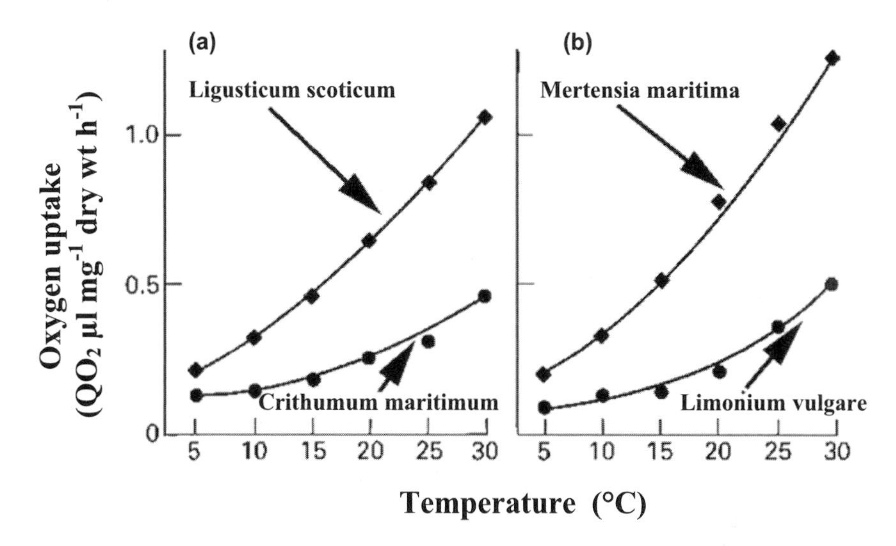

Figure 8 Over-wintering root respiration rate at a range of temperatures (a) *Ligusticum scoticum* , a northern species and *Crithmum maritimum* a southern species (b) *Mertensia maritima* a northern species, and *Limonium vulgare* a southern species (data from Crawford and Palin 1981)

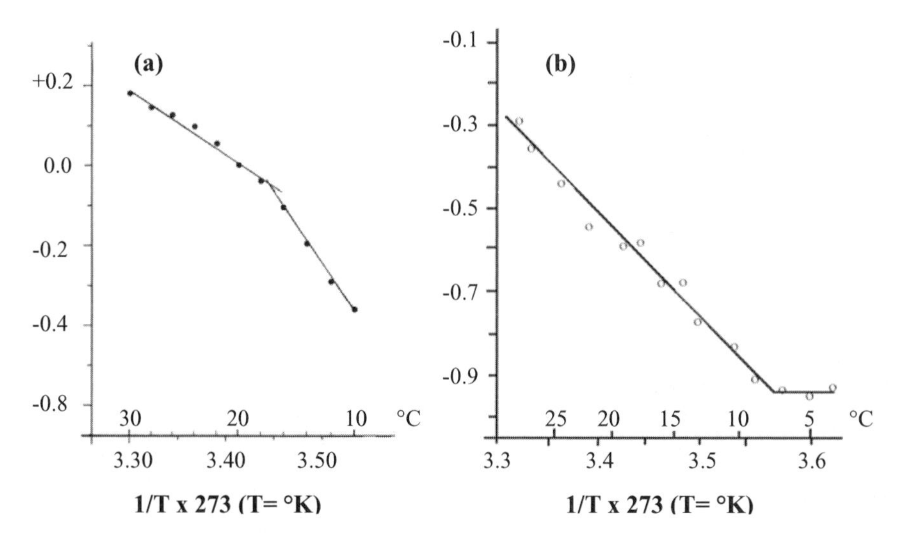

Figure 9 Arrhenius plots of root respiration rate (oxygen uptake - see above) in relation to temperature (a) *Ligusticum scoticum* and (b) *Crithmum maritimum* (data from Crawford and Palin 1981)

In the American study it was apparent that the Arrhenius activation energy does not adapt rapidly to new environments. Within taxa, woody plants that experience broader ranges of temperature during the growing season in their native habitat have smaller activation energies which also varied with growth stage or season. However, in this study no similar relationship was found for estimations made on the young leaves, annuals or herbaceous perennials. For the plants tested, Arrhenius activation energies are high during early spring growth, but changed daily to lower values as the growing season progressed. The shift in Arrhenius activation energies occurs early in the season for southern and low-elevation plants and progressively later for plants from further north or higher elevation and result largely from increases in plant metabolic rates at lower temperatures while little change occurs in the rates at higher temperatures. Altering the temperature dependence of the control of metabolic rate is apparently an important means of response to climate change for woody plants (Criddle et al. 1994).

3 Temperature Variation and the Location of Treelines

Throughout the world trees eventually reach either a polar or else an altitudinal limit for the establishment of closed forest. Many studies have attempted to relate this sometimes very definite line (figure 10) to a range of temperature measurements (Körner 1998; Holtmeier 2000; 2003). Köppen's rule (Köppen 1931) which relates the limit to tree growth as coinciding with the 10°C isotherm has been applied widely, particularly in the Northern hemisphere. In North America, the median July position of the polar front approximates both to the boreal treeline and the 10°C July isotherm and therefore provides a meteorological discontinuity which matches Köppen's rule (Bryson 1966). Whether or not the position of the treeline is caused by the polar front or *vice versa* that the polar front tends to settle on the treeline is not altogether certain. It is possible that the preferred position of the polar front is along the treeline due to the lower albedo of the forest contributing to a greater heating as contrasted with the weaker heating and higher albedo over the adjacent tundra (Pielke and Vidale 1995). A world-wide study looking at numerous thermal indicators found a growing season mean temperature of 6-7°C provided the best generalised indicator of montane treelines from the tropics to the boreal zone (Körner 1999). This modification from a measurement indicating maximum warmth to a temperature mean for the entire growing season reflects a realisation that the latitudinal and altitudinal limits to tree growth are not directly related to the ability to make a net carbon gain. Instead, the treeline

Figure 10 A natural treeline in Patagonia. This well-defined and natural deciduous treeline of the Southern Beech (*Nothofagus pumilo*) in south-western Patagonia extends north-south along the Andes from 36°S to the Tierra del Fuego - a distance of over 2000 km

is more likely to be related to the length of growing season that is needed for the production and development of new tissues (Körner 1999).

Despite the attractive convenience of the concept of a generalised temperature limit for tree growth it is important to take note of the fact that mean temperatures do not exist in nature and therefore should be considered as indicators and not causal factors (Holtmeier 2000; 2003). The same is true for mean soil temperatures. As Holtmeier has argued, it is in the variability of climatic conditions particularly in relation to extreme events, e.g. late and early frosts, drought, snow-rich and snow-poor winters that the controlling effects of temperature exert their ecological influence on limits to tree growth. It is also possible that in some regions of the world the position of the treeline is controlled genetically. It has long been noted when comparisons are made between treelines in the northern and southern hemisphere that the austral timberlines are generally found at lower altitude limits than those of boreal forests. It has sometime been considered that the vast expanse of the oceans in relation to landmass increases the exposure factor in the southern hemisphere to a greater degree at high latitudes than in the northern hemisphere. However, the upper limit for tree growth in the native southern-hemisphere tree species may be due to genetic limitations (see Holtmeier 2000; 2003). In New Zealand, introduced conifers can elevate the tree line by 300 m (Wardle 1986) which suggests that when similar species are compared in the two hemispheres then the altitudinal limitations to tree growth in the austral forests are not as great as has been supposed.

It would appear that despite many attempts, finding a common causal basis for the position of the treeline has proved elusive. This is not surprising given that treelines are frequently only approximate limits to the altitudinal or latitudinal distribution of scattered but upright trees (Sveinbjörnsson 2000). The degree of scatter can vary, and in parts of Russia the development of a mosaic of vegetation at the tundra-taiga interface is particularly noticeable, causing many Russian ecologists to doubt if there is even an approximate treeline that can be mapped with any degree of precision. The basis of the difficulty that forest ecologists have in agreeing on the position of a treeline can be seen in satellite images for North America and Siberia recording the distribution of evergreen vegetation by the Normalised Difference Vegetation Index (NDVI) as seen in May 1998 (figures 11-12). Given these uncertainties in relating thermal indicators to the position of the treeline there is clearly scope for further exploration of this phenomenon. Here again the application of calorimetry may be able to make a contribution to the debate, particularly in assessing the effect of short-term temperature variability during the growing season on 'metabolic use efficiency'. As has been demonstrated, 'metabolic use efficiency' is highly variable both in relation to populations, seasons and the organ being studied and may increase or decrease with rise in temperature but must decrease as temperature variability increases (Hansen

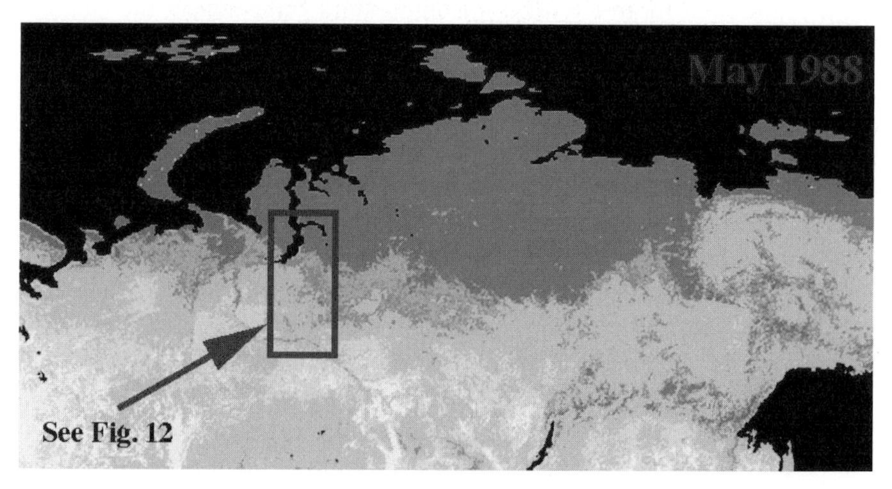

Figure 11 Northern limits to the Eurasian Boreal Forest. Normalised Difference Vegetation Index (NDVI) image recorded in May 1998 (8 km resolution). Colour scale: black = 0; dark grey = 0.48-0.64; light grey = 0.76-0.88

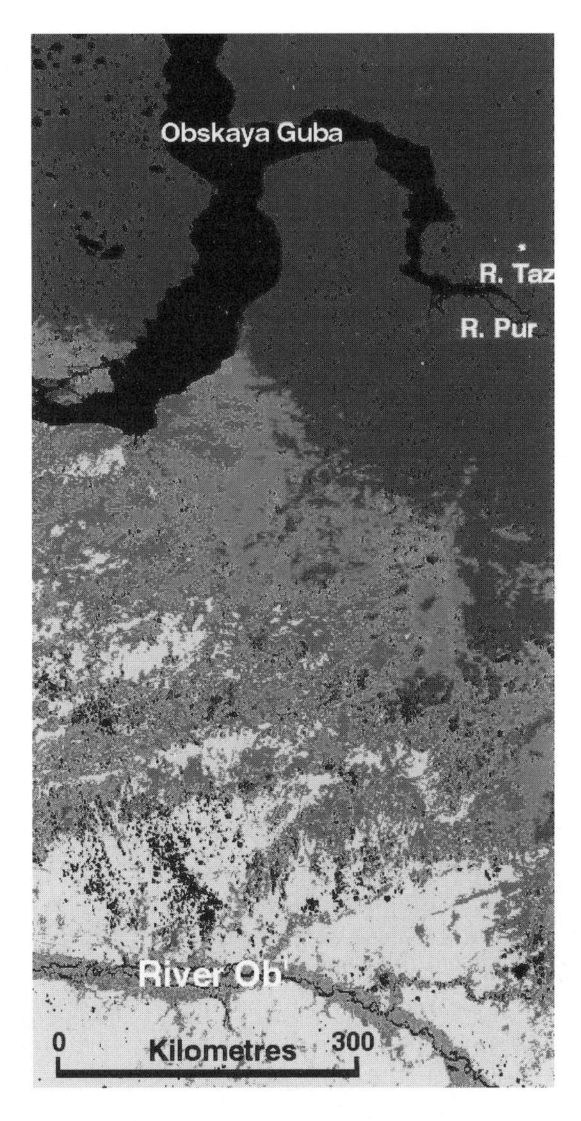

Figure 12 Detail of the transition zone between Forest and Tundra as seen in the normalised vegetation index recorded in May 1998 (1 km resolution). Colour scale black = 0; dark-grey = 0.11-0.25; grey 0.26- 0.40; bright grey = 0.41-0.64. It is considered that this mosaic represents a self-renewing cyclic process taking place over hundreds of years as patches of forest develop on land that dries out after being raised by frost-heave and then reverts again to bog as tree-cover cools the underlying ground

et al. 2002). Thus, as temperature variability (ΔT) increases, even while maintaining a common mean kinetic temperature, it should be expected that growth will decrease. To test this hypothesis a study was made of fifteen year old *Pinus ponderosa* tree plantations growing at four different elevations in the Sierra Nevada Mountains of California. Due to their phenological similarity spring growth was triggered only when daily temperatures reached a similar minimum value which occurred later in populations growing at higher altitudes. Growth therefore occurred at similar temperatures irrespective of altitude or topography. As temperature variability (ΔT) increases with altitude it is possible to compare growth with a number of environmental variables including ΔT. For three of the populations under study (figure 13) no other environmental variable correlated as well with growth rates as (ΔT). The exception at Inskip was due to greater snow fall and branch breakage and seed that was collected from a lower elevation (Hansen et al. 2002).

It follows from the above that there is no reason to expect that there should be a definable threshold temperature that can be related to the position of the treeline. It is more probable that climatic variation together with the occurrence of extreme events combine to create over a period of time a zone above which the probability of tree survival is much reduced. The nature of the habitat (oceanic or continental), the plant communities at the tundra taiga interface (wetland or dryland) and respiratory efficiency as affected by short-term temperature variations during the growing season may all play a role in the reducing tree survival. Thus although the treeline is a widespread and common phenomenon over wide geographic areas the factors controlling its position may vary greatly from one region to another.

It may at first seem counter intuitive, but there is even an argument for suggesting that in oceanic areas climatic warming may lead to a retreat rather than an advance of the treeline. Examination of temperature variations over the past century for Europe and the Arctic from Northern Norway to Siberia suggests that variations in the North Atlantic Oscillation are associated with an increase in oceanicity in certain maritime regions. A southward depression of the treeline in favour of wet heaths, bogs and wetland tundra communities is also observed in several northern oceanic environments. The heightened values currently detected in the North Atlantic Oscillation Index, together with rising winter temperatures, and increased rainfall in many areas in Northern Europe, presents an increasing risk of paludification with adverse consequences for forest regeneration, particularly in areas with oceanic climates. Climatic warming in oceanic areas may increase the area covered by bogs and contrary to general expectations may lead to a retreat rather than an advance in the northern limit of the boreal forest (Crawford et al. 2003).

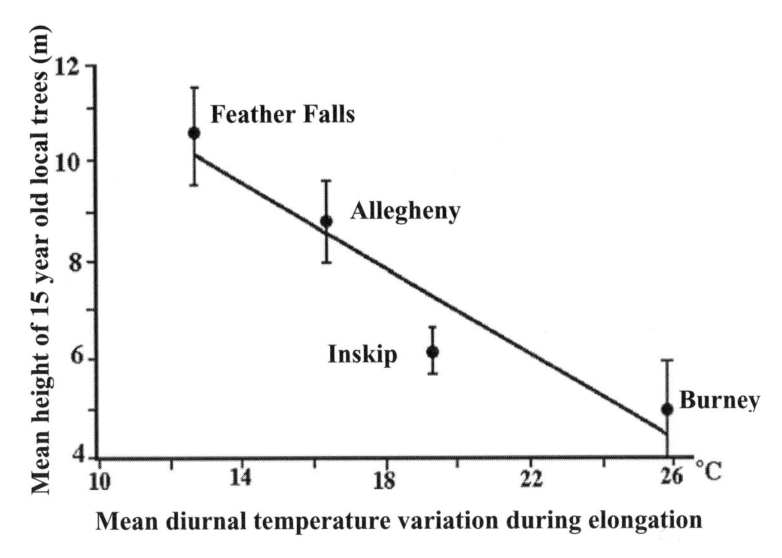

Figure 13 Fifteen year height of local *Pinus ponderosa* trees plotted vs. mean diurnal temperature variation during the shoot elongation period at four different elevations in the Sierra Mountains of California (reproduced with permission from Hansen et al. 2002)

4 Impact of Extreme Temperature Events

The above studies illustrate the wide variety of metabolic responses that can be found in individual plants in relation to the affects of temperature on respiration. If the study of temperature effects is extended to the impact of extreme events on plant populations further examples can be found in plant-life-strategies that exploit the role of ecotypes in adapting species to withstand severe climatic oscillations. At high latitudes cold winters are the norm. An extreme event therefore occurs when warm weather from the south intrudes into polar regions and rain instead of snow falls on frozen ground and turns to ice. Alternatively, warm weather can cause snow to melt which then refreezes as ice. In recent years climatic warming has caused ice-encasement to become a significant hazard to crops and pastures in many northern countries (Bertrand et al. 2001). Ice-encasement of the Tundra is a sporadic stress, disastrous for herbivores such as reindeer, and places much of the low-lying vegetation under a barrier to oxygen diffusion which can last from November until June. Temperatures are low under ice and metabolism will therefore be reduced. However, this does not mean that low-temperature anoxia is not dangerous for plants. Throughout the period under ice there is a total deprivation of oxygen and all aerobic metabolism ceases. After months of encasement in ice, the arrival of spring and snow melt exposes plants to the full oxygen concentration of the air, often within a matter of hours. Plants that have survived months of oxygen-deprivation then face the additional hazard of post-anoxic injury. Tissues prone to this type of injury become soft and spongy, and lose cell constituents. In these cases, oxygen exerts a definite toxic effect and membranes are destroyed irreversibly. The reason for this type of damage is the formation of **ROS** (= reactive oxygen species: superoxide anions $\mathbf{O_2^-}$ and **OH** radicals as well as hydrogen peroxide ($\mathbf{H_2O_2}$)). Ethanol that has accumulated under anoxia is also rapidly oxidised to acetaldehyde by the presence of $\mathbf{H_2O_2}$ and catalase and therefore probably contributes more to membrane damage at the post-anoxic stage than during anoxia (Braendle and Crawford 1999; Bertrand et al. 2001). Post-anoxic injury is avoided, also at a cost which includes provision of anti-oxidants and the maintenance of an enzymatic system capable of detoxifying reactive oxygen species.

Testing arctic plants for anoxia-tolerance reveals that many species are tolerant of prolonged anoxia, even during the growing season (table 1). However, it is probable that only a half to one third of all arctic species are anoxia-tolerant. The genus *Eriophorum* is anoxia tolerant in Arctic but not further south. Similarly, populations of *Saxifraga oppositifolia* sampled in

Table 1 List of species tested in Spitsbergen for anoxia tolerance. Intact plants were kept under total anoxia in anaerobe jars for 7 days at 5°C in the dark. Plants were judged to be anoxia-tolerant if after 48h re-exposure to air they showed no wilting or discolouration of their shoots

ANOXIA TOLERANT (INCLUDING LEAVES)	ANOXIA INTOLERANT
Cardamine nymani	*Cochlearia groenlandica*
Carex misandra	*Equisetum arvense*
Cochlearia groenlandica	*Oxyria digyna*
Eriophorum scheuchzeri	*Polygonum viviparum*
Huperzia selago	*Ranunculus pygmaeus*
Juncus biglumis	*Saxifraga cernua*
Luzula arctica	*S. hieracifolia*
L. arcuata ssp. confusa	
Puccinellia vahliana	
Ranunculus sulphureus	
Saxifraga caespitosa	
S. foliosa	
S. oppositifolia	
Salix polaris (leaves not tolerant)	

Spitsbergen are tolerant of total anoxia while those tested in Scotland die within 4 days of the imposition of total anoxia (Crawford et al. 1994). These variations in the frequency of anoxia tolerance suggest that there must be a hidden disadvantage in having this adaptation. When high latitude plants are kept under prolonged anoxia it is notable that after a certain length of time they suddenly lose their ability to survive oxygen deprivation which suggests that some essential constituent for protection against anoxia has been exhausted (figure 14).

Tolerance of anoxia is therefore expensive, particularly in the use carbohydrate reserves that have to be conserved to support over-wintering anaerobic metabolism, resumption of growth in spring, and provision of the anti-oxidants for defence against post-anoxic injury. In regions with limited resources, it would be advantageous for those species that inhabit areas such as slopes and ridges where ice-encasement is less likely to occur not to use their limited over-wintering reserves for protection against ice-encasement and its associated anoxic and post-anoxic stresses.

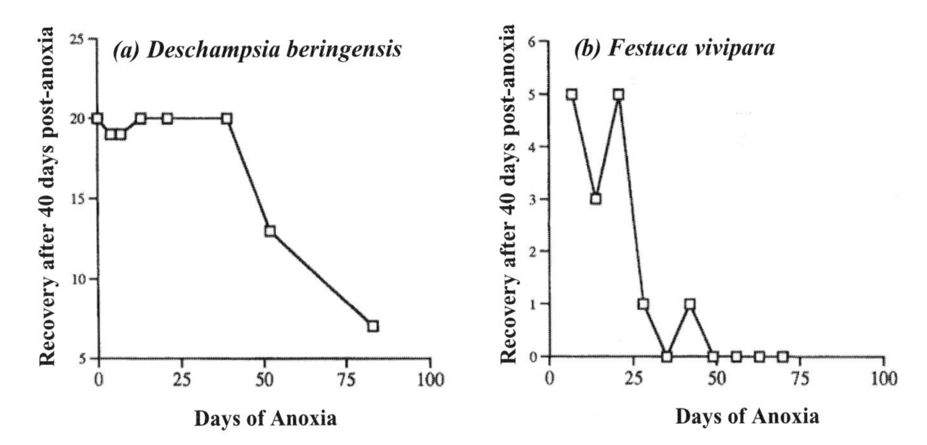

Figure 14 Comparison of anoxia tolerance in grass seedlings of (a) an artic population of *Deschampsia beringensis* from Alaska and (b) a subarctic population of *Festuca vivipara* from Iceland. The seedlings were grown in a cold room at 10°C using seed for *D. beringensis* and clonal pseudo-viviparous plantlets for *F. vivipara*. The y axis records the number of seedlings surviving after allowing over a month for post-anoxic recovery following different lengths of dark anoxic incubation at 5°C

5 Habitat Preferences in Arctic Plant Communities and the Consequences of Incompatible Adaptations

The fact that even within such unproductive environments as the High Arctic, species and populations still exhibit well-defined habitat preferences demonstrates that there must be both positive and negative consequences in the possession of specific adaptations. It has been said that 'adaptation is the first step on the road to extinction'. Adaptation increases habitat specialisation and therefore makes species vulnerable to environmental change. As the Arctic has one of the most variable climatic histories in the entire World there is therefore a risk that some adaptations, although increasing fitness, may prove disadvantageous in the long-term. An unfortunate aspect of both morphological and physiological adaptations is the incompatibility of many adaptations to opposing conditions.

Divisive Effects of Incompatible Adaptations

Plant adaptations to stress can be divided into two categories.

(1) Adaptations that are mutually incompatible.

(2) Adaptations which can co-exist in the same individual plant.

In the first group adaptation to a particular stress may place them at a disadvantage in the unstressed or opposite condition (e.g. drought as opposed to flooding; low soil pH *versus* high soil pH)). Such adaptations are often both morphologically and physiologically incompatible (Crawford 1997; Braendle and Crawford 1999). Consequently, phenotypic plasticity is not available as an option in extending the capacity of these populations to adapt to a range of stresses. Differentiation into genetically distinct ecotypic populations for adaptation to these stresses may be the only choice available for survival. Climatic stresses that are irregular in their impact and yet can occur anywhere throughout the Arctic include: drought, flooding, ice-encasement, cryoturbation and prolonged burial under snow or ice. As these hazards are irregular in occurrence it is probable that only a certain proportion of any population may be adversely affected at any one time. Such a situation allows for the development of balanced polymorphisms where heterogeneous metapopulations can alter their gene frequencies as climatic conditions oscillate. Other potential hazards that are ever-present and ubiquitous include frost, short growing seasons and low growth temperatures. These hazards are a general feature of the arctic environment and are endured by most plants and therefore less likely to discriminate between populations.

In the second group adaptation to one particular stress confers a resistance to other adverse environmental factors. Such groups include heat and drought, as well as frost-hardiness and drought. By possessing adaptations for one particular stress the plant is pre-adapted to survive a related stress when it occurs. Thus plants that can survive long periods of anoxia will be pre-adapted to flooding and ice-encasement if they should occur.

Adaptations that are incompatible within any one individual plant can however co-exist as adjacent populations in neighbouring habitats. In the Arctic, ecotypic variants are frequently found which discriminate between warm, or cold, wet or dry, or early or late sites. The existence of these specialised populations extends the range of habitats that a species can inhabit and can be described as increasing immediate fitness. They can also be considered as increasing long-term fitness by providing a reservoir of genetic variability in neighbouring and interfertile populations which pre-adapts the species as a whole to climate change. This argument for the mutualistic advantages of ecotypic variation has been presented in a number of papers in relation to the divergent forms of *Saxifraga oppositifolia* that are found in arctic habitats (Crawford and Abbott 1994; Crawford 1997).

At high latitudes, *Saxifraga oppositifolia* exists with distinct ecotypes adapted to differences in growing season length (figure 15). There are two extreme types, one tufted with very condensed internodes which is early-flowering and is a cushion-like ecotype of dry, wind-exposed heaths and ridges. The second type is a late-flowering, prostrate ecotype of snow-protected, damp habitats (Teeri 1973). It has been suggested that these subspecies may be differentiated at the tetraploid and diploid levels, respectively, which would promote reproductive isolation. However, the two forms appear to be interfertile and as many intermediates exist there appears no justification for splitting these interbreeding ecotypes into sub-species (Brysting et al. 1996; Kume et al. 1999). The prostrate form is the usual form throughout Scandinavia where only diploid plants (2 = 26) have been found. The tufted arctic form occurs in Svalbard and North America and Greenland where tetraploids and diploids have been observed (Lid and Lid 1994). As yet there is no definitive study which has been able to correlate chromosome number with taxonomy among the many sub-species or ecotypes that occur throughout the circumpolar distribution of *S. oppositifolia* (Webb and Gornall 1989). However, a chloroplast DNA study (Abbott and Brochmann 2003) suggests that the tufted form resulting from shortening of internode length, which can be found in Alaska, Greenland, and Svalbard, has evolved separately in each locality.

In areas with late snow-lie and cold, wet soils, the creeping form exhibits high metabolic rates and rapid shoot production which allows this form to

Figure 15 Two forms of *Saxifraga oppositifolia* with incompatible survival strategies

compensate for ultra-short growing seasons but does not conserve carbohydrate or water for adverse periods. An opposing strategy is evident in the tufted ecotypes living in sites with an earlier resumption of growth, where soils are warmer and drier and the growing season longer. Here, both growth and metabolic rates are lower and result in a greater ability to conserve both carbohydrate and water (Teeri 1973). The existence of opposing strategies for survival in warm and cold habitats, suggests that even in the minimal thermal conditions of the high Arctic a high degree of population diversity gives the species as a whole a wider ecological amplitude. This degree of diversity not only increases the range of sites in which the species can survive but confers an ability to adapt to climate change by altering ecotype frequencies to accommodate climatic fluctuations. This facility may have contributed to the survival of these polymorphic populations of this and other species in the high Arctic during the Last Glacial Maximum. Figure 15 also lists a number of incompatible characters, both morphological and physiological, which have the capacity to adapt *Saxifraga oppositifolia* ecotypes to different environmental conditions.

In a review of regional and local vascular plant diversity in the Arctic (Murray 1997), attention was drawn to the observation that in the Arctic it is only common species that exhibit a broad ecological amplitude which may be due to phenotypic plasticity of the individuals, or else is the result of the species in question being an aggregation of a series of ecotypes. It is this latter possibility that is advanced in this paper as accounting for the outstanding capacity for survival in the Arctic and Subarctic over a very long time of *Saxifraga oppositifolia* and probably other common, widespread, and variable arctic species. The phenomenon by which a species maintains a range of interfertile ecotypes, instead of evolving breeding barriers as a response to competition, has been described as 'suspended speciation'(Murray 1995) and provides a population mechanism for extending the temperature range of the species as a whole. When kept under cultivation under controlled conditions in growth cabinets the two forms differed in the temperature maximum at which they could be kept without dying. The prostrate, creeping form from the wet cold soils died when the temperature in the growth cabinets rose above 6°C while the tufted form from the warmer drier beach ridges tolerated temperatures of up to 10°C (Crawford - unpublished).

6 Mutualism as Mean of Extending Temperature Tolerance

Sub-species variation has been repeatedly noted as present in many arctic species (table 2). A common explanation advanced for ecotypic variation is the need for plants to optimise their use of resources in response to competition. Thus, without implying any environmental variation, competition results in sub-specific ecotypic variation through the '*Red Queen Effect'*. Popularly interpreted, from Lewis Carroll's *Through the Looking-Glass*, this '*effect*' mirrors the need to keep running in order to remain where you are. As originally defined, it represents the environment of a taxon as constantly deteriorating because of the continual evolution of other species, including competitors, predators and parasites (Van Valen 1973). There is however an alternative interpretation for the frequency of ecotypic variation, which does not depend on competition and which has a particular aptness for the High Arctic situation where competition is minimal, and that is the increase in *long-term fitness* that comes from *mutualism*. Selection acting on individuals gives rise to population variations which as a result of competition are

Table 2 Examples of species which maintain a high biodiversity at high latitudes

Species	Observations	References
Arctagrostis latifolia	Three chromosome races within Alaska	(Mitchell 1992)
Armeria maritima	Several sub spp. in Greenland and Arctic America but only 1 sp. on Pacific coast from Vancouver to California	(Lefebvre and Vekemans 1995)
Carex bigelowii	High levels of clonal diversity (genetic diversity) within populations	(Jonsson et al. 1996)
Cerastium arcticum	High levels of genetic diversity at high latitudes with allo-polyploidisation	(Brochmann et al. 1996)
Draba spp.	Allo-polyloidisation prevalent in closely related species with minor variations in habitat preferences	(Brochmann et al. 1992)
Dryas octopetala	Exposed ridge and snow bank populations adapted to different lengths of growing season	(McGraw and Antonovics 1983)
Pedicularis lanata	Breeding system with the highest capacity for outcrossing in the study area and also shows the greatest morphological variation within its geographic distribution	(Philipp 1998)
Pinus sylvestris (arctic populations)	In populations within the Arctic Circle there is strong differentiation between populations with respect to adaptation to the very short growing season	(Savolainen 1996)
Polygonum viviparum	High levels of genetic diversity despite asexual reproduction	(Diggle et al. 1998)
Saxifraga cernua	High levels of genetic diversity despite asexual reproduction	(Gabrielsen and Brochmann 1998)
Saxifraga oppositifolia	Intra-specific variation at high latitude sites	(Abbott et al. 1995; Crawford et al. 1995)
Saxifraga cespitosa	High levels of genetic diversity at high latitudes with allo-polyploidization	(Tollefsrud et al. 1998)
Silene acaulis	Patterns of spatial variation in genetic structure within populations reflected the workings of micro-evolutionary processes	(Abbott et al. 1995 Philipp 1997)
Polygonella spp.	Species within Pleistocene refugia maintain higher levels of biodiversity	(Lewis and Crawford 1995)
Puccinellia phryganodes	High levels of clonal variation within and among populations	(Jefferies and Gottlieb 1983)

associated with specialised habitats. This increases the habitat range occupied by the species as a whole and can be considered as increasing *immediate fitness*.

Mutualism will tend to be most effective where beneficial genes can be exchanged between populations and be readily expressed without masking by competing alleles. This may help to explain why it is that some of the most widespread species in the Arctic are simple diploids (table 3). In a review of regional and local vascular plant diversity in the Arctic (Murray 1997), attention was drawn to the observation that in the Arctic it is only common species that exhibit a broad ecological amplitude which may be due to phenotypic plasticity of the individuals, or else is the result of the species in question being an aggregation of a series of ecotypes. It is this latter possibility that is advanced here as accounting for the outstanding capacity for survival in the Arctic and Subarctic over a very long time of *Saxifraga oppositifolia* and other common and variable arctic species. The phenomenon by which a species maintains a range of interfertile ecotypes, instead of evolving breeding barriers as a response to competition, has been described as 'suspended speciation' (Murray 1997) and can be considered as one of the consequences of mutualism.

Table 3 Examples of autochthonous (ancient, indigenous) species which are widespread in the Arctic as listed by Tolmatchev (1960)

Saxifraga oppositifolia	(2n = 26)
S. hyperborean	(2n = 26)
S. tenuis	(2n = 20)
Ranunculus pygmaeus	(2n = 16)
Dryas octopetala	(2n = 18)
Loiseleuria procumbens	(2n = 24)
Cassiope tetragona	(2n = 26)
Diapensia lapponica	(2n = 12)

7 Polyploidy and Adapting to Climate Change

Using this distinction between long-term and short-term residence in the Arctic it is possible to re-examine other properties that may be associated with survival in the low temperature conditions at high latitudes. One characteristic that deserves a re-examination is polyploidy which is particularly notable in arctic species. Over 80 per cent of the flora of Svalbard is polyploid (Abbott and Brochmann 2003). From such observations it is frequently concluded that polyploid species are better adapted physiologically to cold climates than diploid species. However, as noted above the exceptions to this habit are the autochthonous (ancient and indigenous) species which are usually diploid or occasionally tetraploid (table 3). Consequently, it can be argued that a high level of polyploidy, although common at high latitudes, is not necessarily an obligatory adaptation for the arctic environment. The presence of so many polyploid species in the Arctic could merely be the consequence of plant migrations caused by periods of cooling and warming with the climatic disturbance bringing species into proximity in unusual combinations (Stebbins 1971). The importing of genetic material by migration of polyploids as it might occur when a single long-distance seed dispersal colonised a deglaciated area would automatically carry the entire gene pool of the seed's ancestors (Brochmann and Steen 1999). However, even allowing for the fact that the heterozygosity of polyploids is not necessarily fixed, there will nevertheless be a buffering effect which will always retard the expression of any exchanged genetic information among polyploids. In the case of deleterious genes this may be beneficial but it would also mask the effect of helpful genes.

Most adaptations when closely examined have both positive and negative aspects in relation to survival potential. Species that are drought-tolerant are usually not flood-tolerant. Likewise, floras can be divided ecologically into various contrasting types such as calcicole and calcifuges, glycophytes and halophytes etc. These divergences seems so natural that it is not often questioned why some plants are tolerant of one particular stress yet are intolerant of the converse situation. Plants that are adapted to flooded soils, where iron toxicity is a potential hazard, exclude iron and usually have an oxidising regime at their root surface. This in turn makes them mal-adapted to dry soils where a reducing regime at the root surface is necessary for the solubilisation and absorption of iron. Thus, in relation to the acquisition of iron (an essential element) flooding tolerance is not compatible with drought tolerance.

The question therefore has to be asked if there are similar distinctions between polyploid and diploid species in relation to survival fitness in the Arctic? It has never been demonstrated that polyploid species are more cold-hardy or better adapted to short growing seasons than diploid species. The genetic advantages of polyploidy include restoring fertility to hybrids, and the ability to form new species within the home range of their parents. Polyploidy also has obvious advantages in masking the effects of harmful alleles. More questionable however, is the assertion that polyploids benefit from have a greater store of genetic variability merely because there are more alleles at any one locus. If harmful alleles are masked by polyploidy, then the same must also be true for helpful alleles. In polyploid species mutations are less likely to be expressed and phenotypic variability is likely to be low. In diploid species neighbouring populations with different environmental preferences will have the advantage of being able to profit more readily than polyploids from gene flow, as interchange of alleles from one population to another is less likely to be masked. It is therefore perhaps no coincidence that the ancient (autochthonous) arctic species (see table 3) are mainly diploid (or else have only a low level of polyploidy) as they can adapt readily to climatic change by exchanging readily expressed genes between different ecotypes of the same species. The manner in which mutualism (see above) rather than competition between ecotypes can contribute to species survival will function only if genetic exchange leads to readily expressed adaptations. This is more likely to take place between adjacent diploid than polyploid populations.

Conflicts between adaptations at the physiological, biochemical and morphological level inevitably limit the choice of adaptations that can be expressed by any individual plant. No matter how much genetic information is present within the genome of any individual it is of little use if it cannot be expressed. It follows therefore, that polyploid populations because of their genetic stability are less likely to be able to adapt to changing temperatures (or any other environmental conditions) by exchanging small amounts of genetic material. In short, the adaptive significance of polyploidy, in contributing to physiological fitness in general, or the low temperature regime of the Arctic in particular, has never been proven.

8 Conclusions

In a low thermal environment minor variations in habitat create biologically significant temperature differences. Many of these differences are apparent at a local level and are not amenable to large scale geographic mapping. Plants are highly adaptive in the range of ecotypes that can evolve in relation to minor changes in the habitat and this trait appears well developed even in the species-poor regions of the High Arctic. In the thermally limited, and fluctuating, growing seasons of the polar regions it should be expected that there is a need for maximising metabolic efficiency. Increasing metabolic efficiency creates specialisation and dependence on particular environments as many physiological adaptations, and particularly those associated with temperature, are highly specific to particular thermal regimes. Any one individual plant is therefore limited in the range of adaptations that can be expressed as many physiological and morphological adaptations are mutually incompatible. Consequently, diversity is achieved only if there is genetic flexibility and a population structure that can preserve a range of varying adaptations in adjacent populations. Exchange between ecotypes of genetic information that can readily be expressed will take place most readily between diploid plants. The state of 'suspended speciation' that can be seen in many of the ancient and indigenous (authochthonous) arctic species may reflect a superior ability of diploid species to respond rapidly to temperature environmental changes. The ecotypic richness of the arctic flora should therefore not be viewed as a product of competition but rather as a mutualistic response which increases the genotypic flexibility of the species.

Acknowledgements

I am indebted to Dr Lee Hansen for helpful advice on the role of micro-calorimetry in assessing metabolic efficiency and to Drs A Tobin and RJ Abbott for reading an earlier version of this paper.

References

Abbott RJ, Brochmann C (2003) History and evolution of the arctic flora: in the footsteps of Eric Hultén. Molecular Ecology 12: 299-313

Abbott RJ, Chapman HM, Crawford RMM, Forbes DG (1995) Molecular diversity and derivations of populations of *Silene acaulis* and *Saxifraga oppositifolia*. from the high Arctic and more southerly latitudes. Molecular Ecology 4: 199-207

Arft AM, Walker MD, Gurevitch J, Alatalo JM, Bret-Harte MS, Dale M, Diemer M, Gugerli F, Henry GHR, Jones, MH, Hollister RD, Jonsdottir IS, Laine K, Levesque E, Marion, GM, Molau U, Molgaard P, Nordenhall U, Raszhivin V, Robinson CH, Starr G, Stenström A, Stenström M, Totland O, Turner PL, Walker LJ, Webber PJ, Welker JM and Wookey PA (1999) Responses of tundra plants to experimental warming: Meta-analysis of the international tundra experiment. Ecological Monographs 69: 491-511

Atkinson D (1968) The energy charge of the adenylate pool as a regulatory parameter: interaction with feedback modifiers. Biochemistry 7: 4030-4034

Atkinson TC, Briffa KR and Coope GR (1987) Seasonal temperatures in Britain during the past 22.000 years, reconstructed using beetle remains. Nature 325: 587-592

Barcroft J (1934) Features in the architecture of physiological function Cambridge University Press, Cambridge

Bertrand A, Castonguay Y, Nadeau P, Laberge S, Rochette P, Michaud R, Belanger G and Benmoussa M (2001) Molecular and biochemical responses of perennial forage crops to oxygen deprivation at low temperature. Plant Cell and Environment 24: 1085-1093

Braedle R and Crawford RMM (1999) Plants as amphibians. Perspectives in Plant Evolution and Systematics 2: 56-78

Breidenbach RW, Saxton MJ, Hansen LD and Criddle RS (1997) Heat generation and dissipation in plants: Can the alternative oxidative phosphorylation pathway serve a thermoregulatory role in plant tissues other than specialized organs? Plant Physiology 114: 1137-1140

Brochmann C and Steen SW (1999) Sex and genes in the flora of Svalbard - implication for conservation biology and climate change. In: Nordal I and Razzhivin VY (eds) The species concept in the high north - a panarctic flora initiative, The Norwegian Academy of Science and Letters: Vol. I. Mat-Naturv. Klasse Skrifter, Ny Serie No. 38, Oslo, pp 33-72.

Brochmann C, Gabrielsen TM, Hagen A and Tollefsrud MM (1996) Seed dispersal and molecular phylogeography: glacial survival, tabula rasa or does it really matter? Det Norske Videnskaps-Akademi I. Matematisk-Naturvitenskapelig Klasse Ny Serie 18: 54-68

Brochmann C, Soltis DE and Soltis PS (1992) Electrophoretic relationships and phylogeny of Nordic polyploids in Draba (Brassicaceae). Plant Systematics and Evolution 182: 35-70

Bryson, RA (1966) Air masses, streamlines, and the boreal forest. Geographical Bulletin 8: 228-269

Brysting AK, Gabrielsen TM, Sørlibråten O, Ytrehorn O and Brochmann C (1996) The Purple Saxifrage, *Saxifraga oppositifolia*, in Svalbard: Two taxa or one? Polar Research 15: 93-105

Callaghan TV, Werkman BR and Crawford RMM (2002) The tundra-taiga interface and its dynamics: Concepts and applications. Ambio, 6-14

Cooper EJ, Alsos IG, Hagen D, Smith FM, Coulson S and Hodgkinson I (2004) Plant recruitment in the High Arctic: seedbank and seedling emergence on Svalbard. Journal of Vegetation Sciences 15: 115-124

Crawford RMM (1997) Habitat fragility as an aid to long term survival in arctic vegetation. In: Woodin JS and Marquiss M (eds) Ecology of Arctic Environments, Special Publication No.13 of the British Ecological Society, Blackwell Scientific Ltd Oxford, pp 113-136

Crawford RMM and Abbott RJ (1994) Pre-adaptation of Arctic plants to climate change. Botanica Acta 107: 271-278

Crawford RMM and Palin MA (1981) Root respiration and temperature limits to the north south distribution of four perennial maritime species. Flora 171: 338-354

Crawford RMM, Chapman HM and Hodge H (1994) Anoxia tolerance in high Arctic vegetation. Arctic and Alpine Research 26: 308-312

Crawford RMM, Chapman HM and Smith LC (1995) Adaptation to variation in growing season length in Arctic populations of *Saxifraga oppositifolia* L. Botanical Journal of Scotland 41: 177-192

Crawford RMM, Jeffree CE and Rees WG (2003) Paludification and forest retreat in northern oceanic environments. Annals of Botany 91: 213-226

Criddle RS, Hopkin MS, McArthur ED and Hansen LD (1994) Plant-distribution and the temperature-coefficient of metabolism. Plant Cell and Environment 17:233-243

Dahl E and Mork E (1959) Om sombandet mellom temperatur, ånding og vekst hos gran (*Picea abies* (L.) Karst.). Meddelser Det NorskeSkogforsøksvesen 53: 82-93

Dahl E (1998) The Phytogeography of Northern Europe. Cambridge University Press, Cambridge

Diggle PK, Lower S and Ranker TA (1998) Clonal diversity in alpine populations of *Polygonum viviparum* (Polygonaceae). International Journal of Plant Sciences 159: 606-615

Gabrielsen TM and Brochmann C (1998) Sex after all: high levels of diversity detected in the arctic clonal plant *Saxifraga cernua* using RAPD markers. Molecular Ecology 7: 1701-1708

Gonzalez-Meler MA, Ribas-Cabo M, Giles L and Siedow JN (1999) The effect of growth and measurement temperature on the activity of the alternative respiratory pathway. Plant Physiology 120: 765-772

Hansen LD, Church JN, Matheson S, McCarlie VW, Thygerson T, Criddle RS and Smith BN (2002) Kinetics of plant growth and metabolism. Thermochimica Acta 388: 415-425

Holtmeier F-K (2000) Die Höhengrenze der Gebirgswälder. Arbeiten aus dem Institut für Landschaftsökologie, Band 8. Westfälische Wilhelms-Universität, Münster

Holtmeier F-K (2003) Mountain Timberlines – Ecology, patchiness, and dynamics. Advances in Global Change Research, 14. Kluwer Academic, Dordrecht

Hultén E and Fries M (1986) Atlas of Northern European vascular plants. Koeltz Scientific Books, Königstein

Huntley B and Birks HJB (1983) An Atlas of past and present pollen maps for Europe: 0-13000 years ago. Cambridge Univ. Press

ITEX: www.itex-science.net

Jefferies RL and Gottlieb LD (1983) Genetic variation within and between populations of the asexual plant *Puccinellia x phryganodes*. Canadian Journal of Botany 61: 774-779

Jonsdottir IS Virtanen R and Karnefelt I (1999) Large-scale differentiation and dynamics in tundra plant populations and vegetation. Ambio 28: 230-238

Jonsson BO Jonsdottir IS and Cronberg N (1996) Clonal diversity and allozyme variation in populations of the arctic sedge *Carex bigelowii* (Cyperaceae). Journal of Ecology 84: 449-459

Köppen W (1931) Grundriss der Klimakunde. 2 edn Walter Gruyter, Berlin

Körner C (1998) A re-assessment of high elevation treeline positions and their explanation. Oecologia 115: 445-459

Körner, C (1999) Alpine Plant Life. Springer Verlag, Berlin

Kume A, Nakatsubo T, Bekku Y and Masuzawa T (1999) Ecological significance of different growth forms of purple saxifrage, *Saxifraga oppositifolia* L., in the High Arctic, Ny- Alesund, Svalbard. Arctic Antarctic and Alpine Research 31: 27-33

Lefebvre C and Vekemans X (1995) A numerical taxonomic study of *Armeria maritima* (Plumbaginaceae) in North-America and Greenland. Canadian Journal of Botany 73: 1583-1595

Lewis PO and Crawford DJ (1995) Pleistocene refugium endemics exhibit greater allozymic diversity than widespread congeners in the Genus *Polygonella* (Polygonaceae). American Journal of Botany 82: 141-149

Lid J and Lid DT (1994) Norsk Flora, 6th Edition (ed. Elven R), Det Norske Samlaget: Oslo

Löve A and Löve D (1975) Cytotaxonomical atlas of the arctic flora. J. Cramer, Lehre

McGraw JB and Antonovics J (1983) Experimental ecology of *Dryas octopetala* ecotypes. I. Ecotypic differentiation and life cycle stages of selection. Journal of Ecology 71: 879-897

McNulty AK, Cummins WR and Pellizari A (1988) A field survey of respiration rates in leaves of arctic plants. Arctic 41: 1-5

Meusel H and Jäger EJ (1992) Vergleichende Chorologie der Zentraleuropäischen Flora. Gustav Fischer Verlag, Jena

Mitchell WW(1992) Cytogeographic races of *Arctagrostis latifolia* (Poaceae) in Alaska. Canadian Journal of Botany 70: 80-83

Murray DF (1995) Causes of arctic plant diversity: origin and evolution. In: Chapin FS and Körner C (eds) Arctic and Alpine Biodiversity: Patterns, Causes and Ecosystem Consequences, Springer, Heidelberg, pp 21-32

Murray DF (1997) Regional and local vascular plant diversity in the Arctic. Opera Botanica 132: 9-18

Phillip M (1997) Genetic diversity, breeding system, and population structure in *Silene acaulis* (Caryophyllaceae) in West Greenland. Opera Botanica 132: 89-100

Phillip M (1998) Genetic variation in four species of *Pedicularis* (Scrophulariaceae) within a limited area in West Greenland. Arctic and Alpine Research 30: 396-399

Pielke RA and Vidale PL (1995) The Boreal Forest and the Polar Front. Journal of Geophysical Research-Atmospheres 100: 25755-25758

Pradet A and Bomsel JL (1978) Energy metabolism in plants under anoxia. In: Hook DD and Crawford RMM (eds) Plant life in anaerobic environments, Ann Arbor Science pp 89-118

Preston CD, Pearmann A and Dines TD (eds) (2002) New atlas of the British and Irish flora. Oxford University Press, Oxford

Savolainen O (1996) Pines beyond the polar circle: Adaptation to stress conditions. Euphytica 92: 139-145

Schulze E-D (1982) Plant life forms and their carbon, water and nutrient relations. In: Lange OL, Nobel PS, Osmond CB and Ziegler H (eds) Physiological Plant Ecology II, Vol. New Series 12B, Springer-Verlag, Berlin, pp 615-676

Smith BN, Jones AR and Hansen LD (1999) Growth, respiration rate and efficiency responses to temperature. In: Pessarakli M (ed) Handbook of plant and crop stress (ed), Marcel Dekker, New York, pp 417-440

Stebbins GL (1971) Chromosomal evolution in higher plants. Edward Arnold, London

Sveinbjörnsson B (2000) North American and European treelines: External forces and internal processes controlling position. Ambio 29: 388-395

Teeri JA (1973) Polar desert adaptions of a high Arctic plant species. Science 179: 496-497

Thygerson T, Harris JM, Smith BN, Hansen LD, Pendleton RL and Booth DT (2002) Metabolic response to temperature for six populations of winterfat (*Eurotia lanata*). Thermochimica Acta 394: 211-217

Tollefsrud MM, Bachmann K, Jakobsen KS and Brochmann C (1998) Glacial survival does not matter - II: RAPD phylogeography of Nordic *Saxifraga cespitosa.* Molecular Ecology 7: 1217-1232

Tolmatchev AI (1960) Der autochthone Grundstock der arktischen Flora und ihre Beziehungen zu den Hochgebirgsfloren Nord- und Zentralasiens. Botanisk Tidsskrift 55: 269-276

Van Valen L (1973) A new evolutionary theory. Evolutionary Theory 1: 1-30

Wardle P (1986) Alpine vegetation of New Zealand: a review. In: Barlow B (ed) Flora and fauna of Australasia, ages and origins. CSIRO, Australia

Webb DA and Gornall RJ (1989) Saxifrages of Europe. Christopher Helm, London

Climate Change and High Mountain Vegetation Shifts

Gian-Reto Walther, Sascha Beißner and Richard Pott

Abstract

In the 20th century, the global climate has warmed about 0.6 K. High-mountain areas as well as areas of high latitudes are experiencing even greater increases in temperature especially in the last half century. With changing climatic conditions, the determinants of global, and in particular, altitudinal distribution of plants and plant communities are likely to change and a subsequent reaction of climate sensitive species and ecosystems is expected. The following paper focuses on observed climate-induced changes in the two uppermost altitudinal vegetational ecotones at the treeline and the upper limit of plant life at the alpine-nival transition zone.

1 Introduction

The world is experiencing a period of climate warming with an increase in the global mean surface temperature over the last century by 0.6 ± 0.2 K (Houghton et al. 2001). However, this average increase is neither temporally nor spatially uniform, so that areas with above average warming co-occur with areas with minor or no change or even slight cooling (see e.g. Folland and Karl 2001). In particular, high-mountain areas as well as areas of high latitudes belong to those regions with greatest increase in temperature in the last half century. The increase in average temperature of 0.5 K/decade since 1971 in the European alpine area amounts threefold that of the global trend (cf. OcCC 2002; see also Beniston et al. 1994; 1997). Trends of similar magnitude are also recorded in the Arctic (e.g. Przybylak 2003) and parts of Antarctica, especially on the Antarctic Peninsula (Vaughan et al. 2001; cf. also Turner et al. 2002; Walsh et al. 2002).

It is generally agreed that in many areas climatic conditions shape the ranges and distribution of species and thus the composition of biomes (e.g. Box 1981; 1996; Walter and Breckle 1983ff.; Emanuel et al. 1985; Woodward 1987; Prentice et al. 1992). With a change in the climatic conditions, these determinants are also likely to change and a subsequent reaction of climate sensitive species and ecosystems is expected (e.g. Huntley et al. 1995; Sykes et al. 1996; Iverson and Prasad 1998; Kappelle et al. 1999; Saxe et al. 2001; Theurillat and Guisan 2001). In the following, our attention is drawn to the two uppermost altitudinal vegetational ecotones, i.e. the treeline and the upper limit of plant life. Historical data are compared with the present situation and shifts in vegetation structure and composition are analysed and discussed in the context of recent climate change.

2 Climate Change and Contemporary Treeline Dynamics

The transient from woodland to grassland is visually one of the most striking vegetation boundaries (figure 1). Many research projects have focused treeline ecotones on both, latitudinal and altitudinal gradients (see reviews by e.g. Brockmann-Jerosch 1919; Hermes 1955; Wardle 1974; Tranquillini 1979; Stevens and Fox 1991; Miehe and Miehe 1994; Rochefort et al. 1994; Körner 1998b; Holtmeier 2000; 2003; Sveinbjörnsson 2000; Burga and Perret 2001; Malanson 2001).

Whereas there is general agreement on the involvement of climatic factors in determining treeline positions (here defined according to Körner 1998a), the debate remains controversial on the degree of relevance of the various particular abiotic and biotic parameters driving the limit of tree growth (e.g.

Figure 1 Treeline ecotone on the south-facing slope of Munt la Schera in the Swiss National Park

Körner 1998b; Peterson 1998; Holtmeier 2000; 2003; Malanson 2001; Grace et al. 2002; see also Jobbagy and Jackson 2000). Based on the assumption that growth and reproduction of trees at treeline stands are mainly controlled by temperature, global warming is supposed to rapidly advance the treeline towards higher altitudes resp. latitudes (Grace et al. 2002). Indeed, there is a number of studies from many regions in the world on the potential impact of climate change on the position of the treeline (table 1).

In some areas significant shifts of the treeline have been detected (Meshinev et al. 2000; Kullmann 2002; Penuelas and Boada 2003) and a causal relationship between climate change and the establishment of tree seedlings beyond the forest margin is suggested. In contrast, other studies describe a relatively stable treeline position in the last half century (e.g. Wardle and Coleman 1992; Butler et al. 1994; MacDonald et al. 1998; Cuevas 2000; 2002; Cullen et al. 2001; Masek 2001; Klasner and Fagre 2002). The observed inertia or 'time lag' (cf. Malanson 2001; see also Chapin and Starfield 1997) in the response of treeline position is explained by:

- the relatively poor availability and/or dispersal of seeds of the involved species (e.g. Wardle and Coleman 1992; Lescop-Sinclair and Payette 1995; Masek 2001),
- particular edaphic or aeolian conditions (see e.g. Kupfer and Cairns 1996 and cit. lit., Holtmeier 2000; 2003)
- and/or the importance of disturbance for tree recruitment above the treeline (e.g. Villalba and Veblen 1997; Luckman and Kavanagh 1998; Cullen et al. 2001),

which assign an autogenic component to treeline dynamics, and thus limits their predictability from environmental conditions (Wildi and Schütz 2000). Whereas Germino et al. (2002) outline the importance of herbaceous ground cover ameliorating microclimate and facilitating tree seedling establishment, Magee and Antos (1992) and Wardle and Coleman (1992) highlight the low adaptability of treeline species to spread into and establish in dense grassland vegetation (see also Hobbie and Chapin 1998; Moir et al. 1999).

Instead of a climate induced spatial displacement, several studies reported other more structural features of vegetation change at treeline positions, such as:

- increased growth rates of trees (e.g. Villalba and Veblen 1997; Suarez et al. 1999) unless the warming implies increased water limitation in the regarded ecosystem (cf. Barber et al. 2000; Lloyd and Fastie 2002; see also Miehe 1996; Villalba and Veblen 1997; Moir et al. 1999; Kusnierczyk and Ettl 2002),
- the development of vertical tree stems from pre-established krummholz vegetation (e.g. Lescop-Sinclair and Payette 1995; Luckman and Kavanagh

1998; see also Weisberg and Baker 1995; Hessl and Baker 1997; Dereg and Payette 1998),

- and/or the increase in the density of populations (e.g. Butler et al. 1994; Luckman and Kavanagh 1998; Suarez et al. 1999; Klasner and Fagre 2002).

Table 1 Reports on seedling establishment beyond the present treeline (m = surface meters; alt. m = meters in altitude)

Location	Period	Shift of treeline	Species	References
Chile	since ca. 1850	10 m	*Nothofagus pumilio*	Cuevas (2000), see also Cuevas (2002)
Northwestern Canada	past 150 years	10 – 20 m 2 – 5 alt. m	*Picea glauca*	Szeicz and MacDonald (1995)
Northern Ural mountains, Russia	since 1920	100 – 500 m 20 – 30 alt. m	*Larix sibirica*	Shiyatov (1993, 2000) (cit. in Holtmeier 2000)
New Zealand (South Island)	last 60 years	7 – 9 m 5 – 8 alt.m	*Nothofagus menziesii, N. solandri var. cliffortioides, N. fusca, Prumnopitys ferruginea*	Wardle and Coleman (1992)
Sweden	last 50 years	120 – 375 alt. m	*Betula pubescens ssp. tortuosa* *Sorbus aucuparia, Picea abies, Pinus sylvestris,* *Salix div. spec., Acer platanoides*	Kullman (2002); see also Kullman (2001)
Montseny mountains, Spain	since 1955	70 m	*Fagus sylvatica*	Penuelas and Boada (2003)
Mount Hotham, Victoria, Australia	since 1967	< 15 m	*Eucalyptus pauciflora, E. stellulata*	Wearne and Morgan (2001)
Bulgaria	since 1970	130 (– 340) alt. m	*Pinus peuce*	Meshinev et al. (2000)
Oregon coast range Marys Peak, USA	ca. 30 years	10 m	*Abies procera*	Magee and Antos (1992)
Glacier National Park, Montana, USA	since 1973	mentioned in text	*Abies lasiocarpa, Pinus flexilis,* *Pinus albicaulis, Picea engelmannii, Larix lyallii*	Butler et al. (1994)

The complexity of several interacting factors, whose individual degree of relevance on treeline dynamics depends on the particular local conditions, may explain the great regional and temporal variability in the importance of treeline driving parameters.

In general, the recruitment in treeline forest ecosystems follows a rather episodic than gradual pattern (Szeicz and MacDonald 1995; Hättenschwiler and Körner 1995; Kupfer and Cairns 1996; Cullen et al. 2001; Paulsen et al. 2000). As a consequence, treeline position alternates between periods of relative stasis, when it is unresponsive to climate variation, and periods of rapid change during times, when critical climate parameters exceed some threshold values (Lloyd and Graumlich 1997; Suarez et al. 1999; Paulsen et al. 2000; cf. also Loreau et al. 2001). This inherent resilience of treeline forest to environmental change may imply a temporary disequilibrium relationship with climate (cf. Holtmeier 2000; 2003) and thus, attributes some reservations when using the treeline as criterion for assessing effects of rapid climate change (Körner 1999; Holtmeier 2000; 2003; see also Kupfer and Cairns 1996). Furthermore, although treeline species of the same stand share the same environmental conditions, climate change may induce species-specific and thus, sometimes opposite reactions (see below, but also e.g. Carrer et al. 1998; Luckman and Kavanagh 1998; Motta and Nola 2001; Payette et al. 2001).

3 Climate Change and Alpine Vegetation

The science of alpine vegetation has a long tradition (see e.g. Körner 1999; Körner and Spehn 2002; Nagy et al. 2003). The earliest description of vegetation sequences along altitudinal gradients is mentioned in the probably worlds oldest high mountain monograph by Gessner (1554) 'Descriptio Montis Fracti', Pilatus near Lucern, Switzerland (cit. in Körner 1999). Also one of the longest vegetation monitoring series is derived from a high mountain peak (Piz Linard, Engadine valley, Switzerland) and starts as early as 1835 with the record of Oswald Heer (Heer 1866; see also Pauli et al. 2001a) (cf. table 2). In analogy, an increasing number of high mountain peaks was investigated at the turn of the last century (see e.g. Rübel 1912; Braun 1913). Many of the sites described by the aforementioned authors have been revisited (e.g. Braun-Blanquet 1955; 1957), and thus, an extraordinary well documented database of the spatio-temporal abundance and diversity of high mountain vegetation was established. With regard to global warming, this database has attracted additional attention as it provides a precise description of the situation before the latest warming period.

Table 2 Change in phytodiversity of the summit of Piz Linard (3411 m a.s.l.)

Year	**1835**	**1864**	**1893**	**1911**	**1937**	**1947**	**1992**
Investigator(s)	**Heer**	**Heer**	**Schibler**	**Braun**	**anonymous**	**Braun-Blanquet**	**Grabherr et al.**
Number of species	1	3	4	8	10	10	10
Androsace alpina	+	+	+	+	+	+	+
Ranunculus glacialis		r	+	+	+	1	2
Saxifraga bryoides			r	+	+	1	1
Saxifraga oppositifolia			r	r	+	+	1
Poa laxa				+	+	1	1
Draba fladnizensis				r	+	+	+
Gentiana bavarica				r	r	r	+
Cerastium uniflorum				r	r	r	r
Leucanthemopsis alpina		r			r	r	
Saxifraga exerata					r	r	
Cardamine resedifolia							r
Luzula spicata							r

Note: The descriptive information on the abundance of plant species until 1911 has been transformed in approximates of quantitative cover values (r = very rare, + = rare, 1 = scattered, 2 = common; cf. Braun-Blanquet 1964) in order to make the calculation of multivariate analyses possible

Alpine plants facing climate change may follow in principal three major response strategies: i) to persist at the present location under modified climatic conditions, ii) to migrate to a habitat with more favourable conditions or iii) to die off. The first option again offers different strategies such as the gradual genetic adaptation of populations, phenotypic plasticity or ecological buffering (e.g. to profit from suitable ecological niches) (cf. Theurillat and Guisan 2001; see also Chapin and Körner 1995).

Evidence for the second option, i.e. migration, is given by long-term vegetation comparisons based on the aforementioned descriptions of high mountain vegetation of former times. High mountain peaks in the Bernina area, Engadine valley, Switzerland, investigated at the beginning of the last century by Rübel (1912), were revisited in the 1980ies by Hofer (1992) and again in 2003 by Burga et al. 2004. Nine out of the ten resurveyed summits have shown an increase in species richness, only on Piz Trovat (a summit consisting of rock debris, and therefore difficult to be colonised by plants) a more or less constant number of species has been recorded (figure 2). Overall, species richness increased from a total of 57 (Rübel 1912) to 87 (Hofer 1992) and to 102 vascular plants (Burga et al. 2004) in these almost 100 years between the surveys. Grabherr et al. (1994) presented an extended data

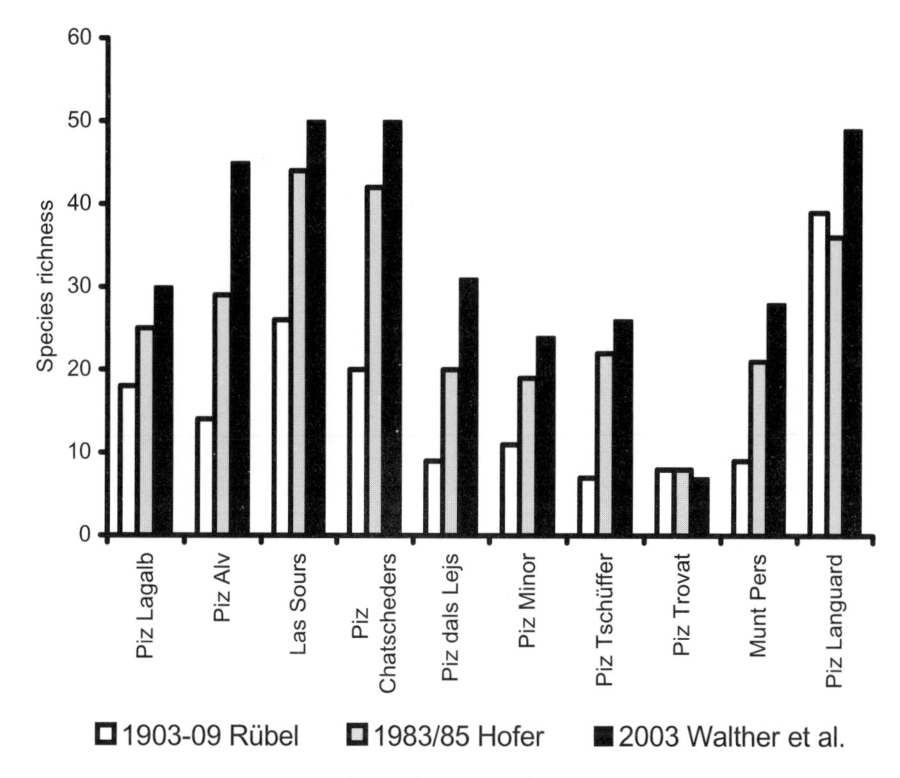

Figure 2 Resurvey of the species richness of 10 high-mountain peaks spanning a period of almost 100 years (Rübel 1912, Hofer 1992, Burga et al. 2004)

set of 26 summits of the eastern Swiss and Austrian Alps (cf. also Grabherr et al. 2001) and concluded that the 'upward movement of the alpine-nival flora is an overall trend'. Using multivariate analysis tools (MULVA-5, Wildi and Orlóci 1996), we calculated the principal coordinate analysis for a subset of these data, i.e. given pairs of the first (1900-1921; data from Rübel 1912; resp. Braun 1913; 1958) and so far latest records (1983-2003; data from Hofer 1992; Grabherr et al. 1994; 2001; Burga et al. 2004) of the summit vegetation surveys of 20 high mountain peaks (figure 3; for details see annex).

The first axis shows a clear separation of two groups discriminating mountains with siliceous (figure 3, left grouping) from those with calcareous substrate (figure 3, right grouping) (correlation with site factor 'substrate type': 0.97). There is also a pronounced shift in uniform direction along the second axis, which correlates best with the number of species (- 0.85) and thus, underlines the directed shift in high-mountain vegetation towards higher species richness. In particular, the increase in species richness is more pronounced on summits of lower altitudes (ca. 2800-3000 m a.s.l.), which is

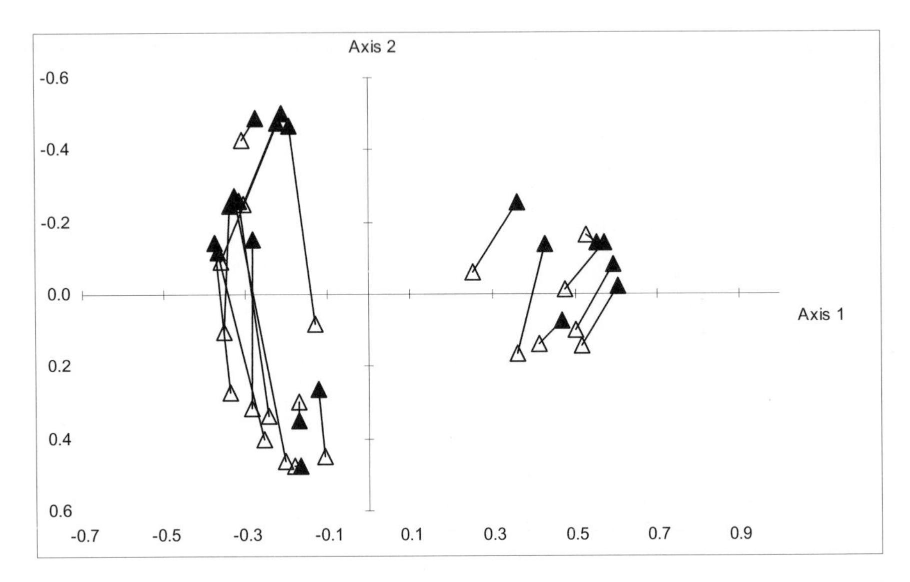

Figure 3 Principal coordinate analysis of first (open triangles) and so far last record (filled triangles) of 20 high mountain peaks of the Engadine valley in Switzerland (data were not transformed for multivariate analysis, for details see text and appendix)

explained by the greater species availability for the filling of potential new ecological niches (Grabherr et al. 1995; Camenisch 2002; see also Keller et al. 2000). Time series with more than one time step reveal different developments. In a first stage, the number of species is increasing (table 2), which is then followed by an increase in the population size of the present species but with stagnating species diversity (table 2 and figure 4). The stagnation in species number may be explained by the dispersal capability of the species, the quality of the available migration routes (cf. Grabherr et al. 1995), but also by the magnitude of the so far experienced climatic change. Overall, the observed upward movement of alpine species lags behind the theoretically calculated shift in vegetation zones based upon the observed change in climatic conditions (Grabherr et al. 1994; cf. also Theurillat and Guisan 2001). An analogue process of increasing species number and frequency has also been reported from Scandinavian mountains. Klanderud and Birks (2003) provide data on increasing species abundance and richness of plants on Norwegian high mountain tops, with climate change as the most likely major driving factor for the changes observed. Based on all these findings, the Global Observation Research Initiative in Alpine Environments (GLORIA) has established a world-wide network to monitor future changes on high mountain summits covering a great variety of longitudinal and latitudinal sites (see e.g. Pauli et al. 2001b; Gottfried et al. 2002).

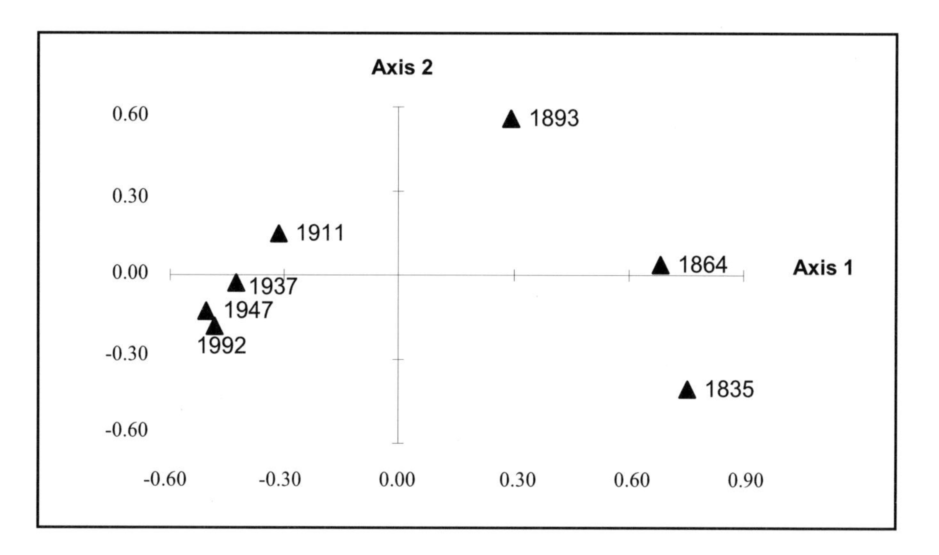

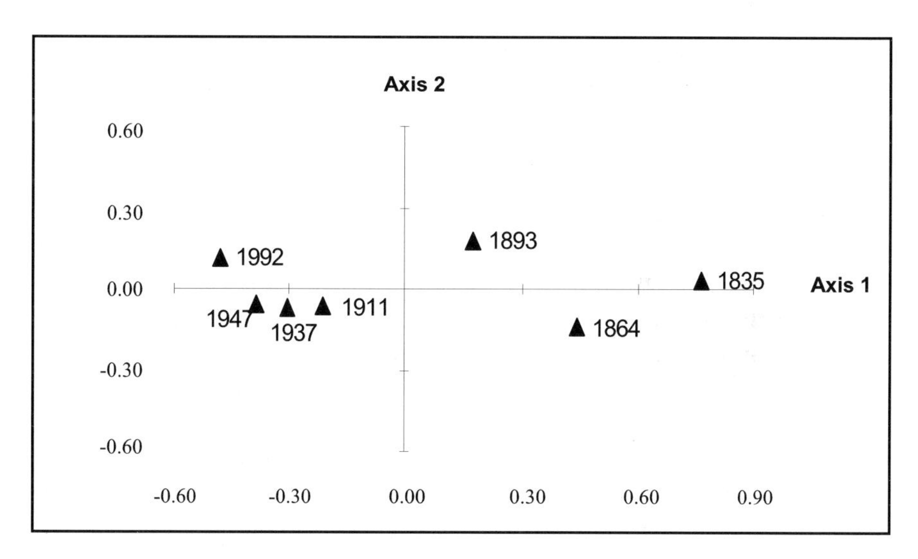

Figure 4 Principal coordinate analyses of approximated quantitative cover values (see table 2) of Piz Linard time series 1835-1992 (correlation of axis 1 with attributes: 'year' = -0.95; 'number of species' = -0.98) (**graph above**: curved because of non-linear data; **graph below**: linearized using 'Flexible Shortest Path Adjustment' (FSPATH) (cf. Wildi and Orlóci 1996)

4 Conclusions

In general, an altitudinal shift always implies a loss in habitat size as a consequence of the diminishing and more fragmented surface area with increasing altitude (cf. Körner 1995; Grabherr 1997; Burga and Perret 1998). Therefore, high-mountain populations are expected to suffer from increased habitat fragmentation (McCarthy et al. 2001) and increased competition with species from lower elevational areas (Camenisch 2002; Burga et al. 2003). In prealpine areas, where species have already reached the top of the mountains and an altitudinal shift towards higher elevational areas is no option, climate change may imply losses in available habitats for the alpine flora unless a cliffy relief (figure 5) provides surrogate habitats for e.g. rock and scree colonising species being able to subsist under warmer climatic conditions.

Another common, general feature of the aforementioned vegetation changes in high mountain ecosystems is given by the fact that the considered species are supposed to shift individually and not as an entity of the present assemblage or vegetation belt, which is comparable to e.g. the late- and post-glacial immigration of stone pine, spruce and fir in the Swiss Alps (Burga and Perret 1998; cf. also Pott et al. 1995; Bauerochse and Katenhusen 1997; Pott 1997; Burga et al. 2003). Specific characteristics of the plants such as reproduction rate, growth strategies, genetic variability, physiological plasticity as well as inter- and intraspecific competition play an important role in the direction and response time of individual species to environmental change and makes it difficult to include the full complexity in models simulating the response of vegetation to climate (cf. Körner 1999; see also Price and Barry 1997). The importance of feedback mechanisms with organisms of other trophic levels for the interpretation of species-specific response patterns to climatic change is exemplified by e.g. Motta and Nola (2001) with sub-alpine forest species in Italy. Whereas *Pinus cembra* showed an increasing trend in the growth rates, the basal area increments of *Larix decidua* of the same stand have oscillated substantially but without similar positive trend. Periodic strong growth reduction of *Larix decidua* has occurred as a consequence of periodic outbreaks of the species-specific larch bud moth (*Zeiraphera diniana*), obviously casting any climatic signal in the growth trends of Larch, but not of the other conifer species (cf. Davis et al. 1998; see also Briones et al. 1998; Holtmeier 2000; 2003; Cherubini et al. 2002).

In parallel with the observed vegetational changes, adaptations in the distribution and composition of the alpine fauna such as birds (D'Oleire-Oltmanns et al. 1995; Järvinen 1995), mammals (Henttonen 1995; Inouye et al. 2000), but also sub-alpine and alpine aquatic systems (e.g. Byron and

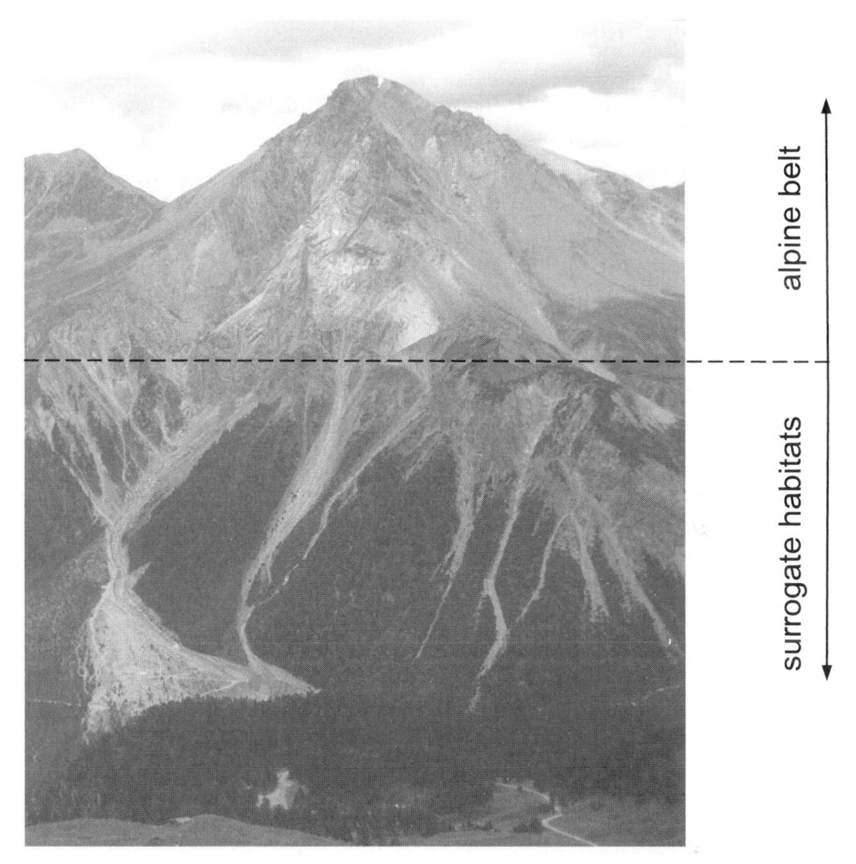

Figure 5 A pronounced relief may provide potential surrogate habitats for alpine species in lower areas than the original alpine belt (Piz Daint, 2968 m a.s.l., Engadine, Switzerland)

Goldman 1990; Sommaruga-Wograth et al. 1997) are expected. Shaw and Harte (2001) postulate, that range shifts and associated changes in species composition have a greater impact on ecosystem processes than the direct influences of the pure physical warming of the system. In any case, the complexity of ecosystems always implies unknown components when assessing the ecological responses to recent climate change (cf. e.g. Walther et al. 2002) and thus, still conceal many uncertainties in the behaviour of high mountain ecosystems under future climate change.

Acknowledgements

Jan Barkowski and Romedi Reinalter assisted at the field survey, which was made in 2003 in collaboration with the Physical Geography Division of the University of Zurich / Switzerland (Prof. C.A. Burga and collaborators). Financial support by the German Research Foundation (DFG Project WA 1523/6-1) is kindly acknowledged.

References

Barber VA, Juday GP, Finney BP (2000) Reduced growth of Alaskan white spruce in the twentieth century from temperature-induced drought stress. Nature 405: 668-673

Bauerochse A, Katenhusen O (1997) Holozäne Landschaftsentwicklung und aktuelle Vegetation im Fimbertal (Val Fenga, Tirol/Graubünden). Phytocoenologia 27(3): 353-453

Beniston M, Rebetez M, Giorni F, Marinucci MR (1994) An analysis of regional climate change in Switzerland. Theoretical and Applied Climatology 49: 135-159

Beniston M, Diaz HF, Bradley RS (1997) Climatic change at high elevation sites: an overview. Climatic Change 36: 233-251

Box EO (1981) Macroclimate and Plant Forms: an Introduction to Predictive Modelling in Phytogeography. Tasks for Vegetation Science 1, Dr W Junk Publishers, The Hague

Box EO (1996) Plant functional types and climate at the global scale. Journal of Vegetation Science 7(3): 309-320

Braun J (1913) Die Vegetationsverhältnisse der Schneestufe in den Rätisch-Lepontischen Alpen. Denkschriften der Schweizerischen Naturforschenden Gesellschaft 48: 1-348

Braun-Blanquet J (1955) Die Vegetation des Piz Languard, ein Massstab für Klimaänderungen. Svensk Botanisk Tidskrift 49(1-2): 1-9

Braun-Blanquet J (1957) Ein Jahrhundert Florenwandel am Piz Linard (3414 m). Bulletin Jardin Botanique Bruxelles, vol jubilaire W. Robyns (Comm. SIGMA 137), pp 221-232

Braun-Blanquet J (1958) Über die obersten Grenzen pflanzlichen Lebens im Gipfelbereich des schweizerischen Nationalparks. Ergebnisse der wissenschaftlichen Untersuchungen des schweizerischen Nationalparks VI(39): 119-142

Braun-Blanquet J (1964) Pflanzensoziologie. Springer, Wien

Briones MJI, Ineson P, Poskitt J (1998) Climate change and *Cognettia sphagnetorum*: effects on carbon dynamics in organic soils. Functional Ecology 12: 528-535

Brockmann-Jerosch H (1919) Baumgrenze und Klimacharakter. Pflanzengeographische Kommission der Schweizerischen Naturforschenden Gesellschaft, Beiträge zur geobotanischen Landesaufnahme 6: 1-255

Burga CA, Perret R (1998) Vegetation und Klima der Schweiz seit dem jüngeren Eiszeitalter. Ott, Thun

Burga CA, Perret R (2001) Monitoring of eastern and southern Swiss alpine timberline ecotones. In: Burga CA, Kratochwil A (eds) Biomonitoring: General and applied aspects on regional and global scales. Tasks for Vegetation Science 35. Kluwer Academic, Dordrecht, pp 179-194

Burga CA, Haeberli W, Krummenacher B, Walther G-R (2003) Abiotische und biotische Dynamik in Gebirgsräumen – Status quo und Zukunftsperspektiven. In: Jeanneret F, Wastl-Walter D, Wiesmann U, Schwyn M (eds) Welt der Alpen – Gebirge der Welt. Jahrbuch der Geographischen Gesellschaft Bern 61. Haupt, Bern, pp 25-37

Burga CA, Walther GR, Beißner S (2004) Florenwandel in der alpinen Stufe des Berninagebiets – ein Klimasignal? Berichte der Reinhold-Tüxen-Gesellschaft 16 (in press)

Butler DR, Hill C, Malanson GP, Cairns DM (1994) Stability of alpine treeline in Glacier National Park, Montana, USA. Phytocoenologia 22(4): 485-500

Byron ER, Goldman CR (1990) The potential effects of global warming on the primary productivity of a sub-alpine lake. Water Resources Bulletin 26(6): 983-989

Camenisch M (2002) Veränderungen der Gipfelflora im Bereich des Schweizerischen Nationalparks: Ein Vergleich über die letzten 80 Jahre. Jahresbericht der Naturforschenden Gesellschaft Graubünden 111: 27-37

Carrer M, Anfodillo T, Urbinati C, Carraro V (1998) High-altitude forest sensitivity to global warming: results from long-term and short-term analyses in the Eastern Italian Alps. In: Beniston M, Innes JL (eds) The impacts of climate variability on forests. Lecture Notes in Earth Sciences 74. Springer, Berlin Heidelberg, pp 171-189

Chapin III FS, Körner C (1995) Patterns, causes, changes, and consequences of biodiversity in arcitc and alpine ecosystems. In: Chapin III FS, Körner C (eds) Arctic and alpine biodiversity. Patterns, causes and ecosystem consequences. Ecological Studies 113, Springer, Berlin, pp 313-320

Chapin III FS, Starfield AM (1997) Time lags and novel ecosystems in response to transient climatic change in arctic Alaska. Climatic Change 35: 449-461

Cherubini P, Fontana G, Rigling D, Dobbertin M, Brang P, Innes JL (2002) Tree-life history prior to death: Two fungal root pathogens affect tree-ring growth differently. Journal of Ecology 90: 839-850

Cuevas JG (2000) Tree recruitment at the *Nothofagus pumilio* alpine timberline in Tierra del Fuego, Chile. Journal of Ecology 88: 840-855

Cuevas JG (2002) Episodic regeneration at the *Nothofagus pumilio* alpine timberline in Tierra del Fuego, Chile. Journal of Ecology 90: 52-60

Cullen LE, Stewart GH, Duncan RP, Palmer JG (2001) Disturbance and climate warming influences on New Zealand *Nothofagus* tree-line population dynamics. Journal of Ecology 89: 1061-1071

Davis AJ, Lawton JH, Shorrocks B, Jenkinson LS (1998) Individualistic species responses invalidate simple physiological models of community dynamics under global environmental change. Journal of Animal Ecology 67(4): 600-612

Dereg D, Payette S (1998) Development of black spruce growth forms at tree line. Plant Ecology 138: 137-147

D'Oleire-Oltmanns W, Mingozzi T, Brendel U (1995) Effects of climate change on birds population. In: Guisan A, Holten JI, Spichiger R, Tessier L (eds) Potential ecological impacts of climate change in the Alps and Fennoscandian Mountains. Editions des Conservatoire et Jardin Botanique de Genève, pp 173-175

Emanuel WR, Shugart HH, Stevenson MP (1985) Climatic change and the broad-scale distribution of terrestrial ecosystems complexes. Climatic Change 7: 29-43

Folland CK, Karl TR (co-ordinating lead authors) (2001) Observed climate variability and change. In: Houghton JT, Ding Y, Griggs DJ, Noguer M, van der Linden PJ, Dai X, Maskell K, Johnson CA (eds) Climate Change 2001: The scientific basis. Cambridge University Press, Cambridge, pp 99-181

Germino MJ, Smith WK, Resor AC (2002) Conifer seedling distribution and survival in an alpine-treeline ecotone. Plant Ecology 162: 157-168

Gottfried M, Pauli H, Hohenwallner D, Reiter K, Grabherr G (2002) GLORIA – The Global Observation Research Initiative in Alpine Environments: Wo stehen wir? Petermanns Geographische Mitteilungen 146: 69-71

Grabherr G (1997) The high-mountain ecosystems of the Alps. In: Wielgolaski FE (ed) Polar and alpine tundra. Ecosystems of the World 3, Elsevier, Amsterdam, pp 97-121

Grabherr G, Gottfried M, Pauli H (1994) Climate effects on mountain plants. Nature 369: 448

Grabherr G, Gottfried M, Gruber A, Pauli H (1995) Patterns and current changes in alpine plant diversity. In: Chapin III FS, Körner C (eds) Arctic and Alpine Biodiversity. Ecological Studies 113, Springer, Berlin Heidelberg, pp 169-181

Grabherr G, Gottfried M, Pauli H (2001) Long-term monitoring of mountain peaks in the Alps. In: Burga CA, Kratochwil A (eds) Biomonitoring: General and applied aspects on regional and global scales. Tasks for Vegetation Science 35, Kluwer Academic, Dordrecht, pp 153-177

Grace J, Berninger F, Nagy L (2002) Impacts of climate change on the tree line. Annals of Botany 90: 537-544

Hättenschwiler S, Körner C (1995) Responses to recent climate warming of *Pinus sylvestris* and *Pinus cembra* within their montane transition zone in the Swiss Alps. Journal of Vegetation Science 6: 357-368

Heer O (1866) Der Piz Linard. Jahrbuch Schweizerischer Alpin Club III, pp 457-471

Henttonen H (1995) Climate change and the ecology of alpine mammals. In: Guisan A, Holten JI, Spichiger R, Tessier L (eds) Potential ecological impacts of climate

change in the Alps and Fennoscandian Mountains. Editions des Conservatoire et Jardin Botanique de Genève, pp 75-78
Hermes K (1955) Die Lage der oberen Waldgrenze in den Gebirgen der Erde und ihr Abstand zur Schneegrenze. Kölner Geographische Arbeiten, Heft 5. Geographisches Institut der Universität Köln
Hessl AE, Baker WL (1997) Spruce-fir growth changes in the forest-tundra ecotone of Rocky Mountain National Park, Colorado, USA. Ecography 20(4): 356-367
Hobbie SE, Chapin III FS (1998) An experimental test of limits to tree establishment in Arctic tundra. Journal of Ecology 86: 449-461
Hofer HR (1992) Veränderungen in der Vegetation von 14 Gipfeln des Berninagebietes zwischen 1905 und 1985. Berichte des Geobotanischen Institutes der Eidg. Technischen Hochschule 58, Stiftung Rübel, Zürich, pp 39-54
Holtmeier F-K (2000) Die Höhengrenze der Gebirgswälder. Arbeiten aus dem Institut für Landschaftsökologie, Band 8. Westfälische Wilhelms-Universität, Münster
Holtmeier F-K (2003) Mountain Timberlines – Ecology, patchiness, and dynamics. Advances in Global Change Research, 14. Kluwer Academic, Dordrecht
Houghton JT, Ding Y, Griggs DJ, Noguer M, van der Linden PJ, Dai X, Maskell K, Johnson CA (eds) (2001) Climate Change 2001: The scientific basis. Contribution of Working Group I to the Third Assessment Report of the Intergovernmental Panel on Climate Change. Cambridge University Press, Cambridge
Huntley B, Berry PM, Cramer W, McDonald AP (1995) Modelling present and potential future ranges of some European higher plants using climate response surfaces. Journal of Biogeography 22(6): 967-1001
Inouye DW, Barr B, Armitage KB, Inouye BD (2000) Climate change is affecting altitudinal migrants and hibernating species. Proceedings of the National Academy of Science of the United States of America 97(4): 1630-1633
Iverson LR, Prasad AM (1998) Predicting abundance of 80 tree species following climate change in the eastern United States. Ecological Monographs 68(4): 465-485
Järvinen A (1995) Effects of climate change on mountain bird populations. In: Guisan A, Holten JI, Spichiger R, Tessier L (eds) Potential ecological impacts of climate change in the Alps and Fennoscandian Mountains. Editions des Conservatoire et Jardin Botanique de Genève, pp 73-74
Jobbagy EG, Jackson RB (2000) Global controls of forest line elevation in the northern and southern hemispheres. Global Ecology and Biogeography 9: 253-268
Kappelle M, van Vuuren MMI, Baas P (1999) Effects of climate change on biodiversity: a review and identification of key research issues. Biodiversity and Conservation 8: 1383-1397
Keller F, Kienast F, Beniston M (2000) Evidence of response of vegetation to environmental change on high-elevation sites in the Swiss Alps. Regional Environmental Change 1(2): 70-77
Klanderud K, Birks HJB (2003) Recent increases in species richness and shifts in altitudinal distributions of Norwegian mountain plants. The Holocene 13(1): 1-6

Klasner FL, Fagre DB (2002) A half century of change in alpine treeline patterns at Glacier National Park, Montana, U.S.A. Arctic, Antarctic and Alpine Research 34(1): 49-56

Körner C (1995) Alpine Plant Diversity: A global survey and functional interpretations. In: Chapin III FS, Körner C (eds) Arctic and alpine biodiversity. Patterns, causes and ecosystem consequences. Ecological Studies 113, Springer, Berlin, pp 45-62

Körner C (1998a) Worldwide positions of alpine treelines and their causes. In: Beniston M, Innes JL (eds) The impacts of climate variability on forests, Lecture Notes in Earth Sciences 74, Springer, Berlin Heidelberg, pp 221-229

Körner C (1998b) A reassessment of high-elevation treeline positions and their explanation. Oecologia 115: 445-459

Körner C (1999) Alpine Plant Life: functional plant ecology of high mountain ecosystems. Springer, Berlin Heidelberg

Körner C, Spehn E (eds) (2002) Mountain Biodiversity: A global assessment. Parthenon, London

Kullman L (2001) 20th century climate warming and tree-limit rise in the southern Scandes of Sweden. Ambio 30(2): 72-80

Kullman L (2002) Rapid recent range-margins rise of tree and shrub species in the Swedish Scandes. Journal of Ecology 90: 68-77

Kupfer JA, Cairns DM (1996) The suitability of montane ecotones as indicators of global climatic change. Progress in Physical Geography 20(3): 253-272

Kusnierczyk ER, Ettl GJ (2002) Growth response of ponderosa pine (*Pinus ponderosa*) to climate in the eastern Cascade Mountains, Washington, USA: Implications for climatic change. Ecoscience 9(4): 544-551

Lescop-Sinclair K, Payette S (1995) Recent advance of the arctic treeline along the eastern coast of Hudson Bay. Journal of Ecology 83: 929-936

Lloyd AH, Fastie CL (2002) Spatial and temporal variability in the growth and climate response of treeline trees in Alaska. Climatic Change 52: 481-509

Lloyd AH, Graumlich LJ (1997) Holocene dynamics of treeline forests in the Sierra Nevada. Ecology 78(4): 1199-1210

Loreau M, Naeem S, Inchausti P, Bengtsson J, Grime JP, Hector A, Hooper DU, Huston MA, Raffaelli D, Schmid B, Tilman D, Wardle DA (2001) Biodiversity and ecosystem functioning: current knowledge and future challenges. Science 294: 804-808

Luckman BH, Kavanagh TA (1998) Documenting the effects of recent climate change at treeline in the Canadian Rockies. In: Beniston M, Innes JL (eds) The impacts of climate variability on forests, Lecture Notes in Earth Sciences 74, Springer, Berlin Heidelberg, pp 121-144

MacDonald GM, Szeicz JM, Claricoates J, Dale KA (1998) Response of the central Canadian treeline to recent climate changes. Annals of the Association of American Geographers 88(2): 183-208

Magee TK, Antos JA (1992) Tree invasion into a mountain-top meadow in the Oregon Coast Range, USA. Journal of Vegetation Science 3: 485-494

Malanson GP (2001) Complex responses to global change at alpine treeline. Physical Geography 22(4): 333-342

Masek JG (2001) Stability of boreal forest stands during recent climate change: evidence from Landsat satellite imagery. Journal of Biogeography 28: 967-976

McCarthy JJ, Canziani OF, Leary NA, Dokken DJ, White KS (eds) (2001) Climate Change 2001: Impacts, Adaptation, and Vulnerability. Contribution of Working Group II to the Third Assessment Report of the Intergovernmental Panel on Climate Change (IPCC). Cambridge University Press, Cambridge

Meshinev T, Apostology I, Koleva E (2000) Influence of warming on timberline rising: a case study of *Pinus peuce* Griseb. in Bulgaria. Phytocoenologia 30 (3-4): 431-438

Miehe G (1996) On the connexion of vegetation dynamics with climatic changes in High Asia. Palaeogeography, Palaeoclimatology, Palaeoecology 120: 5-24

Miehe G, Miehe S (1994) Zur oberen Waldgrenze in tropischen Gebirgen. Phytocoenologia 24: 53-110

Moir WH, Rochelle SG, Schoettle AW (1999) Microscale patterns of tree establishment near upper treeline, Snowy Range, Wyoming, USA. Arctic, Antarctic and Alpine Research 31(4): 379-388

Motta R, Nola P (2001) Growth trends and dynamics in sub-alpine forest stands in the Varaita Valley (Piedmont, Italy) and their relationships with human activities and global change. Journal of Vegetation Science 12: 219-230

Nagy L, Grabherr G, Körner C, Thompson DBA (eds) (2003) Alpine Biodiversity in Europe. Ecological Studies 167. Springer, Berlin Heidelberg

Organe consultatif sur les changements climatiques (OcCC) (2002) Das Klima ändert – auch in der Schweiz. ProClim, Bern

Pauli H, Gottfried M, Grabherr G (2001a) High summits of the Alps in a changing climate. In: Walther G-R, Burga CA, Edwards PJ (eds) 'Fingerprints' of Climate Change – Adapted behaviour and shifting species ranges. Kluwer Academic/ Plenum Publ., New York, pp 139-149

Pauli H, Gottfried M, Hohenwallner D, Hülber K, Reiter K, Grabherr G (eds) (2001b) Global Observation Research Initiative in Alpine Environments (GLORIA). The Multi-Summit Approach. Field Manual – Third Version, Vienna.

Paulsen J, Weber UM, Körner C (2000) Tree growth near treeline: abrupt or gradual reduction with altitude? Arctic, Antarctic and Alpine Research 32(1): 14-20

Payette S, Fortin M-J, Gamache I (2001) The subarctic forest-tundra: the structure of a biome in a changing climate. BioScience 51(9): 709-718

Penuelas J, Boada M (2003) A global change-induced biome shift in the Montseny mountains (NE Spain). Global Change Biology 9: 131-140

Peterson DL (1998) Climate, limiting factors and environmental change in high-altitude forests of Western North America. In: Beniston M, Innes JL (eds) The impacts of climate variability on forests, Lecture Notes in Earth Sciences 74, Springer, Berlin Heidelberg, pp 191-208

Pott R (1997) The Timber Line in Upper Fimbertal. Reports of the DFG 2-3/97: 18-21

Pott R, Hüppe J, Remy D, Bauerochse A, Katenhusen O (1995) Paläoökologische Untersuchungen zu holozänen Waldgrenzschwankungen im oberen Fimbertal (Val Fenga, Silvretta, Ostschweiz). Phytocoenologia 25(3): 363-398

Prentice IC, Cramer W, Harrison SP, Leemans R, Monserud RA, Solomon AM (1992) A global biome model based on plant physiology and dominance, soil properties and climate. Journal of Biogeography 19: 117-134

Price MF, Barry RG (1997) Climate Change. In: Messerli B, Ives JD (eds) Mountains of the World, Pathenon Publisher Group Inc., New York, pp 409-445

Przybylak R (2003) The Climate of the Arctic. Atmospheric and Oceanographic Sciences Library 26. Kluwer Academic / Plenum Publishers, Dordrecht

Rochefort RM, Little RL, Woodward A, Peterson DL (1994) Changes in sub-alpine tree distribution in western North-America: a review of climatic and other causal factors. The Holocene 4: 89-100

Rübel E (1912) Pflanzengeographische Monographie des Berninagebietes. Engelmann, Leipzig

Saxe H, Cannell MGR, Johnson Ø, Ryan MG, Vourlitis G (2001) Tree and forest functioning in response to global warming. New Phytologist 149: 369-400

Shaw MR, Harte J (2001) Control of litter decomposition in a subalpine meadow-sagebrush steppe ecotone under climate change. Ecological Applications 11(4): 1206-1223

Sommaruga-Wograth S, Koinig KA, Schmidt R, Sommaruga R, Tessadri R, Psenner R (1997) Temperature effects on the acidity of remote alpine lakes. Nature 387: 64-67

Stevens GC, Fox JF (1991) The causes of treeline. Annual Review of Ecology and Systematics 22: 177-191

Suarez F, Binkley D, Kaye MW (1999) Expansion of forest stands into tundra in the Noatak National Preserve, northwest Alaska. Ecoscience 6(3): 465-470

Sveinbjörnsson B (2000) North American and European treelines: external forces and internal processes controlling position. Ambio 29(7): 388-395

Sykes MT, Prentice IC, Cramer W (1996) A bioclimatic model for the potential distributions of north European tree species under present and future climates. Journal of Biogeography 23: 203-233

Szeicz JM, MacDonald GM (1995) Recent white spruce dynamics at the subarctic alpine treeline of north-western Canada. Journal of Ecology 83: 873-885

Theurillat J-P, Guisan A (2001) Potential impact of climate change on vegetation in the European Alps: a review. Climatic Change 50: 77-109

Tranquillini W (1979) Physiological Ecology of the Alpine Timberline. Ecological Studies 31. Springer, Berlin Heidelberg

Turner J, King JC, Lachlan-Cope TA, Jones PD (2002) Climate change – Recent temperature trends in the Antarctic. Nature 418(6895): 291

Vaughan DG, Marshall GJ, Connolley WM, King JC, Mulvaney R (2001) Climate Change: Devil in the Detail. Science 293(5536): 1777-1779

Villalba R, Veblen TT (1997) Regional patterns of tree population age structures in northern Patagonia: climatic and disturbance influences. Journal of Ecology 85: 113-124

Walsh JE, Doran PT, Priscu JC, Lyons WB, Fountain AG, McKnight DM, Moorhead DL, Virginia RA, Wall DH, Clow GD, Fritsen CH, McKay CP, Parsons AN (2002) Climate change – Recent temperature trends in the Antarctic. Nature 418(6895): 292

Walter H, Breckle SW (1983ff.) Ökologie der Erde. Band 1-4. Gustav Fischer, Stuttgart

Walther G-R, Post E, Convey P, Menzel A, Parmesan C, Beebee TJC, Fromentin J-M, Hoegh-Guldberg O, Bairlein F (2002) Ecological responses to recent climate change. Nature 416: 389-395

Wardle P (1974) Alpine timberlines. In: Ives JD, Barry RG (eds) Arctic and Alpine Environments, Methuen, London, pp 371-402

Wardle P, Coleman MC (1992) Evidence for rising upper limits of four native New Zealand forest trees. New Zealand Journal of Botany 30: 303-314

Wearne LJ, Morgan JW (2001) Recent forest encroachment into subalpine grasslands near Mount Hotham, Victoria, Australia. Arctic, Antarctic, and Alpine Research 33(3): 369-377

Weisberg PJ, Baker WL (1995) Spatial variation in tree seedling and krummholz growth in the forest-tundra ecotone of Rocky Mountain National Park, Colorado, USA. Arctic and Alpine Research 27(2): 116-129

Wildi O, Orlóci L (1996) Numerical exploration of community patterns. SPB Academic Publishing, New York

Wildi O, Schütz M (2000) Reconstruction of a long-term recovery process from pasture to forest. Community Ecology 1(1): 25-32

Woodward FI (1987) Climate and plant distribution. Cambridge University Press, Cambridge

1 Introduction

Whitebark pine is an important component of high altitude forests of the northern and central Rocky Mountains. It is the only North American member of the stone pines; all classified as genus *Pinus*, subgenus *Strobus*, section *Strobus*, and subsection *Cembrae* (Critchfield and Little 1966; Price et al. 1998; McCaughey and Schmidt 2001). Stone pines are characterized by five needles per fascicle, essentially indehiscent cones, and wingless seeds that are primarily dispersed by two nutcrackers, *Nucifraga columbiana* (Clark's nutcracker, North America) and *Nucifraga caryocatactes* (Europe and Asia) (Hutchins and Lanner 1982; Lanner 1982; 1990; Holtmeier 1999).

The relationship between the Clark's nutcracker and whitebark pine is mutually beneficial (Tomback 1978; 1982; 2001; Lanner 1982; 1996; Tomback and Linhart 1990). Whitebark pine seeds are harvested by Clark's nutcrackers in late summer and early fall and transported to a variety of storage areas throughout the subalpine zone, as well as to areas below and above the current elevational distribution. Seeds are stored in caches of 1-15 seeds, 2-3 cm under the surface. Germination occurs in unretrieved caches (Tomback 1982; Tomback and Linhart 1990).

The influence of the European nutcracker *(Nucifraga caryocatactes*) on the spatial distribution of Swiss stone pines (*Pinus cembra*) has been documented by several studies (Holtmeier 1965; 1966; 1974; 1993; Kuoch and Amiet 1970; Mattes 1978; 1982; 1985). Mattes (1982) and Holtmeier (1993; 1999; 2000; 2003) proposed that the selective site preferences of the nutcracker's caching activity coincide with favorable growing conditions for Swiss stone pine regeneration, causing higher regeneration densities on convex land forms compared to concave slopes and depressions.

Regeneration of Japanese stone pine (*Pinus pumila*) is restricted to open, wind exposed patches near mature pine scrub, despite a more widespread seed distribution by the Japanese nutcracker (*Nucifraga caryocatactes* var. *japonica)* into closed coniferous forest (Hayashida 1994; Kajimoto et al. 1998). Tomback (1982) also reported discrepancies between the most frequent caching environments and sites with the highest seedling recruitment potential for the North American species Clark's nutcracker and whitebark pine, on the eastern slope of the Sierra Nevada.

This study investigated the relationship between seed dispersal, spatial distribution and site characteristics of whitebark pine in the timberline ecotone of the middle Rocky Mountains. In particular, the study focused on regeneration patterns and prevailing site conditions that may limit or promote tree establishment. If regeneration patterns can be sufficiently explained by environmental variables, viable seeds are available on sites with high seedling recruiting potential, no matter whether these sites coincide with the most frequent caching environments of the nutcracker. Contrary results would suggest that seed availability (as the result of the nutcracker's caching and retrieving activities) may be a decisive factor for tree distribution at timberline.

2 Study Sites

The study was conducted in the timberline ecotone of the Beartooth Plateau, Montana and Wyoming, approximately 40 km east of the northeast entrance to Yellowstone National Park (figure 1, photo 1 and 2). The upper timberline of the Beartooth Plateau is located at 2900 m to 3100 m, with occasional stunted trees reaching altitudes up to 3200 m. Study areas were established in two different locations in the upper timberline ecotone. The Tibbs Butte study area is located in Wyoming, on the gentle north to northwest exposed slope of Tibbs Butte, approximately 3 km south of the Gardner Lake trailhead on U.S. Highway 212 (T. 57 N., R. 104 W., southeast quarter of Section 5). The study area is part of an extensive timberline ecotone characterized by Engel-

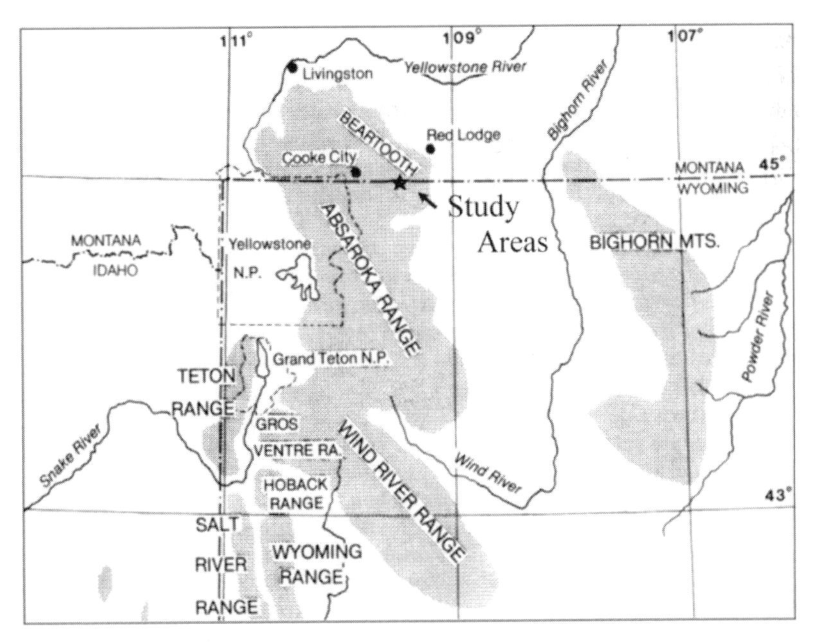

Figure 1 Location of study areas in the middle Rocky Mountains (Arno and Hammerly 1984, modified)

Photo 1 Tibbs Butte study area (marked with arrow), on June 1, 1992. View to SW

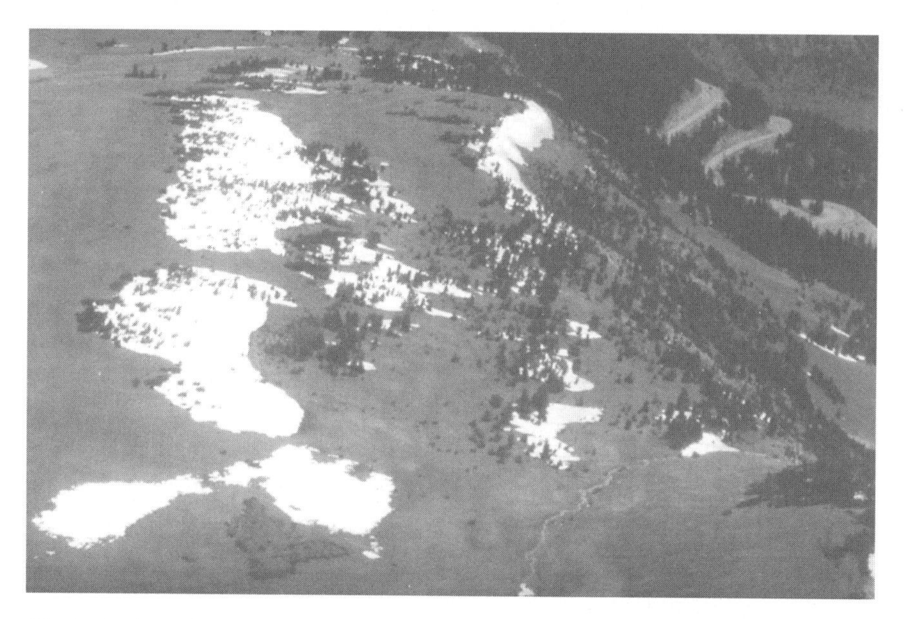

Photo 2 Wyoming Creek study area, on June 1, 1992. View to NW

mann spruce tree islands and groups of whitebark pine (photo 1). From an open subalpine forest at an elevation of 3050 m, the ecotone stretches more than 100 altitudinal meters to the upper-most trees at approximately 3270 m. The Wyoming Creek study area is located in Montana, east of U.S. Highway 212 and approximately 10 km north of Tibbs Butte (T. 9 S., R. 19 E., northeast quarter of Section 32). The area is located leeward of a small ridge, on a bench above the steep slopes of Wyoming Creek (photo 2). On the short, gradual, northeast exposed slope a strip of open whitebark pine woodland developed. Compared to Tibbs Butte, the timberline ecotone is narrow. The whitebark pine woodland reaches an elevation of 2990 m. Only few trees invade the alpine meadows above this altitude.

In all study areas timberline was judged to be caused by climatic rather than by edaphic or orographic factors. No signs of recent disturbance from fire, recreation or grazing were observed.

3 Materials and Methods

In this study, timberline and timberline ecotone are treated as synonyms and describe the transition zone between closed subalpine forest and the most advanced trees of the same species that form the forest below, regardless of their growth form or height. The term 'subalpine' is defined as the altitudinal belt below the timberline ecotone, contrary to suggestions by Löve (1970) who favored the use of 'subalpine' for the timberline ecotone itself.

In the Tibbs Butte and Wyoming Creek study areas, 10 m wide transects were established along altitudinal isoclines. The Tibbs Butte transect, at 3100 m elevation, is 300 m long. It begins at the eastern edge of an open stand of tree islands and continues toward the western end of the northwest exposed slope.

At Wyoming Creek three transects were established at an elevation of 2985 to 2995 m, parallel to each other, with a distance of 25 m between them: the upper transect is 100 m long and runs windward of the woodland, the 85 m long center transect cuts through the woodland, and the lowest 100 m long transect is located on the leeward side of the woodland, where the woodland opens to a treeless meadow. Combined the transects cover 5850 m^2 of timberline ecotone. While the starting points of the transects were subjectively selected, all transects were established perpendicular to the main slope direction and transverse through a random selection of treeless meadows, tree islands and woodland sites.

The locations of all tree individuals, including snags, found within the transect borders were mapped. Additionally, the canopies of trees ≥ 1.5 m tall were recorded in graphs. For all whitebark pine shorter than 1.5 m, the location relative to rocks, logs, trees or other characteristic site features was described. The number of individuals per cache, growth form, diameter of the thickest stem at ground level, height and vitality were recorded. Measurements on whitebark pine clusters referred to the largest and healthiest seedling. Tree islands only partly inside the transect were included in the analysis as long as some of the stems originated from the transect area.

New germinants and seedlings lower than the surrounding herbaceous vegetation can easily be missed while mapping larger areas. Circular subplots of 10 m^2 area were established every 10 m along the transect centerline, amounting to 10 % of the total transect area. In August and September 1993, and again in August 1994, 28 circular plots in the Wyoming Creek transects were carefully searched for new germinants and one- to three-year-old seedlings. The 30 circular plots in the Tibbs Butte transect were surveyed in September 1994. The number of caches, seedlings per cache, vitality, litter depth, and the location relative to rocks, logs, trees, or other microsite characteristics were recorded. Distance and direction from the center pole of the subplot helped relocating germinants and seedlings in September 1995, 1996, and 1997 for determination of survival rates. Each time, the vitality of seedlings was re-evaluated and the diameter and height measured with calipers.

The snow distribution and the patterns of snowmelt on the transects were studied in 1993. The snow depths of the Tibbs Butte transect were measured on May 15, May 30, June 13, June 26, and July 13. The snow depths of the Wyoming Creek transects were mapped on May 18, June 11, June 26, and July 12. Snow depth measurements were taken with a 2.20 m long fiberglass pole in 5 m intervals at the upper boundary, the center, and the lower boundary of the 10 m wide transects. Snowmelt graphs were created using the Geographic Information System IDRISI (Eastman 1995). The IDRISI routine INTERPOL was used to calculate a digital elevation model with 16 cells m^{-2}, from the original 5 m measurements of snow depth. This image was smoothed three times using a mean filter (low pass). The resulting data were classified for the graphic display of snow free areas during spring 1993.

4 Results

Whitebark pine was the only species with abundant regeneration on all studied transects. On the Tibbs Butte transect 20 live spruces and 5 live firs smaller than 1.5 m were mapped; on the Wyoming Creek transects only 3 live spruces and 2 live firs smaller than 1.5 m were found.

The distribution of whitebark pines smaller than 1.5 m was strongly clustered on all mapped transects. Whitebark pine regeneration occurred commonly close to the base of adult trees, adjacent to rocks, and in the vicinity of tree islands. Younger whitebark pines were often growing in groups of several seedlings, originating from the same seed cache. For whitebark pines < 1 cm in diameter, 38 % of regeneration sites contained clusters of two or more individuals. This percentage decreased with increasing diameter. In the diameter class 1- < 2 cm it was only 20 %; at 2- < 3 cm basal diameter the percentage decreased to 7 %. For older saplings it was difficult to distinguish between multiple individuals with intertwined rootstocks or multiple stems from the base. Most whitebark pines ≥1 cm in basal diameter were multi-stemmed. With increasing diameter, the percentage of whitebark pines with multiple stems from the base also increased, from 2 % (diameter class < 1 cm) to 48 % and more (diameter classes ≥ 3 cm). These whitebark pines may have originated from one or several seedlings.

A total of 126 whitebark pine regeneration sites with live seedlings and saplings < 1.5 m were mapped on the Tibbs Butte transect, a density of 4.2 sites 100 m^{-2} (figure 2). Regeneration sites were located close to the main features of this area: 44 % were found near rocks and 47 % were associated with tree islands, most commonly on the leeward side of such structures. At Wyoming Creek, the main feature is the location relative to the open whitebark pine woodland. Live regeneration densities varied strongly between transects, from 4.1 sites 100 m^{-2} on the windward transect, 11.0 sites 100 m^{-2} on the transect leeward the woodland, to 13.5 sites 100 m^{-2} inside the woodland (figure 3). Of the 266 live whitebark pine regeneration sites on all three transects, 52 % were located less than 1 m from the base of a tree (usually a whitebark pine), while another 33 % were found 1- < 5 m from the base of a tree.

Areas inside and leeward of tree islands, close to the base of trees, and in woodland have a higher snow accumulation and melt out later than open and wind exposed sites in the timberline ecotone. Even small rocks or slight depressions can influence snow accumulation and therefore may protect small seedlings from wind and dessication. The tabulation of snowmelt dates and whitebark pine regeneration reveals a nonlinear relationship (table 1). In general, regeneration densities were highest in areas with moderate to long

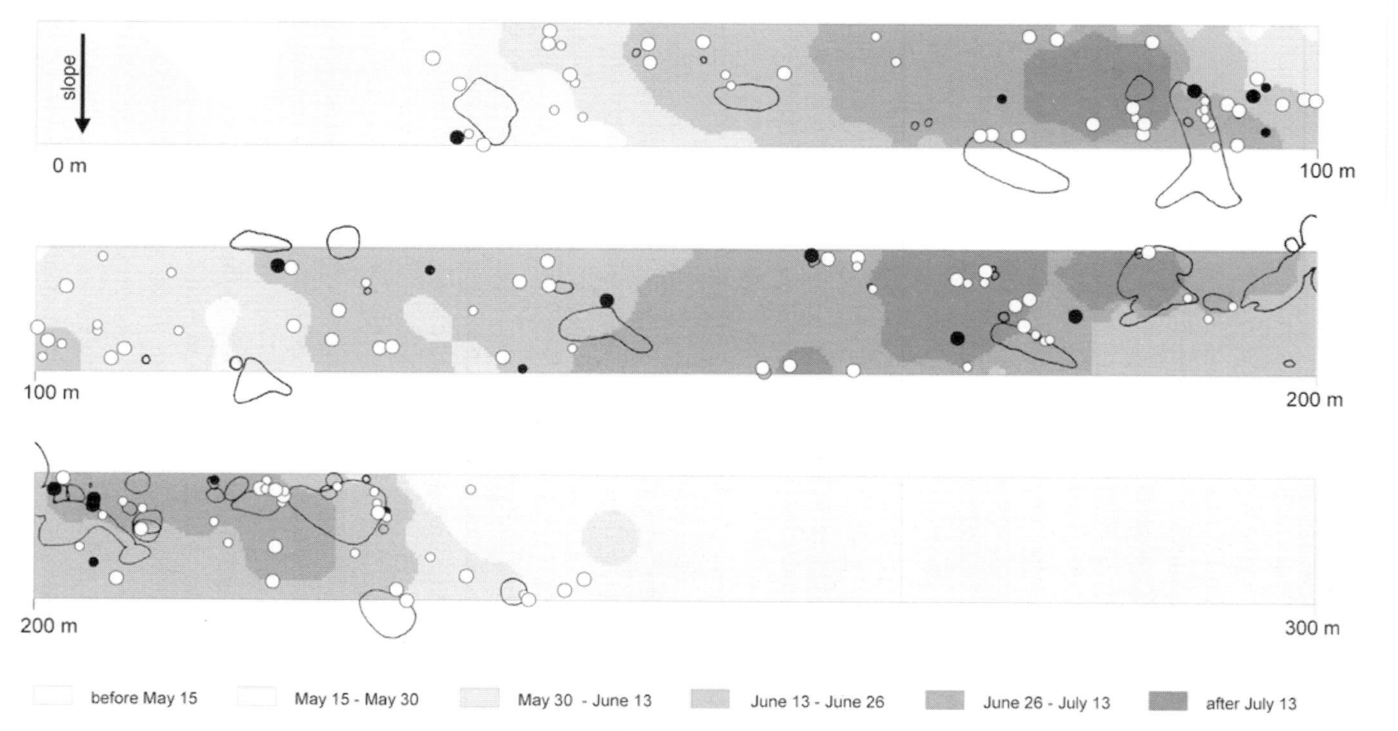

Figure 2 Snowmelt, canopy cover, and whitebark pine regeneration on the Tibbs Butte transect in 1993. Dates of snowmelt are displayed in shades of grey. Black lines show the canopy cover of trees ≥ 1.5 m. Large dots display whitebark pines 0.3 m - < 1.5 m in height, small dots stand for whitebark pines < 0.3 m in height. Dot color indicates the vitality of whitebark pine regeneration, with life whitebark pine = white, and dead whitebark pine = black. The three sections of the diagram represent one continuous transect. The aspect varies from 300°-350°

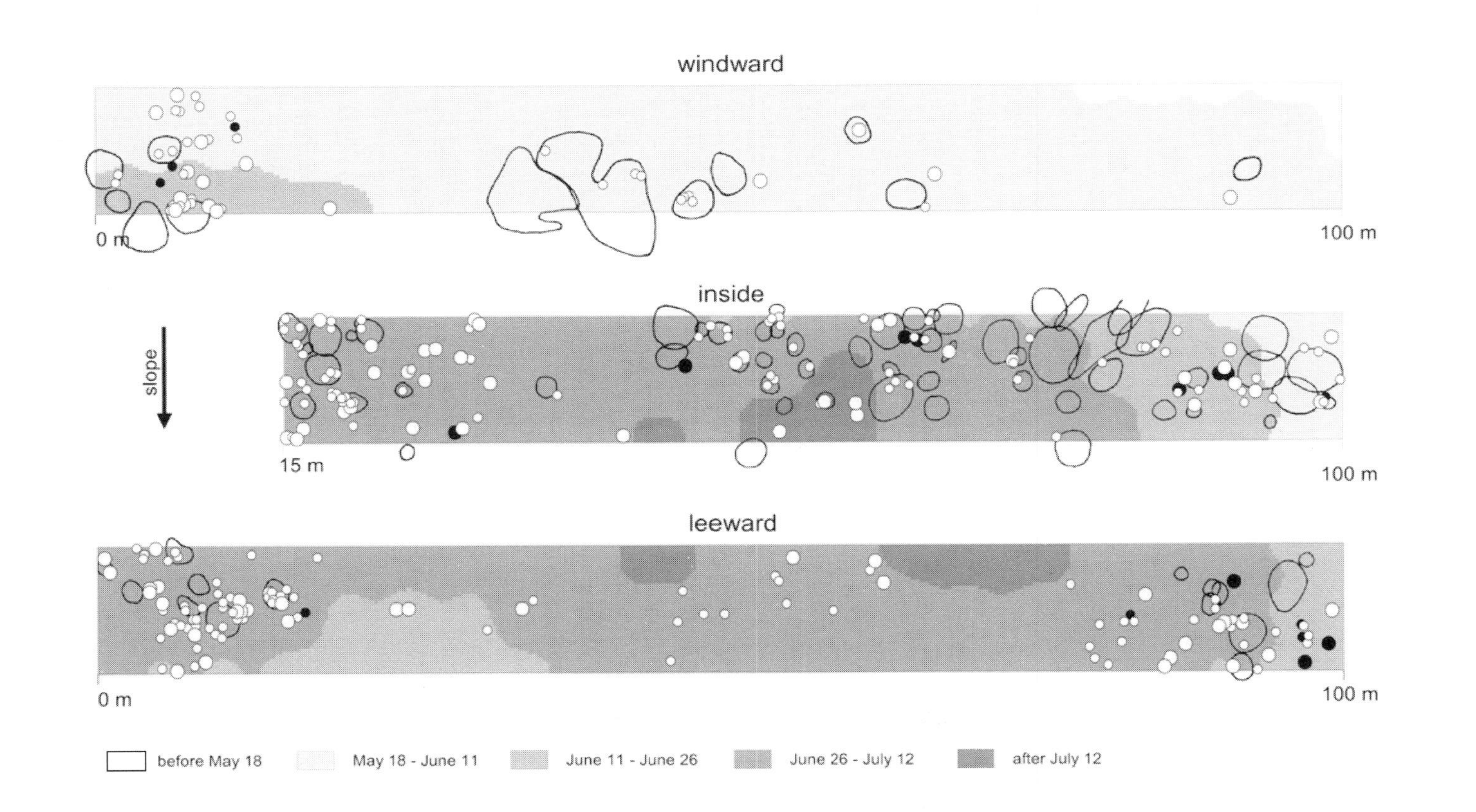

Figure 3 Snowmelt, canopy cover, and whitebark pine regeneration on the Wyoming Creek transects in 1993. See figure 2 for legend. The three sections of the diagram represent three parallel transects. The aspect varies from 30°-60°

Table 1 Density of live whitebark pine regeneration on the Tibbs Butte and Wyoming Creek transects in relation to the time of snowmelt in 1993. Densities were calculated separately for regeneration smaller than 0.3 m in height and regeneration 0.3 m to < 1.5 m in height

Time of snowmelt	**New snow free area**		**Density of regeneration**		**Total regeneration**
Tibbs B. / Wy. Creek	[m²]	[%]	< 0.3 m	0.3-<1.5 m	[%]
Before May 15/18	85	1	0	0	0
May 15/18 – June 13/11	2339	40	1.5	1.2	16
June 13 /11 – June 26	1084	19	3.8	3.5	20
June 26 – July 13/12	1969	34	7.7	4.0	59
After July 13/12	372	6	1.3	3.8	5

snow cover (figures 2 and 3). Few sites on the Tibbs Butte and Wyoming Creek m² transects were snow free before May 15, 1993 and May 18, 1993, respectively. No whitebark pine regeneration was located in the early snow free sites. Thirty-four percent of the transect area melted out between June 26 and July 12/13, 1993. However, 59 % of all regeneration sites with live whitebark pine were found in these areas, a density of 7.9 sites 100 m^{-2} on Tibbs Butte, and 13.4 sites 100 m^{-2} on Wyoming Creek. On July 12/13, 1993, 6 % of the entire transect area was still snow covered. These late snowmelt areas include 5 % of all live whitebark pine regeneration sites.

The seed production of whitebark pine is known to vary significantly between years (Weaver and Forcella 1986; Arno and Hoff 1989). Data on cone production are available from subalpine whitebark pine stands in the Republic Creek drainage south of Cooke City, approximately 40 km east of the study areas. Cone production of whitebark pine in the Republic Creek drainage was good in 1989 and 1991, there were some cones in 1992, and little to no production in 1990 and 1993 (Tomback et al. 2001). This is consistent with observations about whitebark pine seed crops on the Beartooth Plateau between 1991 and 1994. During August and early September 1991, Clark's nutcrackers were seen with filled sublingual pouches near the Tibbs Butte and Wyoming Creek transects. On occasion, nutcrackers were observed caching whitebark pine seeds in the study areas.

Subalpine spruce/fir forests in drainages west and east of Tibbs Butte provide seed sources for Engelmann spruce and subalpine fir. For the Wyoming

Creek transects, the closest spruce and fir seed sources are the subalpine forests in the Wyoming Creek drainage, east of the study area.

All new regeneration located in 1993 and 1994 on the Wyoming Creek and Tibbs Butte transects consisted of whitebark pine seedlings. During this study, not one Engelmann spruce or subalpine fir seedling was located on the transects, neither a new germinant nor a one- to three-year old seedling.

In 1993, 24 seedling clusters with 86 newly germinated whitebark pine seedlings were found in 29 10 m^2 circular plots at Wyoming Creek (10 % of the transect area). The plots also included seven clusters with 16 one- to three-year-old seedlings. In 1994, a new survey of the 10 m^2 plots on the Wyoming Creek transects resulted in five seedling clusters with 16 new germinants. Delayed germination was observed at three clusters from 1993, where seedlings germinating in 1993 were supplemented with three additional seedlings germinating in 1994. On the Tibbs Butte transect, surveys were not completed for 1993. In 1994, only one new seedling and one cluster of three seedlings surviving from 1993 were located in the 30 10 m^2 plots.

Seedling mortality was highest during the first year. On Wyoming Creek, 43 % of 1993 germinants were dead by the end of the summer season. Another 42 % died during the winter of 1993/1994. Seedling mortality dropped to 3 %, 0 %, and 6 % from 1995 to 1997, respectively. This resulted in the survival of three clusters (13 %) with five seedlings (6 %) from 1993 to 1997.

The spatial distribution of whitebark pine germination followed a pattern similar to the distribution of juvenile whitebark pines. Of seedling clusters $\leq$ three years in age, 42 % were located less than one meter from the base of another whitebark pine; 45 % were found in one to three meter distance to a tree base. In 1994, the densities of live seedling clusters germinated between 1991 and 1994 amounted to 0.7 clusters 100 m^{-2} on the Tibbs Butte transect and 6.2 clusters 100 m^{-2} on the Wyoming Creek transects. Cluster densities on Wyoming Creek were two to three times higher on the inside and leeward transects than on the windward transect (figure 4a). In 1993, densities of live whitebark pine clusters germinated between 1991 and 1993 varied between 5.0 clusters 100 m^{-2} on the windward transect to 15.6 clusters 100 m^{-2} inside and 12.0 clusters 100 m^{-2} on the leeward transect. Mortality rates were high on all transects. One year later densities of new germinants and one- to three-year-old seedlings were 3.0 clusters 100 m^{-2} on the wind exposed transect, compared to 6.7 and 9.0 clusters 100 m^{-2} on the inside and leeward transects, respectively. By 1997, only one whitebark pine cluster 100 m^{-2} had survived on the windward transect, compared to 3.3 clusters 100 m^{-2} on the inside and 3.0 clusters 100 m^{-2} on the leeward transect (figure 4b).

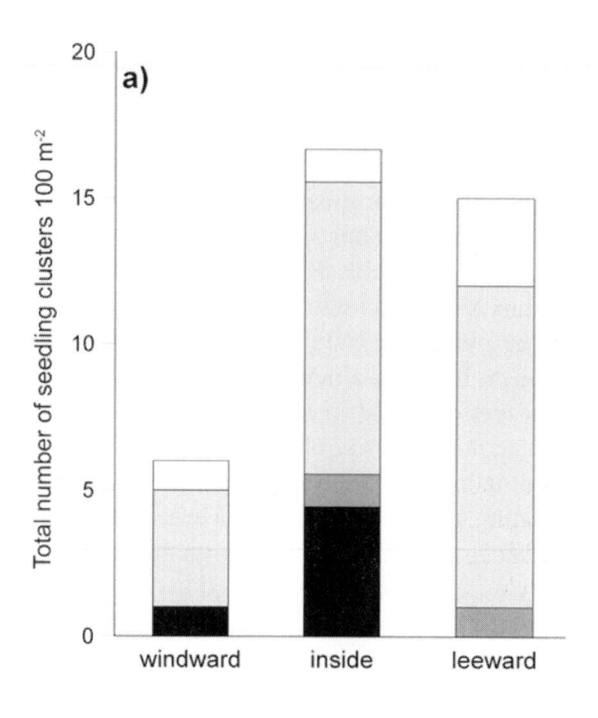

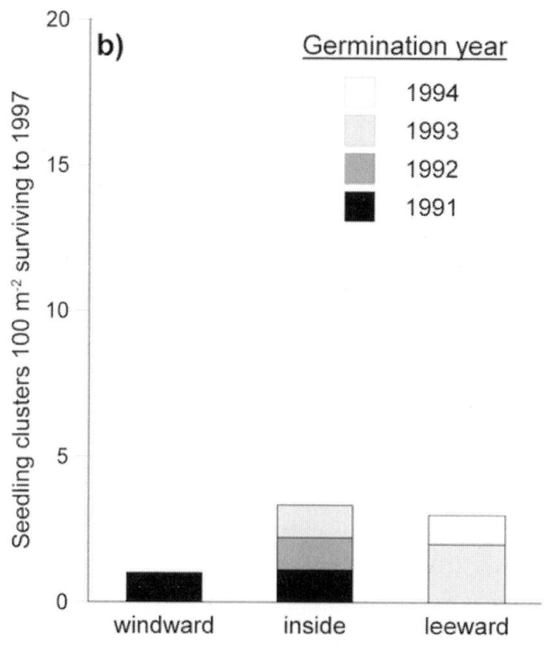

Figure 4 Whitebark pine regeneration densities on the Wyoming Creek transects. a) displays the total densities of seedling clusters ≤ 3 years found in 1993 and 1994 in the 10 m^2-plots; b) shows the densities of seedling clusters surviving to 1997

5 Discussion

The regeneration studies on the transects show that moderate snow cover and intermediate to late snow releases concur with high densities of juvenile whitebark pine ≤ 150 cm in height. Cluster density of seedlings germinated between 1991 to 1994 were higher on the wind protected transects with delayed snowmelt than on the wind exposed transect. In a separate study by the author, the germination and survival of whitebark pine in experimentally seeded sites followed similar trends, with best recruitment in mesic areas with at least moderate snow cover during winter (Mellmann-Brown et al. 2004). In this experiment, suitable site conditions for successful whitebark pine regeneration were indicated by *Salix glauca* and dry *Geum rossii* communities. However, regeneration results of these studies are not consistent with reported caching preferences of the Clark's nutcracker.

According to Tomback (1978; 2001) and Holtmeier (1993), nutcracker caches are frequently established in areas with low snow cover and early snow release. Holtmeier (1993) described higher densities of young limber pines (mostly dispersed by nutcrackers) on wind exposed, convex landforms than on concave surfaces in the timberline ecotone of the Colorado Front Range (also see Wardle 1968). He suggested that these patterns resulted from selective caching preferences of the Clark's nutcracker. Hutchins and Lanner (1982) observed nutcrackers caching 'on all exposures, near a spring, on a streambank, and even in a puddle of water'.

Whether the nutcracker preferably caches seeds on convex slopes, or whether caches are established in almost any exposure, topography, and microsite (Hutchins and Lanner 1982), the spatial distribution of whitebark pine regeneration on the Beartooth Plateau does not appear to be primarily caused by cache distribution pattern. There is no obvious explanation why the nutcracker should prefer areas leeward of tree groups or depressions, where snow tends to linger until midsummer. To the contrary, these sites seem unfavorable for caching, because heavy snow cover restricts access to seeds. There are, however, a number of moisture and temperature related factors that explain patterns of tree regeneration. The importance of moisture for the regeneration of stone pines has been emphasized in studies by Tomback et al. (1993; 2001), Kajimoto et al. (1998), McCaughey and Weaver (1990), and McCaughey (1993; 1994). The germination and survival experiment of whitebark pine at timberline showed higher germination probabilities in areas with late snow release, if temperature conditions were sufficient for germination (Mellmann-Brown et al. 2004). High survival rates were associated with high topographic moisture indices, presence of shade, and low cover percentages of conifer litter. These variables indicate the need for adequate moisture conditions and moderate surface temperatures, two factors that are closely interrelated (Mellmann-Brown et al. 2004). Favorable site conditions were reflected in the vegetational composition of the understory (table 2).

The above results indicate that a sufficient number of whitebark pine seeds are available wherever microsite conditions allow the successful regeneration of whitebark pine, even if these microsites do not coincide with spatial caching preferences of the Clark's nutcracker. Only nutcracker caches that were not retrieved and were established in relatively moist and protected microsites, contribute to the recruitment of whitebark pine in the timberline ecotone. These microsites are more commonly found on concave hillsides, in slight depressions, and adjacent to tree groups than on exposed ridges. In general, concave landforms have ecological disadvantages for tree establishment, i.e., lack of drainage, higher likelihood of avalanches and persisting snow cover late into summer, higher risk of infections with parasitic fungi, and higher probability of nocturnal temperature inversions (Wardle 1968; Senn et al.

Table 2 Regeneration success of seeded whitebark pine in different vegetation types of the timberline ecotone (Mellmann-Brown 2002, modified)

Vegetation type	Site characteristics	Germination	Survival	Growth
Silene acaulis/ Arenaria obtusiloba	wind exposed, dry, low snow cover	poor	none	—
Dryas octopetala	wind eroded soil, low to moderate snow cover	good	poor	very poor
Carex elynoides	dry, low to moderate snow cover	poor	none	—
Dry *Geum rossii*	dry to mesic, moderate snow cover	good	good	fair
Antennaria umbrinella/ Carex phaeocephala	in or near woodland, leeward tree island, high snow cover	very good	poor	poor
Mesic *Geum rossii*	mesic, moderate snow cover, dense herb and grass layer	fair	none	—
Salix glauca	mesic, in slight depressions, moderate to high snow cover	good	very good	good

1994). On the other hand, ecological conditions on convex landforms are not entirely favorable, either. These sites have little to no protection during winter, experience extreme maximum and minimum soil temperatures throughout the year, and may be excessively dry during the latter part of the growing season.

Tree growth on the Beartooth Plateau is generally absent, or at least very stunted, on extreme topographic positions such as pronounced drainages with wetlands and highly exposed ridges. However, many timberline areas on the Beartooth Plateau have little relief, and the danger of creeping snow or avalanches is considerably lower than in rugged mountain ranges. Differences between convex and concave sites are often subtle. The tree regeneration of timberline ecotones with continental climate character may require the additional moisture found in snowdrifts of small depressions or tree groups. This is supported by studies from the Rocky Mountain National Park, where high densities of regeneration coincided with mesic sites parameters or periods of higher precipitation (Weisberg and Baker 1995; Hessl and Baker 1997).

6 Conclusions

Cox concluded for the timberline ecotone on James Peak, Colorado, 'the presence or absence of the tree species appears frequently to be a matter of enough but not too much snow' (Cox 1933, p. 322). This assessment seems well suited for regeneration pattern of whitebark pine on the Beartooth Plateau. In wind exposed gentle terrain (e.g. Tibbs Butte study area), however, trees themselves act as major snow fences, and the question remains how the first trees ever populated these slopes (see Holtmeier 1996; 2000; 2003). A common denominator of 'pioneering' tree groups may be the coarse substrate they tend to grow on (Arno and Hoff 1989; Körner 1998). The protection found between rocks and in small depressions may have been sufficient to allow establishment of isolated tree individuals, which in turn altered the environment around them (Holtmeier 1996). A second, but not exclusive explanation could be the initial establishment of trees under more favorable (in continental areas warmer and moister) climatic conditions. Once established, these 'remnant' trees persist despite moderate climatic variation (Slatyer and Noble 1992; Holtmeier 1995), and may even allow further tree recruitment in sheltered microsites within already established tree clumps.

The above presented relationship between concave landforms and tree recruitment is the result of current environmental conditions. Trends may be reversed in years of high snow accumulation. It is feasible that in years with high and prolonged snow cover slightly convex landforms are more favorable to seedling establishment. Such variable patterns of tree establishment have been shown for the Pacific Northwest, where tree establishment occurred in dry areas when climate conditions were wetter than average, and in wet and cool areas when weather conditions were dryer than average (Fonda and Bliss 1969; Henderson 1973; Woodward et al. 1995). However, the distribution pattern of juvenile trees on the Beartooth Plateau showed similar trends as the distribution pattern of younger whitebark pine seedlings, indicating that critical parameters have not changed significantly during the past 30 - 40 years.

Acknowledgements

Financial support for this study was provided by the Graduiertenförderung Nordrhein-Westfalen, Germany, the Deutscher Akademischer Austausch-dienst, and the USDA Forest Service, Rocky Mountain Research Station, Forestry Sciences Laboratory, in Bozeman, Montana.

Many thanks go to Wyman Schmidt and Ward McCaughey (USDA Forestry Sciences Laboratory, Bozeman, Montana) for logistical support. Kent Houston (USDA Shoshone National Forest, Cody, Wyoming) contributed information and ideas to this study. John Oldemeyer (ret. from the US Fish and Wildlife Service, Fort Collins, Colorado) gave statistical advice and provided useful suggestions to the manuscript.

References

Arno SF, Hammerly RP (1984) Timberline – Mountain and arctic forest frontiers. The Mountaineers, Seattle, Washington

Arno SF, Hoff RJ (1989) Silvics of whitebark pine (*Pinus albicaulis*). USDA Forest Service, Intermountain Research Station, General Technical Report INT-253, Ogden, Utah

Cox CF (1933) Alpine plant succession on James Peak, Colorado. Ecological Monographs 3: 299-372

Critchfield WB, Little EL (1966) Geographic distribution of the pines of the world. USDA Forest Service, Miscellaneous Publication 991, Washington, DC

Eastman JR (1995) IDRISI for Windows, user's guide, version 1.0. Clarks Labs for Cartographic Technology and Geographic analysis, Clarks University, Worchester, Massachusetts

Fonda RW, Bliss LC (1969) Forest vegetation of the montane and subalpine zones, Olympic Mountains, Washington. Ecological Monographs 39: 271-301

Hayashida M (1994) Role of nutcrackers on seed dispersal and establishment of *Pinus pumila* and *Pinus pentaphylla*. In: Schmidt WC, Holtmeier F-K (comps) Proceedings – International workshop on subalpine stone pines and their environment: The status of our knowledge. USDA Intermountain Research Station, General Technical Report INT-GTR-309. Ogden, Utah, pp 159-162

Henderson JA (1973) Composition, distribution and succession of subalpine meadows in Mount Rainier National Park. Ph.D. dissertation, Oregon State University, Corvallis

Hessl AE, Baker WL (1997) Spruce and fir regeneration and climate in the forest-tundra ecotone of Rocky Mountain National Park, Colorado, U.S.A. Arctic and Alpine Research 29: 173-183

Holtmeier F-K (1965) Die Waldgrenze im Oberengadin in ihrer physiognomischen und ökologischen Differenzierung. Dissertation, Mathematisch-Naturwissenschaftliche Fakultät, Universität Bonn, Germany

Holtmeier F-K (1966) Die ökologische Funktion des Tannenhähers im Zirben-Lärchenwald und an der Waldgrenze im Oberengadin. Journal für Ornithologie 4: 337-345

Holtmeier F-K (1974) Geoökologische Beobachtungen und Studien an der subarktischen und alpinen Waldgrenze in vergleichender Sicht (nördliches Fennoskandien/Zentralalpen). Erdwissenschaftliche Forschung 8

Holtmeier F-K (1993) Der Einfluß der generativen und vegetativen Verjüngung auf das Verbreitungsmuster der Bäume und die ökologische Dynamik im Waldgrenzbereich. Beobachtungen und Untersuchungen in Hochgebirgen Nordamerikas und den Alpen. Geoökodynamik 14: 153-182

Holtmeier F-K (1995) Waldgrenzen und Klimaschwankungen. Ökologische Aspekte eines vieldiskutierten Phänomens. Geoökodynamik 16: 1-24

Holtmeier F-K (1996) Die Wirkungen des Windes in der subalpinen und alpinen Stufe der Front Range, Colorado, U.S.A. In: Holtmeier F-K (comp) Beiträge aus den Arbeitsgebieten am Institut für Landschaftsökologie. Arbeiten aus dem Institut für Landschaftsökologie, Westfälische Wilhelms-Universität Münster 1: 19-45

Holtmeier F-K (1999) Tiere als ökologische Faktoren in der Landschaft. Arbeiten aus dem Institut für Landschaftsökologie, Westfälische Wilhelms-Universität Münster 6

Holtmeier F-K (2000) Die Höhengrenze der Gebirgswälder. Arbeiten aus dem Institut für Landschaftsökologie, Westfälische Wilhelms-Universität Münster 8

Holtmeier F-K (2003) Mountain Timberlines – Ecology, patchiness, and dynamics. Advances in Global Change Research, 14. Kluwer Academic, Dordrecht

Hutchins HH, Lanner RM (1982) The central role of Clark's nutcracker in the dispersal and establishment of whitebark pine. Oecologia 55: 192-201

Kajimoto T, Onodera H, Ikeda D, Daimaru H, Seki T (1998) Seedling establishment of subalpine stone pine (*Pinus pumila*) by nutcracker (*Nucifraga*) seed dispersal on Mt. Yumori, Northern Japan. Arctic and Alpine Research 30: 408-417

Körner Ch (1998) A re-assessment of high elevation treeline positions and their explanation. Oecologia 115: 445-459

Kuoch R, Amiet R (1970) Die Verjüngung im Bereich der oberen Waldgrenze der Alpen. Mitteilungen der schweizerischen Anstalt für das forstliche Versuchswesen 46: 159-328

Lanner RM (1982) Adaptations of whitebark pine for seed dispersal by Clark's nutcracker. Canadian Journal of Forest Research 12: 391-402

Lanner RM (1990) Biology, taxonomy, evolution and geography of stone pines of the world. In: Schmidt WC, McDonald KJ (comps) Proceedings – Symposium on whitebark pine ecosystems: Ecology and management of a high-mountain resource. USDA Forest Service, Intermountain Research Station, General Technical Report INT-270. Ogden, Utah, pp 14-24

Lanner RM (1996) Made for each other: A symbiosis of birds and pines. Oxford University Press, New York

Löve D (1970) Subarctic and subalpine: Where and what? Arctic and Alpine Research 2: 63-73

Mattes H (1978) Der Tannenhäher (*Nucifraga caryocatactes* L) im Engadin - Studien zu seiner Ökologie und Funktion im Arvenwald. Münsterische Geographische Arbeiten 2

Mattes H (1982) Die Lebensgemeinschaft von Tannenhäher und Arve. Eidgenössische Anstalt für das forstliche Versuchswesen, Berichte 241

Mattes H (1985) The role of animals in cembran pine forest regeneration. In: Turner H and Tranquillini W (eds) Establishment and tending of subalpine forest: Research and management. Proceedings 3rd IUFRO workshop P1.07-00, Eidgenössische Anstalt für das forstliche Versuchswesen, Berichte 270: 195-205

McCaughey WW (1993) Delayed germination and seedling emergence of *Pinus albicaulis* in a high-elevation clearcut in Montana, U.S.A. In: Edwards DG (comps) Proceedings – Symposium: Seed dormancy and barriers to germination. Forestry Canada, IUFRO project group P2.04-00. Victoria, British Columbia, Canada, pp 67-72

McCaughey WW (1994) The regeneration process of whitebark pine. In: Schmidt WC, Holtmeier F-K (comps) Proceedings – International workshop on subalpine stone pines and their environment: The status of our knowledge. USDA Intermountain Research Station, General Technical Report INT-GTR-309. Ogden, Utah, pp 179-187

McCaughey WW, Schmidt WC (2001) Taxonomy, distribution and history. In: Tomback DF, Arno SF, RE Keane (eds) Whitebark pine communities: Ecology and restoration. Island Press, Washington, DC, pp 29-40

McCaughey WW, Weaver T (1990) Biotic and microsite factors affecting whitebark pine establishment. In: Schmidt WC, McDonald KJ (comps) Proceedings - Symposium on whitebark pine ecosystems: Ecology and management of a high-mountain resource. USDA Forest Service, Intermountain Research Station, General Technical Report INT-270, Ogden, Utah, pp 140-150

Mellmann-Brown S (2002) The regeneration of whitebark pine in the timberline ecotone of the Beartooth Plateau, Montana and Wyoming. Ph.D. dissertation, Westfälische Wilhelms Universität, Münster, Germany

Mellmann-Brown S, McCaughey WW, Holtmeier F-K (2004) Germination and survival of *Pinus albicaulis* in the timberline ecotone of the Beartooth Plateau, Montana and Wyoming. In preparation.

Price RA, Liston A, Strauss SH (1998) Phylogeny and systematics of *Pinus*. In: Richardson DM (ed) Ecology and biogeography of Pinus. Cambridge University Press, New York, pp 49-68

Senn J, Schönenberger W, Wasem U (1994) Survival and growth of planted Cembran pines at the alpine timberline. In: Schmidt WC, Holtmeier F-K (comps) Proceedings – International workshop on subalpine stone pines and their environment: The status of our knowledge. USDA Intermountain Research Station, General Technical Report INT-GTR-309. Ogden, Utah, pp 105-109

Slatyer RO, Noble IR (1992) Dynamics of montane treelines. In: Hansen AJ, di Castri F (eds) Landscape boundaries. Consequences for biotic diversity and ecological flows. Ecological Studies 92. Springer, Berlin, pp 346-359

Tomback DF (1978) Foraging strategies of Clark's nutcracker. Living Bird 16:123-161

Tomback DF (1982) Dispersal of whitebark pine seeds by Clark's nutcracker: A mutualism hypothesis. Journal of Animal Ecology 51: 451-467

Tomback DF (2001) Clark's nutcracker: Agent of regeneration. In: Tomback DF, Arno, SF, Keane RE (eds) Whitebark pine communities: Ecology and Restoration. Island Press, Washington, DC, pp 89-104

Tomback DF, Linhart YB (1990) The evolution of bird-dispersed pines. Evolutionary Ecology 4: 185-219

Tomback DF, Sund SK, LA Hoffmann (1993) Post-fire regeneration of *Pinus albicaulis*: Height-age relationships, age structure, and microsite characteristics. Canadian Journal of Forest Research 23: 113-119

Tomback DF, Anderies AJ, Carsey KS, Powell ML, Mellmann-Brown S (2001) Delayed seed germination in whitebark pine and regeneration patterns following the Yellowstone fires. Ecology 82: 2587-2600

Wardle P (1968) Engelmann spruce (*Picea engelmannii* Engel.) at its upper limit on the Front Range, Colorado. Ecology 49: 483-495

Weaver T, Forcella F (1986) Cone production in *Pinus albicaulis* forests. In: Shearer RC (comp) Proceedings – Conifer tree seed in the Inland Mountain West Symposium. USDA Forest Service, Intermountain Research Station, General Technical Report INT-203. Ogden, Utah, pp 68-76

Weisberg PJ, Baker WL (1995) Spatial variation in the tree regeneration in the forest-tundra ecotone, Rocky Mountain National Park, Colorado. Canadian Journal of Forest Research 25: 1326-1339

Woodward A, Schreiner EG, Silsbee DG (1995) Climate, geography, and tree establishment in subalpine meadows of the Olympic Mountains, Washington, U.S.A. Arctic and Alpine Research 27: 217-225

Structure and the Composition of Species in Timberline Ecotones of the Southern Andes

William Pollmann and Renate Hildebrand

Abstract

The subalpine southern Andean timberline is characterized by deciduous *Nothofagus pumilio* forests, which change with increasing altitude to *Nothofagus* krummholz, built of deciduous *N. pumilio* and *N. antarctica.* The current study examines the structure and species composition in timberline ecotones of the southern Andes to assess whether the deciduous timberline in the Andes results as convergent structure of, for example, *Betula*-timberlines of the Northern Hemisphere.

Through a Braun-Blanquet phytosociological approach, we show how the characteristic structure and combination of timberline vegetation in the southern Andes vary in latitudes from 33°S to 55°S and altitudes between 2000 m and 600 m. Timberline composition was distinct at sites and included a variety of assemblages ranging from northern Azaro-Nothofagetalia communities to assemblages of the Adenocaulo-Nothofagetalia and Violo-Nothofagetalia in the South.

Krummholz generally occurred in a ten to some hundred meter wide ecotone. Four distinctive *N. pumilio* growth forms were identified; (1) 'elfin woods' where trees become more and more stunted as they approach the woodland limit readily observed in previous studies; (2) 'cornice-like' growth form characterized by branches close to the ground; (3) restricted to the lee-ward side, a single-stemmed habit occurred that was characterized by stems aligned downhill before curving up into vertical alignment; and (4) on the wind-exposed side, growth forms were characterized by single-stemmed habit and long branches running uphill. These growth form changes were apparently controlled by changes in abiotic factors and climate rather than genetically determined.

In conclusion, the recognition of the deciduous foliage and deformed habit of timberline trees in the southern Andes is explained by extreme climate. Comparison of same latitudes revealed that maximum altitudes of southern Andean timberline were generally higher than those of evergreen *Nothofagus* limits in New Zealand but lower than timberlines of Holarctic mountain ranges.

These results serve to emphasize that understanding structure and physiognomy of the southern Andean timberline and its apparent stability may require attention to ecophysiological adaptations and responses of timberline *Nothofagus* to high-mountain environment.

1 Introduction

Latitudinal and altitudinal timberlines are ecotones between continuous forests of erect trees and adjacent non-forested vegetation (Holtmeier 1985; 2003). At high altitudes, the transition between these two communities includes a 'krummholz' forest, trees of low stature with decurrent and deformed stems (Daubenmire 1954; Wardle 1965; Tranquillini 1979; Grace 1989; Armand 1992; Körner 1998; Miehe and Miehe 2000; Holtmeier 2003). Krummholz growth forms are genetically determined in some species, but are more frequently a phenotypic response to climatic stress (J. Wardle 1974; P. Wardle1998; Holtmeier 2003). Holtmeier (1981) differentiated between environmentally determined and genetically determined dwarf trees, applying the term 'krummholz' to the latter only (e.g. *Pinus mugo* var. *prostrata*, *Alnus viridis*). However, in many studies krummholz is used in a wider sense, applying this term to all stunted and dwarfed trees at upper limit of forest margin (cf. Tranquillini 1979; Holtmeier 2003).

At mountain timberlines, vegetation grades from continuous forest of erect trees to non-forested vegetation. The transition zone between forest and non-forested vegetation, with the upper and lower boundaries delimited by 'timberline' and 'treeline' is termed 'timberline ecotone' (Daubenmire 1954; Wardle 1965; Ellenberg 1966; Tranquillini 1979; Holtmeier 1985; 1989; 1995), and is a major ecological boundary (Armand 1992; Holtmeier 2003). Timberline is the boundary between trees with erect and krummholz growth forms. Above timberline, the density of trees decreases and forest cover becomes less continuous as the krummholz forest grades into non-forested vegetation.

Depending on the forest type, environmental conditions and site history, the structure and extent of krummholz forests and growth forms vary markedly (e.g. Holtmeier 1973; 1989; 2003). For example, the lichen-spruce forest limit in Canada is extended for several kilometers (Lavoie and Payette 1994), whereas the *Nothofagus* krummholz in South America and New Zealand is narrow, limited to only ten to some hundred meters of crippled trees or 'elfin woods' (Wardle 1965; 1998; Ellenberg 1966).

The timberline ecotone in the southern Andes is still for the most part in its natural state compared, for example, with the Northern Hemisphere (the Alps; Ellenberg 1966; 1996). Above ca. 1000 m in the southern Andes at 40°S, the broad-leaved evergreen trees are absent and, thus, monotypic *Nothofagus* form subalpine forests. Accordingly, the timberline ecotone is generally dominated by krummholz forms of *N. pumilio* with some *N. antarctica* (Daniels and Veblen 2003a). With increasing altitude, *N. pumilio* becomes more and more stunted as it approaches the woodland limit but even at timberline it forms dense stands (figure 1). The south-Andean *N. pumilio* trees serve as a good example for an upper woodland limit in which case timberline and treeline might be nearly equivalent. At undisturbed sites, the width of the ecotone, measured as the distance from timberline to treeline, was limited to a few meters. Where trees have been cut and grazing practiced ('anthropo-zoogenic impacts'), timberline ecotone was commonly >100 m wide (Ellenberg 1966; Veblen et al. 1977). In addition, the Andean forest landscape is dynamic and frequently disturbed (Veblen et al. 1996). Several studies observed that timberline in the southern Andes is also influenced by episodic catastrophic disturbances, such as earthquake-triggered mass movements, volcanic eruptions, and snow avalanches (Veblen et al. 1977; 1981; Veblen and Ashton 1978; 1979; Veblen 1979; 1989a; 1989b). Although large-scale, infrequent disturbances commonly influence *Nothofagus* forests in the southern Andes (Pollmann and Veblen in press), undisturbed *N. pumilio* forest-to-timberline ecotones as described by Ellenberg (1966) can be observed in various areas throughout the Andean range.

Nothofagus pumilio timberlines in the southern Andes differ from northern temperate timberlines and also from New Zealand *Nothofagus* timberlines in

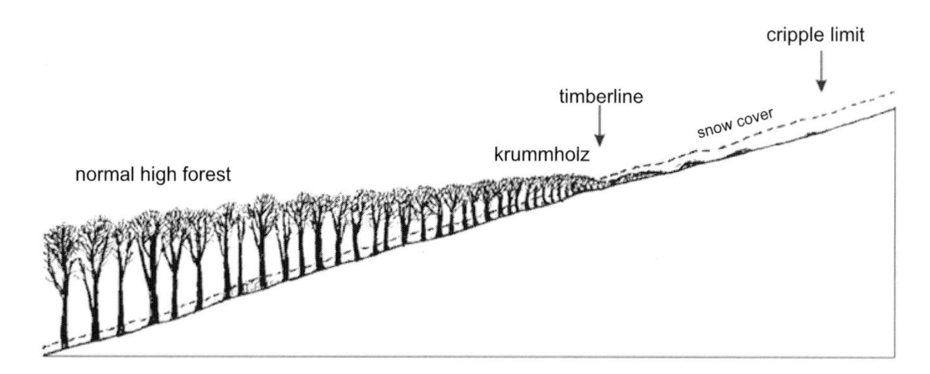

Figure 1 The upper woodland limit in the Andes near 40° S. Under natural conditions, southern beech (*Nothofagus pumilio*) becomes more and more stunted as it approaches the limit but even here it forms dense stands. After Ellenberg (1966)

important ways. *Nothofagus pumilio* is a broadleaf, deciduous species with different growth forms and vegetation structure at timberline relative to the treelines of mostly evergreen conifers in the Northern Hemisphere (Ellenberg 1996; Holtmeier 2003), and evergreen *Nothofagus* treeline in New Zealand, where for example, mountain beech (*N. solandri* var. *cliffortioides*) and silver beech (*N. menziesii*) form timberline (Ogden et al. 1996).

Studies in the Northern Hemisphere and in New Zealand have shown that a large variety of growth forms occur in the timberline ecotone and that they are caused by wind and snow (Holtmeier 1973; 1980; Norton and Schönenberger 1984) as well as biotic and other factors (Holtmeier 2003). Although several different growth forms have been classified for *Nothofagus* timberlines in New Zealand (Wardle 1963; 1965; 1971; Norton and Schönenberger 1984) less has been made to describe the main forms present in Andean *Nothofagus* krummholz (Daniels and Veblen 2003a). Treeline elevations were higher east of the Andes, reflecting the more continental climate in Argentina than in Chile, plus regional impacts of volcanic eruptions (Daniels and Veblen 2003b).

The purpose of the present investigation is to examine the structure and species composition in timberline ecotones of the Chilean Andes between 33° and 55°S. Multi-scale (from broad- to fine scale, sensu Whiters and Meentemeyer 1999) analysis of vegetation data, such as distribution, structure, composition and competition, and abiotic factors, such as snow and wind were used to characterize the structure of *N. pumilio* timberlines within latitudinal and altitudinal gradients. The study integrates a regional scale analysis of distribution and species composition of timberline vegetation with an intensive spatial analysis at stand and single tree scale of timberline structures related to snow cover severity and wind deformation. Comparisons on growth forms are made between the Andean *Nothofagus* timberline and the evergreen *Nothofagus* timberlines in New Zealand and Tasmania (P. Wardle 1973; 1974; Norton and Schönenberger 1984; Ogden et al. 1996; Read and Brown 1996), and northern deciduous timberlines (Skottsberg 1910; 1916; Kalela 1941a; 1941b; Godley 1960; Oberdorfer 1960; Haemet-Ahti 1963; 1986; Wardle 1965; Haemet-Ahti and Ahti 1969). The specific objectives are (1) to document the structure of *N. pumilio* timberline ecotone as a result of abiotic factors, (2) to examine the species composition of high-elevated timberline ecotone (between 35° and 55°S), and (3) to assess whether the deciduous timberline of the southern Andes results as convergent structure of subarctic and subalpine *Betula*-timberlines, and submeridional *Fagus*-timberlines of the Northern Hemisphere.

2 Study Area

2.1 Climate and Forest Differentiation

The stands sampled are located in the distribution area of *Nothofagus pumilio* krummholz between 35° and 55°S in the Andean Range (figure 2). Two well-marked climatic gradients are observed in the southern Andes, reflected by the natural forest vegetation. There is a steep north-to-south gradient of vegetation types from Mediterranean type forests in the Central Chilean forest region to rainforests in the Valdivian and Patagonian forest region of south-central and southern Chile, to the Magellanic forest region at Tierra del Fuego (Veblen et al. 1995; 1996; Seibert 1996; Pollmann 2001; Hildebrand-Vogel 2002). This environmental gradient is one of the most striking vegetation changes reflecting steep rainfall differences from ca. 1000 mm per year in Central Chile to ca. 10000 mm per year at Tierra del Fuego (Taljaard 1972; Miller 1976, Prohaska 1976). Rain and snow are concentrated in fall and winter, when over 75 % of the annual precipitation occurs (Miller 1976). In addition, there is also an important west-east moisture gradient. Precipitation increases with elevation, reaching a maximum at Andean peaks. The precipitation on the mountain peaks is usually at least three to four times higher than the amount over the central Valley at the same latitude (Miller 1976). Mean annual air temperature decreases from north to south, from 14.9°C in Talca to 6.0°C at Navarino Island (Amigo and Ramírez 1998). Wardle (1998) compared matched climate stations at timberline in the southern Andes and in New Zealand, and confirmed that annual temperature ranges were similar in the two regions, both near sea and at inland localities, where annual ranges were wider. However, the Andean stations appeared to be colder, even after temperatures have been calibrated to adjust for altitudinal and latitudinal differences between the matched stations. The relative coldness of the Andean stations may probably result from the proximity of the cold Peruvian current and from clear and cold nights during the summer dry season. Monthly mean maximum and minimum air temperatures from December through April average 1°C and 2.5°C cooler at the Andean stations; and severe summer frosts appeared. Subalpine *N. pumilio* timberline is covered by snow for approximately nine months each year, from April to December. In contrast to the regular northern hemispheric feature, the *N. pumilio* timberline ecotone lacks permafrost at all (Troll 1948; 1960; Godley 1960; Miehe and Miehe 2000).

Forest vegetation in montane and subalpine elevation zones of southern Chile and adjacent Argentina have been extensively described, for example,

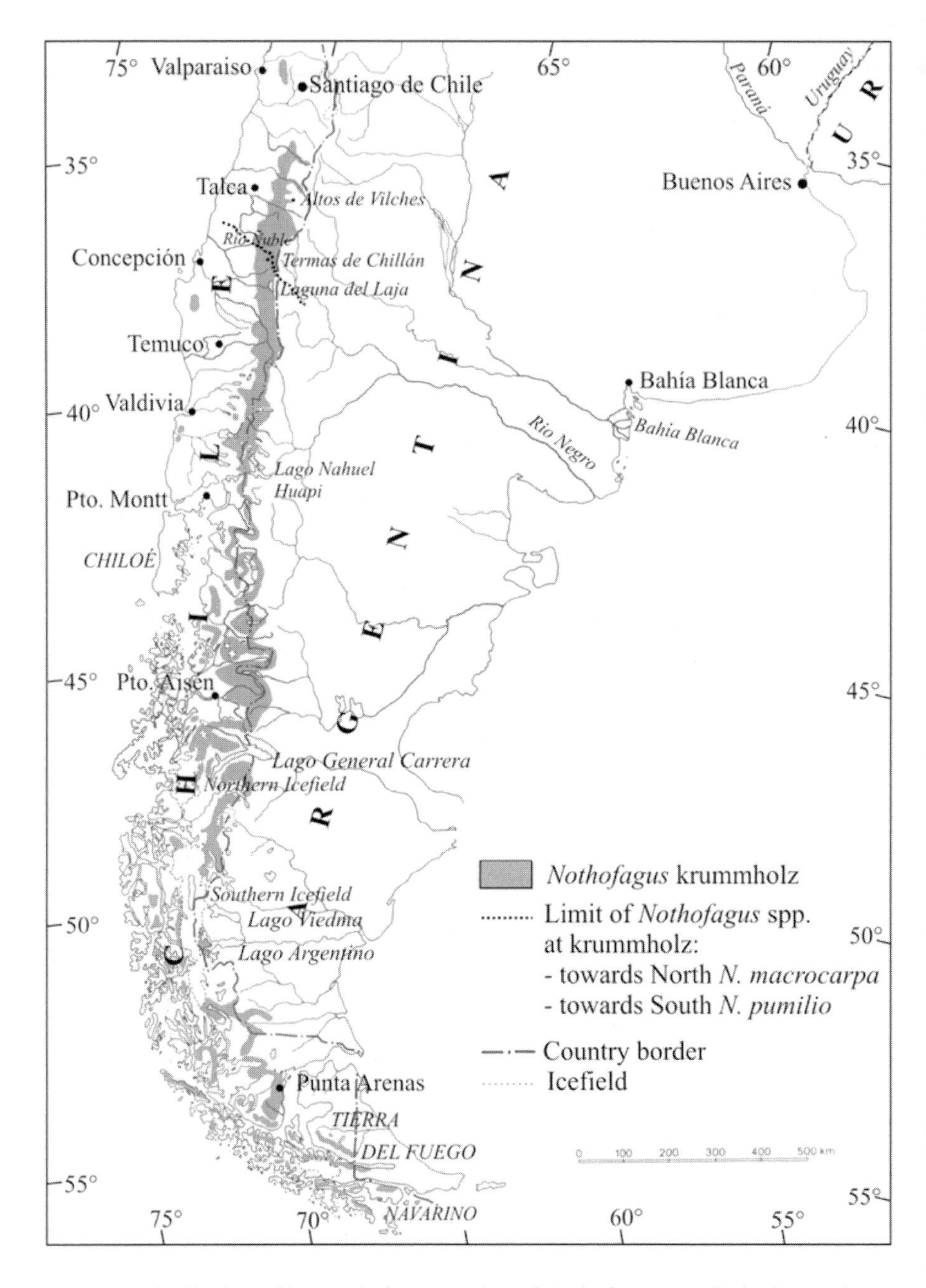

Figure 2 Distribution of krummholz vegetation of *Nothofagus pumilio* in the southern Andes

by Oberdorfer (1960), Gajardo (1994), Seibert (1996), Eskuche (1999) and Pollmann (2001). Southern Hemisphere climatic conditions and peculiarities of South American vegetation history led to forest types characterized by combinations of evergreen and deciduous elements, which are distributed naturally according to climatic features, and reflect these climatic gradients (Hildebrand-Vogel 2002).

2.2 High-elevation Forests in the Southern Andes

Within the area of approximately 2000 km from north-to-south (35°S - 55°S), *N. pumilio* occurs at both cooler and dryer sites than its congeners, *N. dombeyi*, *N. betuloides*, and *N. alpina.* Thus, mixed deciduous-evergreen *Nothofagus* forests give way to monotypic *N. pumilio* forests at higher elevations and in more easterly locations in the rain-shadow of the Andes (Kozdon 1958; Eskuche 1973; Hildebrand-Vogel et al. 1990; 1998; Finckh 1996; Daniels and Veblen 2003a; Pollmann and Veblen in press). Although *N. pumilio* forms almost pure subalpine forests at timberline, it may also be associated with other species, such as *Araucaria araucana* and *N. antarctica.* There is evidence that mixed stands at timberline may be due to fire-induced changes in forest composition (Veblen et al. 1996). The failure of regeneration is most conspicuous for *N. pumilio* on drier sites at high elevations (e.g. Northern Patagonia), where Veblen et al. (1996) observed a lack of tree seedlings in numerous burns more than forty years old, likely due to fire-induced edaphic changes or simply to opportunistic prior establishment of a dense cover of herbaceous plants, or strong desiccating winds (Holtmeier 2003).

Between 37°30′S and 39°30′S, *Araucaria araucana* is a component of the *N. pumilio* forests of relatively dry habitats (cf. Finckh and Paulsch 1995). *Araucaria*, superbly adapted to fire by its thick bark, protected terminal buds, and ability to resprout after being burned (Veblen 1982; Burns 1993), appears to be one of the longest-lived South American tree species, occasionally attaining ages of more than 1000 years.

Disturbances such as fire, volcanism, snow avalanches, and windthrow have an important influence on forest vegetation of the southern Andes (cf. Veblen et al. 1981; Pollmann and Veblen in press), and some timberlines have been depressed and greatly affected, for example, by volcanism (Veblen et al. 1977; Veblen and Ashton 1978; 1979; Veblen 1979). However, recent studies suggested that old-growth *N. pumilio* forests are in a steady state (sensu Whittaker 1975) in that the dominant species is regenerating primarily by treefall gap dynamics (Rebertus and Veblen 1993; Rebertus et al. 1993; Veblen et al. 1996).

2.3 Timberline in the Southern Andes

In the southern Andes, *N. pumilio* occurs at timberline from 35°30′ to 55°S (figure 2), ranging from 2000 m in the North (Talca, 7th Region) to ca. 600 m in the South (Isla Navarino, 12th Region), above which andine grass- or shrublands occur. Other tree species making up timberline stands are *Nothofagus macrocarpa*, to be found only in the northernmost part of the forest region (33°-35°30′S) and *Nothofagus antarctica*, appearing together with or instead of *N. pumilio* through the whole forest region from 35°30′ to 55°S. Non-climatic factors such as soil moisture, nutrient availability, competition, and disturbance also influence the timberline ecotone (e.g. Hall et al. 1992; Holtmeier and Broll 1992; Slatyer and Noble 1992; Noble 1993; Holtmeier 1994; Lloyd and Graumlich 1997).

Several traits make *N. pumilio* well suited to grow at high elevation: (1) it is deciduous and is one of the most tolerant of low temperatures among tree species in southern South America (Sakai et al. 1981; Alberdi et al. 1985; Alberdi 1987; 1995). However, in case of frequent disturbances *N. antarctica* with its tendency to grow shrublike seems to be advantageous. (2) *Nothofagus pumilio* produces new leaves earlier in the spring, probably giving *N. pumilio* a competitive advantage over *N. antarctica* (Veblen et al. 1977; Lechowicz 1984). There is morphological evidence that juvenile krummholz trees grow to erect adult trees when conditions are favorable (Veblen et al. 1977; Hildebrand-Vogel et al. 1990). (3) Phenotypic plasticity likely benefits its long-term survival at timberline. (4) *Nothofagus pumilio* is a prolific seed producer (Rusch 1993; Veblen et al. 1996; Cuevas 2000). Although seed remains viable only one year after production, small seedlings can be extremely abundant and persistent for many years (Veblen et al. 1981; Cuevas and Arroyo 1999; Cuevas 2000). (5) Finally, vegetative reproduction of *N. pumilio* from basal shoots and layering occurs, but it is not the dominant form of reproduction (Veblen et al. 1996).

3 Methods

3.1 Analysis of Vegetation Data

Field work was carried out from the spring 1995 to the summer of 1998; 51 phytosociological relevés according to the Braun-Blanquet approach were obtained (Braun-Blanquet 1964; Dierschke 1994), at sites with krummholz vegetation, selected on the basis of physiognomy, structure and species dominance (e.g. *N. pumilio, N. antarctica*). The plot size was based on the concept of the minimum area of relatively uniform stands (Westhoff and van der Maarel 1973), and usually ranged from 80 to 100 m^2 for the majority of the krummholz stands (cf. Ramírez et al. 1997). The abundance/dominance values followed the 6-grade scale of Braun-Blanquet (1964), modified by Reichelt and Wilmanns (1973). The environmental data collected for each site include altitude, slope, aspect, depth of snow cover, and in addition, the altitudinal range of the timberline ecotone between 35° and 55°S. The species nomenclature follows Marticorena and Quezada (1985) and Marticorena and Rodríguez (1995; 2001), with the exceptions from Landrum (1988; 1999), Matthei (1997), and Vázquez and Rodríguez (1999); nomenclature of syntaxonomy follows Hildebrand (in press). The characterization of homogenous plant communities is one of the principle aims in phytosociology and landscape ecology (e.g. Mueller-Dombois and Ellenberg 1974). Syntaxonomical sorting was used to detect and characterize preliminary vegetation types in the data matrix (Bergmeier et al. 1990; Dierschke 1994).

3.2 Analysis of Krummholz Structure

In each sample plot, shrub height and length of creeping trunk of *Nothofagus* timberline trees were measured. Spatial distribution and vertical structure of *Nothofagus* krummholz vegetation were assessed by measuring and mapping branching patterns. In three transects of approximately 5 m in length, detailed stand-scale data were collected within the timberline ecotone. Within each study area, growth forms were examined to compare northern (at ca. 37°S, 1950 m, Termas de Chillán) and southern (at ca. 55°S, 600 m, Navarino Island) *N. pumilio* krummholz. Timberline profiles easily permit comparisons between growth forms of krummholz in the southern Andes and in other parts of the world (cf. Norton and Schönenberger 1984; Holtmeier 2003).

4 Results

4.1 Timberline Vegetation

Nothofagus krummholz and species composition fit within the framework of regional forest syntaxa. In dependence of length of vegetation period at timberline undergrowth features vary between well developed (e.g. covering > 95%) and nearly missing. Abundant are character species of the Nothofagetea pumilionis class, such as *Maytenus disticha, Rubus geoides* and *Berberis montana* together with the class character species of Andean heathlands, *Pernettya pumila.*

In the northernmost part of the forest region, three *Nothofagus* species (*N. macrocarpa, N. antarctica, N. pumilio*) occur at timberline (table 1). Single-

Table 1 Species composition and classification in the southern Andes between 35°S and 37°S

Class I: Nothofagetea pumilionis — Class II: Nothofagetea antarcticae
Order *1*: Adenocaulo-Nothofagetalia pumilionis
Order *2*: Azaro alpinae-Nothofagetalia pumilionis

Unit	I.1	I.1	I.1	I.1	I.1	I.1	I.1	I.2	I.2	I.2	I.2	II	II	II	II	II	II
Number	1	2	3	4	5	6	7	8	9	10	11	12	13	14	15	16	17
Authors	*	*	*	*	*	*	*	S	S	S	S	S	S	S	S	S	S
Elevation (10 m a.s.l.)	179	175	190	190	190	167	167	-	-	-	-	-	-	-	-	-	-
Aspect	SE	SE	.	SE	SE	S	SE	-	-	-	-	-	-	-	-	-	-
Slope angle (°)	44	20	.	32	20	10	5	-	-	-	-	-	-	-	-	-	-
Structure																	
Cover tree layer [%]	75	80	85	85	85	90	85	-	-	-	-	-	-	-	-	-	-
Height tree layer [m]	3	5	1	2	3	10	6	-	-	-	-	-	-	-	-	-	-
Cover shrub layer [%]	5	30	.	.	.	70	70	-	-	-	-	-	-	-	-	-	-
Height shrub layer [m]	<1	1	.	.	.	2	2	-	-	-	-	-	-	-	-	-	-
Cover herb layer [%]	<1	<1	<1	40	2	25	15	-	-	-	-	-	-	-	-	-	-
Number of species	7	8	2	2	5	12	19	12	10	13	7	13	16	14	15	17	28
Adenocaulo-Nothofagetetalia pumilionis																	
c. c *Nothofagus pumilio*	5	5	5	5	5	5	5	5	5	5	5	5	5	2a	2a	.	.
c. c *Maytenus disticha*	1	2a	.	3	1	.	.	1	1	1	1	1	.	.	.	.	.
d. o *Pernettya pumila*	.	.	.	.	.	+	+	+	+	+	+	.	.	.	.	.	+
d. o *Myoschilos oblonga*	.	1	.	.	.	.	.	+	+	+	+	.	.	.	.	.	.
d. o *Berberis montana*	r	r	.	.	+	+	1	.	.	.	.	.	.	.	.	.	.
d. o *Escallonia alpina*	.	.	.	.	.	.	1	.	+	.	.	.	.	.	.	.	+
c. al *Chusquea montana*	1	1	.	.	.	.	.	.	.	.	.	.	.	.	.	.	.
d. al *Valeriana laxiflora*	+	r	.	.	.	.	.	.	.	.	.	.	.	.	.	.	.
d. o *Codonorchis lessonii*	.	.	.	.	.	+	1	.	.	.	.	.	.	.	.	.	.
c. al *Chusquea culeou*	.	.	.	.	.	+	+	.	.	.	.	.	.	.	.	.	.
Azaro alpinae-Nothofagetalia pumilionis																	
c. o *Azara alpina*	1	1	.	.	.	+	+	+	+	+	+	.	.	.	+	+	+
c. o *Perezia nutans*	.	.	.	.	.	.	r	.	.	.	.	.	.	+	.	.	.
Berberis rotundifolia	.	.	.	.	.	+	+	+	+	+	+	+	+	+	.	.	+
Festuca acanthophylla	.	.	.	.	.	.	1	+	+	+	.	+	+	+	+	.	+
Senecio polygaloides	.	.	r	.	r	.	r	.	.	+	.	.	.	.	+	+	+
Berberis microphylla	.	.	.	.	.	.	.	+	+	+	+	.	+	.	+	+	+
Nothofagetea antarcticae																	
c. c *Nothofagus antarctica*	.	.	.	.	.	.	.	.	.	.	.	1	1	4	2a	5	4
Nothofagus macrocarpa	.	.	.	.	.	.	.	.	.	.	.	1	.	1	+	.	+
Alstroemeria spathulata	.	.	.	.	.	.	.	.	.	.	.	+	+	+	+	.	.
Acaena pinnatifida	.	.	.	.	.	.	.	.	.	.	.	.	+	.	+	+	+
Senecio eruciformis	.	.	.	.	.	.	.	.	.	.	.	+	.	.	.	+	+
Gamochaeta stachydifolia	.	.	.	.	.	.	.	.	.	.	.	.	.	+	.	+	+
Astragalus berterii	.	.	.	.	.	.	.	.	.	.	.	+	.	+	.	.	.
Dioscorea brachybothrya	.	.	.	.	.	.	.	.	.	.	.	.	+	.	+	.	.
Acaena splendens	.	.	.	.	.	.	.	.	.	.	.	.	+	.	+	.	.
Others																	
Ribes magellanicum	+	+	.	.	+	+	+	.	.	.	.	.	+	.	.	+	+
Proto-Usnea spp.	2m	2m	.	.	.	2m	2m	.	.	.	.	.	.	.	.	.	.
Misodendrum linearifolium	.	.	.	.	.	2m	2m	.	.	.	.	.	.	.	.	.	.
Misodendrum oblongifolium	.	.	.	.	.	2m	2m	.	.	.	.	.	.	.	.	.	.
Osmorhiza chilensis	.	.	.	.	.	2a	+	.	.	.	.	.	.	.	.	.	+
Coniza spp.	.	.	.	.	.	2a	1	.	.	.	.	.	.	.	.	.	.

Table 1 (continued)

Unit	I.1	I.1	I.1	I.1	I.1	I.1	I.1	I.2	I.2	I.2	I.2	II	II	II	II	II	II
Number	1	2	3	4	5	6	7	8	9	10	11	12	13	14	15	16	17
Rumex acetosella	.	.	.	.	.	.	.	.	.	+	.	+	+	+	+	+	1
Baccharis retinoides	.	.	.	.	.	.	.	+	.	+	.	.	+	+	+	+	+
Euphorbia peplis	.	.	.	.	.	.	.	+	.	.	.	+	+	.	+	+	.
Berberis empetrifolia	.	.	.	.	.	.	.	+	.	.	.	.	+	.	+	+	+
Acaena ovalifolia	.	.	.	.	.	+	.	.	.	+	.	.	+	.	.	.	+
Carex trichodes	.	.	.	.	.	.	r	.	.	.	.	+	.	.	.	.	.
Cynanchum nummularifolium	.	.	.	.	.	.	.	+	.	.	.	.	.	.	.	.	+

single species (number/abundance):

Chloraea alpina 7/+
Galium cotinoides 7/r
Geranium spp. 7/r
Poa spp. 7/r
Sisyrinchium junceum 7/r
Gastridium ventricosum 9/+
Chaetanthera chilensis 12/+
Rosa rubiginosa 13/+
Trisetum spp. 14/+
Hieracium chilense 14/+
Bromus hordeaceus 17/+
Maytenus chubutensis 17/+
Gaultheria phillyreifolia 17/+
Stachys grandidentata 17/+

Abbreviations: *c.* = character species; *d.* = differential species; c = class; o = order; al = alliance.
Authors: S = San Martín et al. (1991); * = this study.
Location of number: 1: 36°54´ S, 71°23´ W; 2: 36°54´ S, 71°23´ W; 3: 36°54´ S, 71°23´ W; 4: 36°54´ S, 71°23´ W; 5: 36°54´ S, 71°23´ W; 6: 35°34´ S, 70°40´ W; 7: 35°34´ S, 70°40´ W

species *N. pumilio* krummholz, characterized by the occurrence of *Maytenus disticha, Myoschilos oblonga, Berberis montana* and *Codonorchis lessonii*, belongs to the Azaro alpinae-Nothofagetalia order of the Nothofagetea pumilionis class (table 1). However, multiple-species stands with these three *Nothofagus* species, and monospecific *N. antarctica* stands are assigned to Nothofagetea antarcticae class, differentiated by the occurrence of the *Nothofagus* species together with *Acaena pinnatifida, Alstroemeria spathulata*, and *Maytenus chubutensis* (table 1).

In the central and southern part of the forest region, *N. pumilio* trees dominate the krummholz vegetation (table 2), leaving only small sites with special history to *N. antarctica.* Thus, multiple-species stands are rarely found, and are interpreted as results of the successional replacement of early *N. antarctica* by later *N. pumilio.* Three communities differentiated in Table 2 characterize the Valdivian forest region (1/1, 1/2, 2/3), further three the Northern Patagonian (3/1, 3/2, 4/1) and the last one the Magellanic forest region (II/1). Stands assigned to Quinchamalio-Nothofagion (table 2: I.1) and Drimydo-Nothofagion alliance (table 2: I.2/1, I.2/2), typical for the driest sites of the Valdivian forest region, comprise species such as *Ribes valdivianum, Myoschilos oblonga, Araucaria araucana* and *Ovidia andina.* Moreover and exclusively, stands of comparatively wet habitats, characterized by *Valeriana lapathifolia* (Valeriano-Nothofagetum, table 2: I.2/3), enable the cryophilous *Drimys andina* to reach as high as timberline. The Northern Patagonian forest region in parallel manner has three typical species compositions (classified as three associations, table 2) of drier sites along

Table 2 Species composition and classification of timberline vegetation in the southern Andes between 37°S and 55°S

Class: Nothofagetea pumilionis
- Order I: Adenocaulo-Nothofagetalia
  - Quinchamalio-Nothofagion pumilionis
    - *1/1* Festuco scabriusculae-Nothofagetum pumilionis
  - Drimydo-Nothofagion pumilionis
    - *2/1* Drimydo-Nothofagetum pumilionis
    - *2/2* Species-poor form of the Drimydo-Nothofagetum
    - *2/3* Valeriano lapathifoliae-Nothofagetum pumilionis
  - Chiliotricho diffusi-Nothofagion pumilionis
    - *3/1* Poo alopecuri-Nothofagetum pumilionis
    - *3/2* Empetro rubri-Nothofagetum pumilionis
  - Senecioni acanthifolii-Nothofagion pumilionis
    - *4/1* Senecioni-Nothofagetum pumilionis

	Unit	1/1	2/1	2/1	2/1	2/2	2/2	2/2	2/2	2/2	2/2	2/2	2/2	2/3	2/3	2/3	2/3
	Number	1	2	3	4	5	6	7	8	9	10	11	12	13	14	15	16
	Author(s)	Fi	*	*	Fi	Fr	Fi	Fi	Esk	Fi	Fi	Fi	Fi	Fl	*	*	*
	Elevation (10 m a.s.l.)	158	158	164	151	-	158	152	164	149	152	135	136	132	122	120	116
	Aspect	NW	SE	S	SW	-	NW	S	-	-	SE	SE	-	N	S	SW	NW
	Slope angle (°)	5	5	30	2	-	10	2	-	-	2	5	0	40	25	25	25
	Structure																
	Cover tree layer [%]	30	70	90	.	.	.	.	.	.	.	60	90	70	80	75	70
	Height tree layer [m]	.	3	3	.	.	.	.	.	.	.	.	.	5	4	8	8
	Cover shrub layer [%]	50	20	30	70	.	80	70	.	50	70	50	20	80	30	30	.
	Height shrub layer [m]	.	0.7	0.7	.	.	.	.	.	.	.	.	.	1.0	1.0	1.0	.
	Cover herb layer [%]	<5	5	5	<10	.	<10	<10	.	<10	<10	<10	<10	30	80	80	80
	Number of species	10	9	9	10	10	9	9	9	10	8	9	11	14	24	25	16
Nothofagetea pumilionis																	
c. c	*Nothofagus pumilio*	4	4	5	5	5	5	4	5	2	4	5	5	4	5	4	4
c. c	*Maytenus disticha*	1	2b	3	2	2	+	3	1	3	3	2	2	3	2b	2b	2a
c. c	*Pernettya pumila*	1	.	.	1	1	.	1	+	.	1	3	1	.	+	r	.
c. c	*Rubus geoides*	.	.	.	1	1	.	1	1	.	.	.	.	+	.	r	+
c. c	*Berberis montana*	+	.	.	+	.	+	+	+	+	+	.	+	.	+	+	.
d. c	*Escallonia alpina*	.	.	.	.	+	.	.	.	.	.	.	.	1	3	3	2a
c. c	*Viola reichei*	.	.	.	.	+	.	.	.	.	.	.	.	+	2b	2b	+
c. c	*Codonorchis lessonii*	.	.	.	.	.	.	.	.	.	.	1	1	.	.	+	.
c. c	*Adenocaulon chilense*	1	+	.	.	.	.	.	.	.	.	.	.	+	.	.	+
c. c	*Gavilea lutea*	1	.	.	.	.	.	.	.	.	.	.	.	.	.	.	.
d. c	*Blechnum penna-marina*	.	1	.	.	.	1	.	.	.	.	.	.	+	.	+	.
Adenocaulo-Nothofagetalia																	
	Berberis serrato-dentata	.	.	.	+	1	.	.	.	.	.	.	.	+	r	+	2a
	Macrachaenium gracile	.	.	.	.	.	.	.	.	.	.	.	.	.	2b	1	1
	Lycopodium magellanicum	.	.	.	.	.	.	.	.	.	.	.	.	+	r	1	+
Myoschilo-Nothofagenalia																	
d.	*Ribes valdivianum*	1	+	1	1	.	1	2	1	1	1	1	1	.	.	.	.
d.	*Araucaria araucana*	.	.	r	.	.	.	+	.	+	+	2	+	.	.	.	.
d.	*Ovidia andina*	.	.	.	2	.	.	1	.	+	+	.	.	.	.	.	.
d.	*Myoschilos oblonga*	.	.	+	+	1	1	1	.	+	2	1	1	+	.	.	+
d.	*Pernettya myrtilloides*	.	+	.	.	.	.	.	.	.	.	2	.	.	.	+	4
d.	*Perezia pedicularifolia*	.	+	r	.	.	.	.	.	.	.	.	.	.	+	+	.
Quinchamalio-Nothofagion pumilionis																	
c. al	*Quinchamalium chilense*	+	+	.	.	.	.	.	.	.	.	+	.	.	.	.	.
Festuco scabriusculae-Nothofagetum																	
c. as	*Festuca scabriuscula*	1	.	.	.	.	.	.	.	.	.	.	.	.	.	.	.

Table 2 (continued)

Class: Nothofagetea pumilionis
Order II: Violo magellanicae-Nothofagetalia pumilionis
II/1 Violo magellanicae-Nothofagion pumilionis

3/1	3/1	3/1	3/2	3/2	3/2	3/2	3/2	3/2	3/2	3/2	3/2	3/2	3/2	3/2	3/2	3/2	3/2	3/2	3/2	3/2	3/2	4/1	II/1	II/1	II/1
17	18	19	20	21	22	23	24	25	26	27	28	29	30	31	32	33	34	35	36	37	38	39	40	41	42
*	*	*	*	*	*	*	*	*	*	*	*	*	*	*	*	*	*	*	*	*	*	*	*	Do	*
106	107	118	100	104	111	117	121	121	121	119	98	113	112	115	125	112	116	104	110	104	108	131	52	56	61
S	S	S	N	NW	NW	W	W	NW	W	S	SW	S	S	-	W	W	NW	SW	W	-	NW	E	NE	NE	NW
40	10	5	36	15	24	0	0	0	40	30	45	3	10	-	32	28	40	5	40	-	-	20	15	10	13
70	70	.	80	80	.	80	80	80	80	.	70	70	70	80	75	70	75	70	70	75	5	70	75	.	.
9	9	.	7	6	.	7	7	7	5	1.7	8	7	7	6	7	8	6	7	5	6	4	10	9	.	.
.	.	80	.	.	80	.	1	3	10	70	10	.	.	.	.	.	.	.	.	.	.	.	.	95	85
.	.	3.0	.	.	3.0	.	0.7	0.7	1.5	.	.	.	.	.	.	.	.	.	.	.	.	.	.	3.0	0.7
75	80	20	95	70	70	30	70	80	65	60	60	80	85	30	90	85	80	30	80	60	85	30	70	2	30
10	7	6	13	7	6	12	12	10	9	4	10	6	7	10	9	10	9	6	4	4	5	12	11	3	3
4	5	5	5	5	5	5	5	5	5	4	4	4	4	5	5	4	5	4	4	5	4	5	5	5	5
3	2b	.	2b	.	.	2	2	2	2b	.	2b	.	2a	+	2a	2a	1	.	.	.	.	.	.	.	.
.	.	.	.	.	.	1	1	2	.	r	.	.	.	.	.	.	.	.	.	.	.	1	.	.	2b
.	.	r	r	1	2a	1	+	.	+	.	.	.	.	+	.	1	.	.	.	.	.	.	+	.	.
.	.	.	.	.	.	.	.	.	1	.	.	.	.	.	.	.	1	.	.	+	.	.	.	.	.
.	2a	.	.	.	.	.	.	.	.	.	2a	r	2a	.	r	+	.	.	.	.	.	.	.	.	.
.	.	.	1	.	.	.	.	.	.	.	r	.	.	+	.	+	+	.	.	.	.	.	.	.	.
.	.	r	.	.	.	1	.	.	.	.	.	.	.	+	.	.	.	+	.	.	.	.	+	.	.
.	.	.	.	.	.	.	.	.	.	.	.	.	.	.	.	.	.	.	.	.	.	.	r	.	.
.	.	.	.	.	.	+	+	+	.	.	.	r	.	.	r	.	.	+	.	.	.	.	.	.	.
.	.	.	r	.	.	.	.	.	.	.	.	.	.	.	.	.	.	.	.	.	.	.	.	.	.
														.	.	.	.	.	.	.	.				
.	.	.	.	.	.	.	.	.	.	.	+	.	.	.	.	.	.	.	.	.	.	.	.	.	.
1	.	.	1	.	.	.	.	.	.	.	1	.	.	.	.	.	.	.	.	.	.	.	.	.	.
.	.	.	.	r	.	.	.	.	.	.	.	.	.	.	.	.	.	.	.	.	.	.	.	.	.
.	.	.	.	.	.	.	.	.	.	.	.	.	.	.	1	.	+	.	.	.	.	.	.	.	.
.	.	.	.	.	.	.	.	.	.	.	.	.	.	.	.	.	.	.	.	.	.	.	.	.	.
.	.	.	.	.	.	.	.	.	.	.	.	.	.	.	.	.	.	.	.	.	.	.	.	.	.
.	.	.	.	.	.	.	.	.	.	.	.	.	.	.	.	.	.	.	.	.	.	.	.	.	.
.	.	.	.	.	.	.	.	.	2a	.	.	.	.	.	.	.	1	.	.	.	.	.	.	.	.
.	.	.	.	.	.	.	.	.	.	.	.	.	.	.	.	.	.	.	.	.	.	.	.	.	.
r	+	.	.	.	.	.	.	.	.	.	.	.	.	.	.	.	.	+	+	.	+	.	.	.	.
.	.	.	.	.	.	.	.	.	.	.	.	.	.	.	.	.	.	.	.	.	.	.	.	.	.

Table 2 (continued)

	Unit	1/1	2/1	2/1	2/1	2/2	2/2	2/2	2/2	2/2	2/2	2/2	2/2	2/3	2/3	2/3	2/3
	Number	1	2	3	4	5	6	7	8	9	10	11	12	13	14	15	16
Drimydo-Nothofagion																	
c. al	*Drimys andina*	.	.	.	+	.	.	.	.	.	.	.	.	3	+	1	+
Drimydo-Nothofagetum																	
c. o	*Perezia prenanthoides*	.	1	.	.	.	.	.	.	.	.	.	.	.	.	.	.
c. o	*Valeriana laxiflora*	.	.	1	.	.	.	.	.	.	.	.	.	.	.	.	.
d. o	*Vicia nigricans*	.	.	+	.	.	.	.	.	.	.	.	.	.	.	.	.
Valeriano-Nothofagetum																	
d. as	*Valeriana lapathifolia*	.	.	.	.	.	.	.	.	.	.	.	.	2	4	2a	.
	Hypochoeris arenaria	.	.	.	.	.	.	.	.	.	.	.	.	.	r	r	.
	Nothofagus betuloides	.	.	.	.	.	.	.	.	.	.	.	.	.	.	+	+
Chiliotricho diffusi-Nothofagion																	
c. al	*Chiliotrichum diffusum*	.	.	.	.	.	.	.	.	.	.	.	.	.	.	.	.
d. al	*Proto-Usnea* spp.	+	.	.	.	.	.	.	.	.	.	.	.	.	.	.	2m
d. al	*Pernettya mucronata*	.	.	.	.	.	.	.	.	.	.	.	.	.	.	.	.
d. al	*Festuca magellanica*	.	.	.	.	.	.	.	.	.	.	.	.	.	.	.	.
Poo alopecuri-Nothofagetum																	
d. as	*Poa alopecurus*	.	.	.	.	.	.	.	.	.	.	.	.	.	r	r	.
d. as	*Poa rigidifolia*	.	.	.	.	.	.	.	.	.	.	.	.	.	+	+	.
Empetro-Nothofagetum																	
d. as	*Empetrum rubrum*	.	.	.	.	.	.	.	.	.	.	.	.	.	+	r	+
Senecioni acanthifolii-Nothofagion																	
c. al	*Schizeilema ranunculus*	.	.	.	.	.	.	.	.	.	.	.	.	.	.	.	.
d. al	*Ourisia breviflora*	.	.	.	.	.	.	.	.	.	.	.	.	.	.	.	.
d. al	*Ortachne rariflora*	.	.	.	.	.	.	.	.	.	.	.	.	.	.	.	.
d. al	*Oxalis magellanica*	.	.	.	.	.	.	.	.	.	.	.	.	.	.	.	.
	Perezia magellanica	.	.	.	.	.	.	.	.	.	.	.	.	.	.	.	.
	Marsippospermum philippi	.	.	.	.	.	.	.	.	.	.	.	.	.	.	.	.
	Deyeuxia suka	.	.	.	.	.	.	.	.	.	.	.	.	.	.	.	.
	Acaena antarctica	.	.	.	.	.	.	.	.	.	.	.	.	.	.	.	.
	Caltha appendiculata	.	.	.	.	.	.	.	.	.	.	.	.	.	1	1	.
c. al	*Senecio acanthifolius*	.	.	.	.	.	.	.	.	.	.	.	.	.	r	.	.
d. c	*Gunnera magellanica*	.	.	.	.	.	.	.	.	.	.	.	.	.	2a	4	.
Violo magellanicae-Nothofagetalia and **Violo magellanicae-Nothofagion**																	
c. o	*Cardamine glacialis*	.	.	.	.	.	.	.	.	.	.	.	.	.	.	.	.
d. o	*Acaena magellanica*	.	.	.	.	.	.	.	.	.	.	.	.	.	.	.	.
d. o	*Dysopsis glechomoides*	.	.	.	.	.	.	.	.	.	.	.	.	.	.	.	.
d. o	*Ranunculus peduncularis*	.	.	.	.	.	.	.	.	.	.	.	.	.	.	.	.
Others																	
	Ribes magellanicum	.	.	.	.	1	.	.	.	2	.	.	+	+	.	.	.
	Misodendrum spp.	.	.	.	.	.	+	.	.	.	.	.	.	.	.	.	2m
	Lagenifera hirsuta	.	.	.	.	1	.	.	.	.	.	.	.	.	+	1	.
	Acaena ovalifolia	.	.	.	.	.	.	.	.	.	.	.	.	.	+	1	.
	Nothofagus antarctica	.	.	.	.	.	.	.	.	.	.	.	.	.	2a	.	.

Abbreviations: *c.* = character species; *d.* = differential species; c = class; o = order; al = alliance;

Authors: Do = Dollenz (1983) ; Esk = Eskuche (1973); Fi = Finckh (1996); Fl = Flores (1998); Fr = Freiberg (1985); * = this study.

Locations of relevés:

1: 39°29´ S, 71°45´ W	7: 39°28´ S, 71°50 W	13: 41°27´ S, 72°02´ W
2: 38°30´ S, 71°20´ W	8: without coordinates	14: 41°55´ S, 71°40´ W
3: 38°30´ S, 71°20´ W	9: 39°28´ S, 71°50´ W	15: 41°55´ S, 71°40´ W
4: 39°28´ S, 71°51´ W	10: 39°28´ S, 71°50´ W	16: 43°50´ S, 72°00´ W
5: without coordinates	11: 39°27´ S, 71°51´ W	17: 45°30´ S, 72°00´ W
6: 39°35´ S, 71°32 W	12: 39°27´ S, 71°51´ W	18: 45°30´ S, 72°00´ W

Table 2 (continued)

3/1	3/1	3/1	3/2	3/2	3/2	3/2	3/2	3/2	3/2	3/2	3/2	3/2	3/2	3/2	3/2	3/2	3/2	3/2	3/2	3/2	3/2	4/1	II/1	II/1	II/1
17	18	19	20	21	22	23	24	25	26	27	28	29	30	31	32	33	34	35	36	37	38	39	40	41	42
.	.	.	.	.	.	.	.	.	.	.	.	.	.	.	.	.	.	.	.	.	.	.	.	.	.
.	.	.	.	.	.	.	.	.	.	.	.	.	.	.	.	.	.	.	.	.	.	.	.	.	.
.	.	.	.	.	.	.	.	.	.	.	.	.	.	.	.	.	.	.	.	.	.	.	.	.	.
.	.	.	.	.	.	.	.	.	.	.	.	.	.	.	.	.	.	.	.	.	.	.	.	.	.
.	.	.	.	.	.	.	.	.	.	.	.	.	.	.	.	.	.	.	.	.	.	.	.	.	.
.	.	.	.	.	.	.	.	.	.	.	.	.	.	.	.	.	.	.	.	.	.	.	.	.	.
.	.	.	.	.	.	.	.	.	.	.	.	.	.	.	.	.	.	.	.	.	.	.	.	.	.
1	1	.	+	.	.	+	+	1	2b	.	2a	r	r	.	1	1	2a	.	r	.	.	.	.	.	.
2a	2a	.	.	.	.	.	.	.	2m	.	2a	2a	2a	.	.	.	.	.	1	.	2m	.	.	.	.
2b	3	r	r	2b	2b	+	+	+	.	2b	2a	2a	2b	1	.	+	.	2b	.	.	3	.	.	.	.
r	.	.	.	.	.	+	+	+	.	.	.	.	.	.	.	.	.	.	.	.	.	.	.	.	.
r	r	r	r	.	.	1	1	+	.	.	1	.	r	.	.	.	.	.	.	.	.	.	.	.	.
r	.	.	.	.	.	.	.	.	+	.	.	.	.	.	+	1	+	.	.	r	.	.	.	.	.
														.	.	.	.	.	.	.	.				
+	.	r	4	4	4	2	3	4	4	3	1	2b	3	2a	5	5	5	+	5	4	4	.	.	.	3
.	.	.	r	+	+	.	.	.	.	.	.	.	.	.	.	.	.	.	.	.	.	+	.	.	.
.	.	.	.	.	.	.	.	.	.	.	.	.	.	.	.	.	.	.	.	.	.	+	.	.	.
.	.	.	.	.	.	.	.	.	.	.	.	.	.	.	.	.	.	.	.	.	.	2	.	.	.
.	.	.	.	.	.	.	.	.	.	.	.	.	.	.	.	.	.	.	.	.	.	+	.	.	.
.	.	.	.	.	.	.	.	.	.	.	.	.	.	.	.	.	.	.	.	.	.	2a	.	.	.
.	.	.	.	.	.	.	.	.	.	.	.	.	.	.	.	.	.	.	.	.	.	1	.	.	.
.	.	.	.	.	.	.	.	.	.	.	.	.	.	.	.	.	.	.	.	.	.	+	.	.	.
.	.	.	.	.	.	.	.	.	.	.	.	.	.	.	.	.	.	.	.	.	.	+	.	.	.
.	.	.	.	.	.	.	.	.	.	.	.	.	.	.	.	.	.	.	.	.	.	2a	.	.	.
.	.	.	1	.	r	.	.	.	.	.	.	.	.	.	.	.	.	.	.	.	.	.	4	+	.
.	.	.	.	.	.	.	.	.	.	.	.	.	.	1	.	.	.	.	.	.	.	1	2	+	.
.	.	.	.	.	.	.	.	.	.	.	.	.	.	.	.	.	.	.	.	.	.	.	1	.	.
.	.	.	.	.	.	.	.	.	.	.	.	.	.	.	.	.	.	.	.	.	.	.	3	.	.
.	.	.	.	.	.	.	.	.	.	.	.	.	.	.	.	.	.	.	.	.	.	.	2	.	.
.	.	.	.	.	.	.	.	.	.	.	.	.	.	.	.	.	.	.	.	.	.	.	2	.	.
.	.	.	.	.	.	+	+	+	.	.	.	.	.	.	.	.	.	.	.	.	.	.	.	.	.
2m	2m	.	.	2m	.	.	.	.	2m	.	2m	2m	2m	.	.	.	.	.	.	.	.	.	.	.	.
.	.	.	+	+	.	.	.	.	.	.	.	.	.	.	.	.	.	.	.	.	.	.	.	.	.
.	.	.	.	.	.	.	.	.	.	.	.	.	.	.	.	.	.	.	.	.	.	.	.	.	.
.	.	.	.	.	.	.	.	.	.	.	.	.	.	.	.	.	.	.	.	.	.	.	.	.	.

as = association.

Locations of relevés: (cont.)

19: 45°45´ S, 71°47´ W
20: 48°28´ S, 72°29´ W
21: 48°28´ S, 72°29´ W
22: 48°28´ S, 72°29´ W
23: 45°25´ S, 71°50´ W
24: 45°25´ S, 71°50´ W
25: 45°25´ S, 71°50´ W
26: 46°05´ S, 72°05´ W
27: 45°30´ S, 72°00´ W
28: 45°30´ S, 72°00´ W
29: 45°30´ S, 72°00´ W
30: 45°30´ S, 72°00´ W
31: 45°37´ S, 72°15´ W
32: 45°37´ S, 72°15´ W
33: 45°37´ S, 72°15´ W
34: 45°38´ S, 72°13´ W
35: 47°11´ S, 72°29´ W
36: 47°10´ S, 72°31´ W
37: 47°11´ S, 72°29´ W
38: 47°12´ S, 72°31´ W
39: 48°28´ S, 72°29´ W
40: 54°55´ S, 67°35´ W
41: 53°30´ S, 72°00´ W
42: 54°55´ S, 67°35´ W

with one of the wet habitats: the first (Poo-Nothofagetum) and the second (Empetro-Nothofagetum) belong to Chiliotricho-Nothofagion (table 2: I.3), the third (Senecioni-Nothofagetum) to the Senecioni-Nothofagion alliance (table 2: I.4). Dominant dwarf shrub undergrowth of *Empetrum rubrum* differentiates the Empetro-Nothofagetum association of nutrient poor sites (table 2: I.3/2) from Poo-Nothofagetum (table 2: I.3/1) with *Poa alopecurus*. The Senecioni-Nothofagetum association proofs its vicinity to wetland vegetation by the occurrence of *Ourisia*, *Marsippospermum*, and *Caltha* among others. Already Magellanic forest vegetation is poor in species, but even poorer are Magellanic timberline stands (table 2: II.1). We distinguish three types of krummholz vegetation in the southern Andes: (1) a tall forb krummholz, characterized by *Senecio acanthifolius* and *Gunnera magellanica* (table 2: 40), (2) timberline shrubs with ericaceous undergrowth of *Empetrum rubrum* and *Pernettya pumila* (table 2: 42), and (3) a monospecific *Nothofagus* (i.e., *N. pumilio*) krummholz that practically lacks further species (such as shrubs and herbs) in the understory (table 1: 5).

4.2 Timberline Elevation, Extension, and Structural Differentiation

In the southern Andes, the upper timberline drops gradually to the higher latitudes, from 2000 m at 35°S to approximately 600 m at 55°S (figure 3). At these latitudes, *N. pumilio* krummholz generally occurs in a ten to some hundred meter wide ecotone, decreasing slightly from the northern limit to the south (figure 3). In south-central Chile (around 35°S), the upper *N. pumilio* timberline reaches its highest position (~ 2000 m) in Alto de Vilches (San Martín et al. 1991; Lara et al. 2001). While in the northern part of the forest region there is a marked difference in the altitudinal position of the timberline in dependence of exposition, in its southern part (e.g. Navarino Island), the altitude of timberline is independent of the slope aspect. On remote, the altitudinal position of the southernmost *N. pumilio* timberline seems mainly controlled by factors such as wind and snow rather than by temperature.

At the alpine timberline in the southern Andes, four distinctive *N. pumilio* growth forms were identified. Generally, *N. pumilio* becomes more and more stunted as it approaches the woodland limit (figure 1; ‘elfin wood’, cf. Ellenberg 1966; Holtmeier 2003). Thus, the structure of the *Nothofagus* forest changes from single-stemmed, erect trees to multiple-stemmed trees with naturally coppiced growth forms at timberline. At many sites, presumably those less affected by strong winds, dense stands of naturally coppiced *N. pumilio* form the timberline (figure 1). Extreme deformations of *N. pumilio* krummholz found at the limits of the distribution area serve to address the

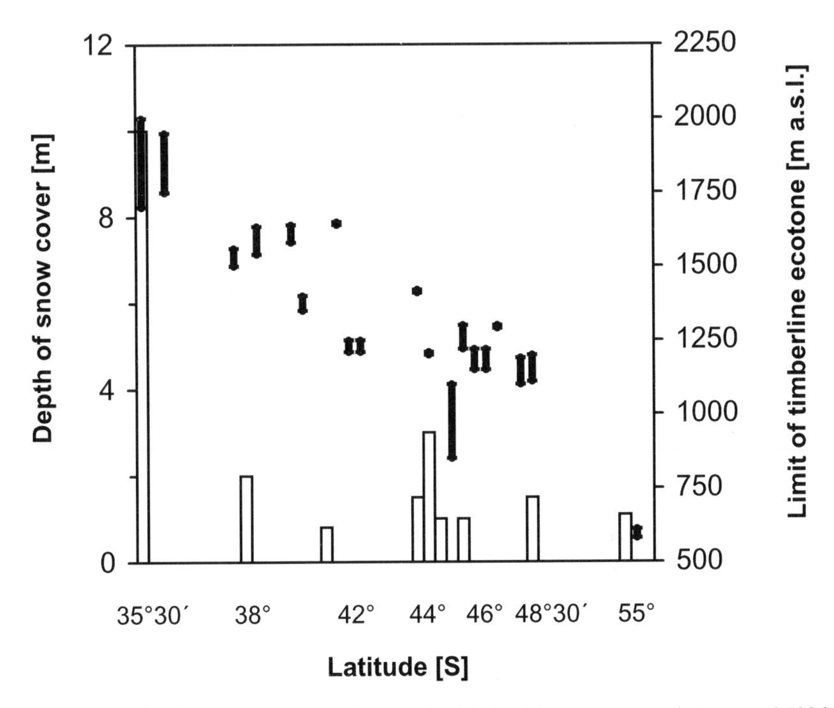

Figure 3 Depth of snow cover (columns) in timberline ecotones between 35°30´ and 55°S of the southern Andes. The vertical lines show altitudinal ranges of the timberline ecotone (right-hand y-axis), between 35°30´S and 55°S

predominant factors driving the growth forms observed in the Andes (figure 4). At around 37°S (figure 4a), well developed timberline shrubs are restricted to the shade-aspect slopes (which are leeward to the strong northerlies) while the sun-side slopes are severely eroded and bare of forest and/or timberline vegetation. At Navarino Island of Tierra del Fuego archipelago at the southernmost end of the forest region, for example (figures 4b, 4c), timberline altitudinal position is hardly influenced by slope aspect but predominantly by wind exposure.

Timberline structure in the northern forest region (figure 4a) is characterized by a compact krummholz with branches close to the ground on the leeward side, and some *Maytenus disticha* shrubs in the understory. Dead buds and lacking foliage is a distinctive feature providing abundant evidence of winter death (frost, desiccation, abrasion). A second tree deformation and special timberline structure have been frequently observed, for example at Navarino Island: on leeward sides, there is typically a growth form with a gnarled bole, characterized by a single-stemmed habit and stems aligned approximately downhill, before curving up into vertical alignment (figure 4b). Often the layering trunks, creeping approximately 3 to 4 m above the ground

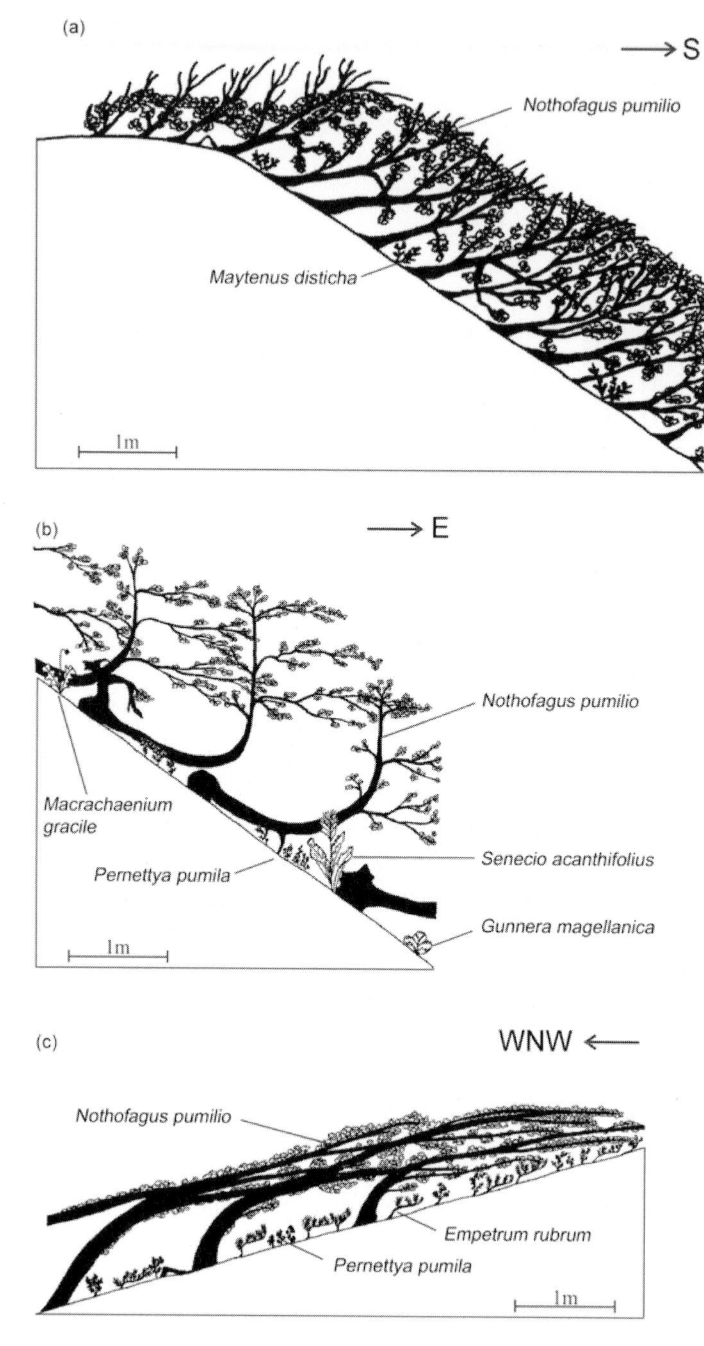

Figure 4 Growth forms at timberline under influence of wind and snow in the southern Andes: (a) Termas de Chillán, 37°S, 1950 m; (b) Cerro la Bandera, Navarino Island, 55°S, 590 m; and (c) Navarino Island, 55°S, 600 m

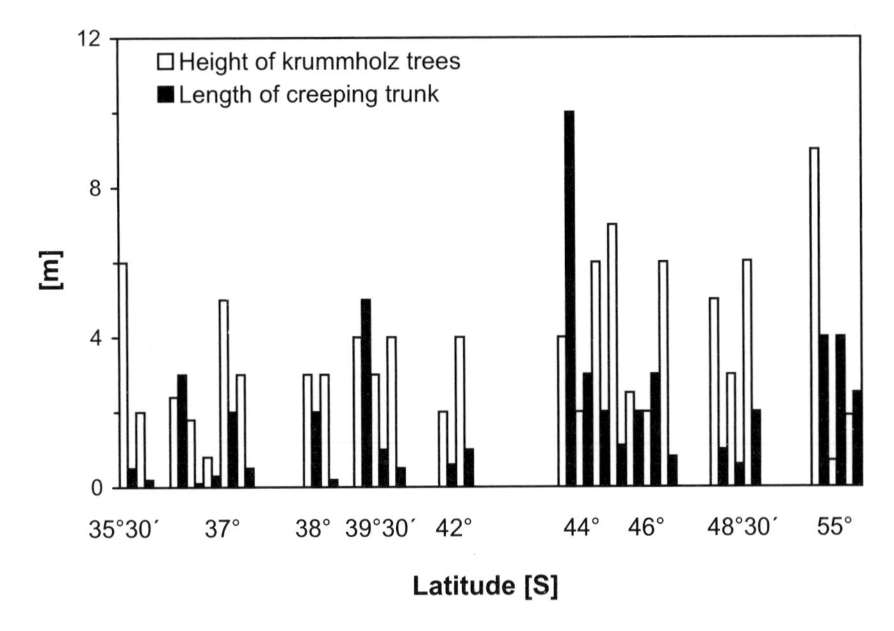

Figure 5 Height of *Nothofagus* krummholz and length of creeping trunk of *N. pumilio* timberline trees in the southern Andes (35°30´ S to 55°S)

(exceptional > 8 m), are even longer than vertical alignment of krummholz trees (figure 5). The understory is often dominated by herbs such as *Macrachaenium gracile*, *Senecio acanthifolius*, and *Gunnera magellanica.* This growth form occurs at Tierra del Fuego timberlines (e.g. Navarino Island) where locally enormous amounts of wet snow cover occur. Finally, an extreme situation can exclusively be observed in the south of Andean timberlines on windward sides (figure 4c); growth forms are characterized as fan-shaped bushes running uphill for several meters. Branches arise from the uphill side of the very large boles. The top surface of the big branches is absolutely free of ramification obviously due to the abrasive forces of ice and soil particles driven by strong winds. These trees form stands associated with dwarf shrubs (e.g. *Pernettya pumila, Empetrum rubum*) in the understory.

5 Discussion

5.1 Phytosociology and Habitat Types at Andean Timberline

Bibliographic research on *N. pumilio* krummholz vegetation between 35°S and 55°S allowed us to conclude that 45 of the relevés (88 %; table 3) correspond to vegetation types that could be considered zonal communities

Table 3 Synsystematical overview of timberline vegetation in the southern Andes

	Table 1			*Table 2*							
	I Nothofagetea antarcticae II Nothofagetea pumilionis II/1 Azaro alpinae-Nothofagetalia pumilionis II/2 Adenocaulo-Nothofagetalia pumilionis II/2/1 Quinchamalio-Nothofagion II/2/2 Drimydo-Nothofagion II/2/3 Chiliotricho diffusi-Nothofagion II/2/4 Senecioni acanthifolii-Nothofagion II/3 Violo magellanicae-Nothofagetalia										
	35°-37° S			37°- 55° S							
Unit	I	II/1	II/2	2/1	2/2	2/2	2/2	2/3	2/3	2/4	II/3
Number	1	2	3	4	5	6	7	8	9	10	11
Number of relevés	6	4	7	1	3	8	4	3	19	1	3
Number of species	17	11	8	10	6	9	20	8	8	12	6
Syntaxonomy / Species											
Nothofagetea antarcticae class											
Nothofagus antarctica	V	.	.	.	.	.	1	.	.	.	.
Acaena pinnatifida	IV	.	.	.	.	.	.	.	.	.	.
Alstroemeria spathulata	IV	.	.	.	.	.	.	.	.	.	.
Gamochaeta stachydifolia	III	.	.	.	.	.	.	.	.	.	.
Senecio eruciformis	III	.	.	.	.	.	.	.	.	.	.
Nothofagus macrocarpa	IV	.	.	.	.	.	.	.	.	.	.
Azaro alpinae-Nothofagetalia order											
Azara alpina	II	4	III	.	.	.	.	.	.	.	.
Perezia nutans	I	.	I	.	.	.	.	.	.	.	.
Berberis rotundifolia	IV	4	II	.	.	.	.	.	.	.	.
Festuca acanthophylla	V	3	I	.	.	.	.	.	.	.	.
Senecio polygaloides	III	1	II	.	.	.	.	.	.	.	.
Berberis microphylla	IV	4	.	.	.	.	.	.	.	.	.
Adenocaulo-Nothofagetalia order											
Ribes valdivianum	.	.	.	1	3	IV	.	.	I	.	.
Araucaria araucana	.	.	.	.	1	III	.	.	.	.	.
Ovidia andina	.	.	.	.	1	II	.	.	.	.	.
Myoschilos oblonga	.	4	I	.	2	V	2	.	.	.	.
Pernettya myrtilloides	.	.	.	.	1	I	2	.	I	.	.
Perezia pedicularifolia	.	.	.	.	2	.	2	.	.	.	.
Berberis serrato-dentata	.	.	.	.	1	I	4	.	+	.	.
Macrachaenium gracile	.	.	.	.	.	.	3	1	I	.	.
Lycopodium magellanicum	.	.	.	.	.	.	4	.	+	.	.
Quinchamalio-Nothofagion alliance											
Quinchamalium chilense	.	.	.	1	1	I	.	2	I	.	.
Festuca scabriuscula	.	.	.	1	.	.	.	.	.	.	.
Drimydo-Nothofagion alliance											
Drimys andina	.	.	.	.	1	.	4	.	.	.	.
Perezia prenanthoides	.	.	.	.	1	.	.	.	.	.	.
Valeriana laxiflora	.	.	II	.	1	.	.	.	.	.	.
Vicia nigricans	.	.	.	.	1	.	.	.	.	.	.
Valeriana lapathifolia	.	.	.	.	.	.	3	.	.	.	.
Hypochoeris arenaria	.	.	.	.	.	.	2	.	.	.	.
Nothofagus betuloides	.	.	.	.	.	.	2	.	.	.	.
Lagenifera hirsuta	.	.	.	.	.	I	2	.	I	.	.

Table 3 (continued)

Unit	I	II.1	II.2	2/1	2/2	2/2	2/2	2/3	2/3	2/4	II.3
Number	1	2	3	4	5	6	7	8	9	10	11
Chiliotricho diffusi-Nothofagion alliance											
Proto-Usnea spp.	.	.	III	+	.	.	1	2	II	.	.
Pernettya mucronata	.	.	.	.	.	.	.	3	V	.	.
Chiliotrichum diffusum	.	.	.	.	.	.	.	2	IV	.	.
Festuca magellanica	.	.	.	.	.	.	.	1	I	.	.
Poa alopecurus	.	.	.	.	.	.	2	3	II	.	.
Poa rigidifolia	.	.	.	.	.	.	2	1	II	.	.
Empetrum rubrum	.	.	.	.	.	.	3	2	V	.	1
Senecioni acanthifolii-Nothofagion	.	.	.								
Schizeilema ranunculus	.	.	.	.	.	.	.	.	I	+	.
Ourisia breviflora	.	.	.	.	.	.	.	.	.	+	.
Ortachne rariflora	.	.	.	.	.	.	.	.	.	2m	.
Oxalis magellanica	.	.	.	.	.	.	.	.	.	+	.
Perezia magellanica	.	.	.	.	.	.	.	.	.	2a	.
Marsippospermum philippi	.	.	.	.	.	.	.	.	.	1	.
Deyeuxia suka	.	.	.	.	.	.	.	.	.	+	.
Acaena antarctica	.	.	.	.	.	.	.	.	.	+	.
Caltha appendiculata	.	.	.	.	.	.	2	.	.	2a	.
Senecio acanthifolius	.	.	.	.	.	.	1	.	I	.	2
Gunnera magellanica	.	.	.	.	.	.	2	.	+	1	2
Violo magellanicae-Nothofagetalia order											
Cardamine glacialis	.	.	.	.	.	.	.	.	.	.	1
Acaena magellanica	.	.	.	.	.	.	.	.	.	.	1
Dysopsis glechomoides	.	.	.	.	.	.	.	.	.	.	1
Ranunculus peduncularis	.	.	.	.	.	.	.	.	.	.	1
Nothofagetea pumilionis class											
Nothofagus pumilio	IV	4	V	4	3	V	4	3	V	5	3
Maytenus disticha	I	4	III	1	3	V	4	2	III	.	.
Pernettya pumila	I	4	II	1	1	VI	2	.	II	1	1
Rubus geoides	.	.	.	.	1	III	3	1	III	.	1
Berberis montana	.	.	IV	+	1	IV	2	.	I	.	.
Escallonia alpina	I	1	I	.	.	I	4	1	II	.	.
Viola reichei	.	.	.	.	.	I	4	.	I	.	.
Codonorchis lessonii	.	.	II	.	.	II	1	1	I	.	1
Adenocaulon chilense	.	.	.	1	1	.	2	.	.	.	1
Gavilea lutea	.	.	.	1	.	.	.	.	II	.	.
Blechnum penna-marina	.	.	.	.	1	I	2	.	+	.	.
Others											
Ribes magellanicum	III	.	IV	.	.	II	1	.	I	.	.
Acaena ovalifolia	II	1	I	.	.	.	2	.	.	.	.
Misodendrum spp.	.	.	II	.	.	I	1	2	II	.	.
Osmorhiza chilensis	I	.	II	.	.	.	.	.	.	.	.
Chusquea montana	.	.	II	.	.	.	.	.	.	.	.
Chusquea culeou	.	.	II	.	.	.	.	.	.	.	.
Rumex acetosella	V	1	.	.	.	.	.	.	.	.	.
Euphorbia empetrifolia	IV	1	.	.	.	.	.	.	.	.	.
Baccharis retinoides	V	2	.	.	.	.	.	.	.	.	.
Berberis empetrifolia	IV	1	.	.	.	.	.	.	.	.	.

(Oberdorfer 1960; Eskuche 1973; Hildebrand-Vogel et al. 1990; Finckh 1996) that are 'persistent' (sensu Veblen 1992) at subalpine elevations, and will not become extinct during a given time period, or, if they will do (e.g. by disturbance; Veblen et al. 1981), communities will recolonize the area within the time span required for one turnover of population. Generally, the understories of *Nothofagus* krummholz of the southern Andean timberline ecotones are often dominated by species that originate from adjacent forest regions (i.e., Central Chilean forest region 30°-38°S, Valdivian forest region 38°-42°S, Northern Patagonian forest region 42°-48°S, and Magellanic forest region 48°-55°S). Floristic comparison with zonal high-elevation forests reveals that krummholz can be easily included into these synsystematical units (Hildebrand in press), instead of being treated as an independent association as proposed by Eskuche (1973). *Maytenus disticha*, a character species of the Nothofagetea pumilionis class, becomes absent only within Magellanic Violo-Nothofagion and Southern Patagonian Chilitricho-Nothofagion communities, feature in accordance with findings from Tierra del Fuego (M.T.K Arroyo pers. comm.). We interpret this as the typical community type of relative wind-exposed, dry and cold habitats with sparse snow cover.

Among South American beech species, *Nothofagus antarctica* occurs in the widest range of habitat types: areas of poor drainage at low to high elevations, exposed sites of unstable substrate at timberline, topographic depressions subject to cold air drainage, steep slopes with shallow soils, and dry sites near the Patagonian steppe ecotone (Veblen et al. 1996). It is a typical representative of sites that are too harsh for most of the other tree species (Seibert 1996). The results of this study imply that *N. antarctica* may occur in association with *N. pumilio* at timberline, but most commonly *N. pumilio* forms extensive krummholz in single-species stands (table 3). The frequent occurrence of *N. antarctica* at the northern *Nothofagus* timberline may be an indicator of a reduced growth period due to long lasting snow cover that *N. antarctica* probably takes advantage over *N. pumilio* because it produces new leaves later in the spring (Veblen et al. 1977). Along the ecotone from rainforests to Patagonian steppe (west-to-east topographic gradient), *N. antarctica* resprouts vigorously after burning (Veblen et al. 1996). Likewise, in the Central Chilean forest and in the northern part of the Valdivian rainforest, *N. macrocarpa* and *N. obliqua* both resprout after being burned. Thus, given the failure of *N. pumilio* regeneration after fire for several decades (Veblen et al. 1996), influence of fire may change species composition from monotypic *N. pumilio* stands towards multiple-species stands, also at timberline. However in conclusion, at the regional scale, *N. pumilio* presence and dominance at Andean timberline may partly reflect its life history differences (tolerance to low temperatures, leaf production early in spring, phenotypic plasticity, prolific seed producer, and reproductive biology) in comparison to most of the associates in the subalpine zone. Given these competitive traits for regeneration and growth at timberline, deciduous *N. pumilio* trees also establish, at the local scale, in topographic depressions subject to cold air drainage.

Ecology of the timberline is primarily determined by the key factors limiting tree growth at higher elevations: drought in the northern part of the forest area, and low temperatures respective shortage of the growing season in its southern part. Shortage of water supply is accompanied by depletion of soil material and therefore of nutrients on windward sites; however, the leeward sites have long lasting snow cover, resulting in better water balance conditions, but consisting of a shorter vegetation period. The species composition of the understory in *N. pumilio* krummholz stands apparently reflects different site properties as follows:

- •Sparse undergrowth dominated by shrubs such as *Maytenus* and/or *Myoschilos* indicates dry habitats, relatively warm and nutrient-rich;
- •Undergrowth dominated by dwarf shrubs (i.e., Ericaceae) indicates dry habitats that are poor in nutrients;
- •Understory with herbs and/or tall forbs indicates wet habitats, usually relatively cold and rich in nutrients; however, in the northern part of the forest region wet site conditions are extremely rare;
- •Understory of Ericaceae together with wetland herbs and grasses indicates wet habitats, but relatively cold sites and poor in nutrients.

5.2 Growth Forms at Andean Timberline

The most frequent tree deformation found at timberline that stems and branches run downhill for several meters before curving vertical alignment is generally attributed to the forces exerted by sliding snow, snow creeping, and settling snow cover resulting in mechanical damages of the trees (Eskuche 1973). In southern Chile with predominant western and northern winds, windward sites are at the same time sun-exposed and leeward slopes are shady, resulting in pronounced differences at exposed ridges. While leeward tremendous snow masses pile up that protect trees from abrasion and frost desiccation (figures 4a, 4b). Adjacent windward sites are treeless or have prostrate growing fan-shaped trees running uphill at shallow slopes, coating the slopes with a compact dwarf thicket less than 1 m in height (figure 4c).

Growth forms with branches close to the ground (termed 'cushion') have been described for *N. solandri* var. *cliffortioides* trees from exposed spur crests at the New Zealand timberline (Norton and Schönenberger 1984); in addition, Holtmeier (1980) describes a similar growth form (termed 'cornice-like') also for gymnosperm trees at timberline in Colorado, North America. Growth forms of *N. solandri* var. *cliffortioides* trees (termed 'decurrent') with long branches running downhill for several meters (P. Wardle 1963, J. Wardle 1970, Norton and Schönenberger 1984) closely conform to those described for *N. pumilio* in this study (figures 4a, 4b, and 5).

5.3 Reproductive Biology and Stability at Timberline

At alpine timberlines, it has been suggested that sexual reproduction is probably of much less importance to trees than at lower altitudes (Álvarez-Uría and Díaz González 2001; Cuevas 2002). In general, vegetative reproduction seems to be important in allowing trees to grow at timberline, especially at periods unfavorable to seedling growth (e.g. short growth period, freeze-thaw events in spring). However, contrary to accounts of limited seed production at altitudinal timberline in the Northern Hemisphere (Tranquillini 1979; Holtmeier 1994) and in New Zealand (Allen and Platt 1990), seed availability does not appear to prevent *N. pumilio* seedling establishment at the timberline in the southern Andes (Daniels and Veblen 2003b). *Nothofagus pumilio* growing in subalpine forests produces prolific seed, although production varies between sites and years (Rusch 1993). However, seedling age structures indicated that at least some viable seed was available annually although seedling frequency was low in some years (Barrera et al. 2000). In addition, variations in the density of small seedlings at multi-scales (Daniels and Veblen 2004) suggest that soil moisture is important for *N. pumilio* seedling establishment and survival at altitudinal treeline, implying interactions between temperature and precipitation, and disturbance-induced changes of site conditions that affect seedling survival as well as establishment. However, given that in Central Chile mammalian fauna eat *Nothofagus* seed it is likely that they also consume tree seeds in southern Chile (Veblen et al. 1996). Although birds have been observed eating *Nothofagus* seed, there are no quantitative data on the role of *Nothofagus* seed in birds' diets. The majority of tree seeds identified in the diets of birds in southern Chile are indigestible seeds (Armesto and Rozzi 1989). Thus, birds may even contribute to seed dispersion; however, the degree to which birds utilize *Nothofagus* seed is unknown.

In the southern Andes, it is likely that generative regeneration is important in allowing *N. pumilio* trees to persist at timberline. However, layering also occurs on *N. pumilio*, and can probably prolong life of individual trees; sometimes adventitious roots are formed where branches touched the ground (figures 4), but usually new stems relied on the roots of the original treefall (Veblen et al. 1996). Repeated dieback and subsequent epicormic branching from old boles may prolong the life of single trees, at least at the base. *Nothofagus pumilio* trees rarely exceed 300 to 400 years of age (Veblen et al. 1996), although trees up to 625 years old occur (Pollmann and Veblen in press). Additional tree-ring research on forest dynamics, establishment and seedling survivorship/mortality at timberline may clarify age structure and may give advice on longevity and persistence in the southern Andes.

Given the significant spatial and temporal variation of timberline related to disturbance, and given differences in regional, local, and micro-climates, as well as climate variability over time (Veblen et al. 1996; 1999; Villalba and Veblen 1997a; 1997b; 1998; Daniels and Veblen 2003b), the probable occurrence of old *N. pumilio* trees and adequate seedling establishment in the krummholz ecotone suggest that the present timberline margin may have been stable for several centuries. *Nothofagus pumilio* was not observed above the present-day timberline, although depression of timberline was observed after volcanism (Veblen et al. 1977; Veblen and Ashton 1978). At regional scale, treeline elevations were higher in Argentina than in Chile, corresponding to differences in temperature and continentality (Daniels and Veblen 2003b). Tree-ring based temperature reconstructions for Northern Patagonia indicate that prior to the 1970ies, temperatures were generally below average since ca. 1820 (Villalba et al. 1997a). Given predominately cool temperatures, treeline elevations in the southern Andes may have been limited by low temperature over the long-term and at large scales. In this context, a warmer and more continental climate in Argentina explains the higher treeline elevations at mid-latitudes in inland localities relative to western localities (Wardle 1998).

In the Northern Hemisphere, crippled timberline trees aged at 800 years and more have been interpreted (Ives and Hansen-Bristow 1983) as implying considerable stability of timberline; new trees are recruited continuously, the old trees having persisted through past adverse periods. It is likely that similarly in *N. pumilio* forests, persistence of established trees occurred through past adverse periods; recruitment of new individuals occurred regularly. In contrast to previous observations (McQueen 1976), seedling establishment thus may have an important role in maintaining timberline stability and it is unlikely, that the position of the timberline margin has changed in the last several hundred years and has probably been relatively stable for longer periods (Daniels and Veblen 2004). Contradictory to Northern Hemisphere gymnosperm forests and New Zealand *N. solandri* forests, generative regeneration is of high importance in *N. pumilio* forests (Cuevas 2000). However, vegetative reproduction does occur occasionally, in line with observations by Veblen et al. (1996) who mentioned that *N. pumilio* sometimes produces basal shoots when damaged. *Nothofagus pumilio* seedlings more than 10 cm tall were abundant in undisturbed understories and can rapidly response to the creation of canopy gaps, in line with previous observations (Veblen et al. 1981; Rebertus and Veblen 1993). However, instabilities both in seedling establishment and tree radial growth of *N. pumilio* were illustrated, over time and through space (Daniels and Veblen 2003b). These spatial and temporal instabilities in vegetation - climate relationships exemplify the need of exercise caution when conducting historical reconstructions of climate - vegetation relationships that may be unstable through time. Understanding these connections contributes to knowledge of the responses of *N. pumilio* forests in the southern Andes to past, present, and future climate variation (Villalba and Veblen 1998, Daniels and Veblen 2004). When relationships are unstable, short-term intensive analyses may provide a more reliable understanding of timberline dynamics in relation to climate change rather than alternative research approaches that assume stability.

5.4 Southern Andean, New Zealand, Tasmanian and Northern Hemisphere timberlines

In New Zealand and Tasmania (Australia), *Nothofagus* species also form timberline. The evergreen *N. solandri* commonly occurs at timberline in New Zealand, and is, despite of its contrasting foliage seasonality, ecologically comparable to *N. pumilio* (McQueen 1976; Hämat-Ahti 1986; Ogden et al. 1996; Wardle 1998). However, in the southern Andes, there is much more tendency than in New Zealand for *Nothofagus* to form stunted growth or krummholz (Wardle 1973; 1981; 1998). In Tasmania the mountains do not raise much above timberline and the upper forest limit is irregular and fragmented (Read and Brown 1996). Species of *Eucalyptus* and *Athrotaxis* commonly occur at timberline; the dwarf deciduous *N. gunnii* occurs near timberline on exposed sites and is capable of surviving frequent snow and ice storms (Jackson 1973; Haemet-Ahti 1986).

Regarding the climatic situation, the timberline of southern South America and New Zealand exhibit only few similarities to those of the Northern Hemisphere (e.g. snow cover, similar physiological response of tree growth to climatic influences during winter; cf. Troll 1973; Wardle 1978; Holtmeier 2003). The comparison of southern Andean and New Zealand climates at timberline (Wardle 1981; 1998) indicated that the southern Andes are significantly cooler than New Zealand at comparable latitudes and altitudes in winter, summer, and through the year. Subalpine forests in the southern Andes occupy a belt that is above the forest limit in New Zealand, and that is subject to snow of much greater depth and duration, and to winds that are stronger and colder. Evergreen *Nothofagus* from New Zealand treeline is less cold-resistant than deciduous *Nothofagus* from the Andes, and requires longer growing seasons than Northern Hemisphere timberline trees (Wardle 1998). Thus, structure, deciduous foliage, and adapted growth forms of trees described from the southern Andes may be explained by considering the extreme climatic conditions (Wardle 1981; Holtmeier 2003). However, krummholz and leaf phenology is not only a morphological response to the adverse climate but also reflects a life form with functional advantages (Barrera et al. 2000).

From a worldwide point of view, most vegetation of timberline is evergreen or mixed deciduous/evergreen in the non-seasonal lower latitudes (P. Wardle 1974; Holtmeier 2003), but there is hardly any general rule to explain the occurrence of different tree species at timberlines (Wardle 1981; Körner 1998). Thus, timberline built by deciduous broad-leaved trees seems to be a special feature of mid- and high-latitudinal mountain ranges, where heavy snowfall occurs, in combination with wind storms. Some timberlines of high-oceanic to sub-oceanic areas in meridional, temperate and boreal zones are built by deciduous trees. For example, in northern Eurasia (northern Fennoscandia) and southwestern Europe (Spain), deciduous broad-leaved *Betula* trees occur at timberline, as well as in oceanic southeastern Alaska and adjacent Canada (Rikli 1946; Blüthgen 1960; Haemet-Ahti 1963; 1987; Armand 1992; Nakhutsrishvili 1999; Álvarez-Uría and Díaz González 2001; Holtmeier 2003). Comparisons of forest line elevation between hemispheres support that under similar thermal conditions there are no altitudinal differences between forest lines of the two hemispheres. Several sites in Europe under the same type of oceanic climates that occur in the southern hemisphere had forest lines occupied by broadleaf species, even when conifers were present in the region (Jobbágy and Jackson 2000). Forest lines tend to be dominated by broadleaf species, then by evergreen conifers, and finally by deciduous conifers as thermal seasonality increases. This assessment agrees with the observation that the tundra - taiga ecotone in the arctic is occupied by *Betula* spp. in the most oceanic portion of the arctic belt and by conifers in the rest of the area (Hustich 1983). Thus, Andean *N. pumilio* may be comparable both with the Fennoscandian *Betula tortuosa* and the Cantabrian *B. celtiberica* because of comparable physiognomy and physiological responses. In oceanic meridional southern Europe the deciduous forest tree *Fagus sylvatica* also occurs up to timberline (Nikolovski 1970; Schreiber 1998), behaving similar to *N. pumilio* in austral and antarctic southern Chile, which in subalpine forests also becomes more and more stunted as it approaches the limit (Ellenberg 1966). In line with previous observations (Haemet-Ahti 1986), we can state the predominance of *N. pumilio* over a variety of timberline habitats in the southern Andes. Concerning the deciduous foliage of *Nothofagus* timberline, our observations suggest that the time required to produce new leaves probably makes deciduous *N. pumilio* more sensitive and competitive to the length of the growing season than are evergreen species, for example in New Zealand. In general, broadleaf deciduous trees seem to be more prone to moisture stress (in high-oceanic regions) than are evergreen trees, in which case low temperature may not be the dominant climatic factor limiting *N. pumilio* and *Betula* spp. at timberline. Thus, deciduous *N. pumilio* timberline in the Andes includes convergence to both the *Betula*-timberline in high-oceanic submeridional and boreal latitudes, and the *Fagus*-timberline of the oceanic meridional-montane Mediterranean zone. However, the altitudinal position of the European timberline exceeds the position of this limit in the southern Andes by about 400 m (southern Italy, 38°N) to 550 m (Cantabrian Mountains, 43°N).

6 Conclusions

Species composition and structure of southern Andean timberline found at latitudes between 35°S and 55°S and altitudes between 2000 m and 600 m are uncommon and distinctive. Although all timberline ecotones are dominated by *Nothofagus pumilio*, main differences reflect understory vegetation originating from adjacent forest regions. The appearance of similar growth forms at southern timberline give evidence that these growth forms are determined by extreme environments that develop comparable stunted forms in dependence from sites, rather than by genetic differentiation or species constitution. The recognition of main differences of southern Andean timberline compared to New Zealand alpine treelines such as the deciduous foliage of timberline trees may be explained by more extreme climatic conditions (temperature *and* precipitation) in South America. By structure and physiognomy, the *Nothofagus pumilio* timberline appears rather more similar to northern hemispheric *Betula* and *Fagus* timberlines than to those of the related *Nothofagus* species in New Zealand; its altitudinal position is fairly intermediate – higher than in respective New Zealand and lower than in European mountains.

Acknowledgements

We are grateful to L. Flores, R. Godoy, and A. Vogel for cooperation in the field and to M. Alberdi, M.T.K Arroyo, R. Godoy, C. Ramírez, and J. San Martín for their contributions to project collaboration. The Corporación Nacional Forestal (Chile) provided local logistical support. This study is part of research work supported by the German National Science Foundation (DFG, grants Hi 535/2 and Po 703/1) and the German Academic Exchange Service

(DAAD, grant D/02/00809). We thank the referees for their constructive comments.

References

Alberdi ML (1987) Ecofisiología de especies chilenas del genero *Nothofagus*. Bosque 8: 77-84

Alberdi ML (1995) Ecofisiología de especies leñosas de los bosques higrófilos templados de Chile: Resistencia a la sequía y bajas temperaturas. In: Armesto JJ, Villagrán C, Arroyo MTK (eds) Ecología de los bosques nativos de Chile, Edit Univers, Santiago, pp 279-299

Alberdi ML, Romero M, Ríos D, Wenzel H (1985) Altitudinal gradients of seasonal frost resistance in *Nothofagus* communities of southern Chile. Acta Œcol 6: 21-30

Allen RB, Platt KH (1990) Annual seedfall variation in *Nothofagus solandri* (Fagaceae), Canterbury, New Zealand. Oikos 57: 119-206

Álvarez-Uría P, Díaz González TE (2001) Recruitment limitation of celtiberic birch near the treeline in the Cantabrian Mountains (NW Spain) an experimental study. Verh Ges Ökol 31: 336

Amigo J, Ramírez C (1998) A bioclimatic classification of Chile: woodland communities in the temperate zone. Plant Ecol 136: 9-26

Armand AD (1992) Sharp and gradual mountain timberlines as a result of species interaction. In: Hansen AJ, di Castri F (eds) Landscape boundaries consequences for biotic diversity and ecological flows. Ecol Stud 92: 360-378. Springer , New York

Armesto JJ, Rozzi R (1989) Seed disperal syndromes in the rain forest of Chiloé: evidence for the importance of biotic dispersal in a temperate rain forest. J Biogeog 16: 219-226

Barrera MD, Frangi JL, Richter LL, Perdomo MH, Pinedo LB (2000) Structural and functional changes in *Nothofagus pumilio* forests along an altitudinal gradient in Tierra del Fuego, Argentina. J Veg Sci 11: 179-188

Bergmeier E, Härdtle W, Mierwald U, Nowak B, Peppler C (1990) Vorschläge zur syntaxonomischen Arbeitsweise in der Pflanzensoziologie. Kieler Notizen Pflanzenkde Schleswig-Holstein Hamburg 20: 92-103

Blüthgen J (1960) Der skandinavische Fjällbirkenwald als Landschaftsform. Peterm Geogr Mitt 2/3: 119-144

Braun-Blanquet J (1964) Pflanzensoziologie. Grundzüge der Vegetationskunde. 3rd ed., Springer Wien

Burns BR (1993) Fire-induced dynamics of *Araucaria araucana-Nothofagus antarctica* forest in southern Andes. J Biogeog 20: 669-685

Cuevas JG (2000) Tree recruitment at the *Nothofagus pumilio* alpine timberline in Tierra del Fuego, Chile. J Ecol 88: 840-855

Cuevas JG (2002) Episodic regeneration at the *Nothofagus pumilio* alpine timberline in Tierra del Fuego, Chile. J Ecol 90: 52-60
Cuevas JG, Arroyo MTK (1999) Ausencia de banco semillas persistent en *Nothofagus pumilio* (Fagaceae) en Tierra del Fuego, Chile. Rev Chil Hist Nat 72: 73-82
Daniels LD, Veblen TT (2003a) Altitudinal treelines of the southern Andes near 40° S. For Chron 79: 237-241
Daniels LD, Veblen TT (2003b) Regional and local effects of disturbance and climate on altitudinal treelines in northern Patagonia. J Veg Sci 14:733-742
Daniels LD, Veblen TT (2004) Spatiotemporal influences of climate on altitudinal treeline in Northern Patagonia. Ecol 85: 1284-1296
Daubenmire R (1954) Alpine timberlines in the Americas and their interpretation. Butler Univ Bot Stud 11:119-136
Dierschke H (1994) Pflanzensoziologie. Ulmer, Stuttgart
Dollenz O (1983) Fitosociología de la Reserva Forestal „El Parrillar" Peninsula de Brunswick, Magallanes. Ans Inst Pat 14:109-118
Ellenberg H (1966) Leben und Kampf an den Baumgrenzen der Erde. Naturw Rundschau 19:133-139
Ellenberg H (1996) Vegetation Mitteleuropas mit den Alpen. 5th ed, Ulmer, Stuttgart
Eskuche U (1973) Estudios fitosociológicos en el Norte de Patagonia I Investigación de algunos factores de ambiente en comunidades de bosque y de chaparral. Phytocoenologia 1: 64-113
Eskuche U (1999) Estudios fitosociológicos en el norte de la Patagonia II Los bosques del Nothofagion dombeyi. Phytocoenologia 29: 177-252
Finckh M (1996) Die Wälder des Villarrica-Nationalparks (Südchile). Lebensgemeinschaften als Grundlage für ein Schutzkonzept. Diss Bot 259: 1-181
Finckh M, Paulsch A (1995) *Araucaria araucana.* Die ökologische Strategie einer Reliktkonifere. Flora 190: 365-382
Flores L (1998) Los bosques de *Nothofagus pumilio* (Poepp. et Endl.) Krasser, entre los 40° y 43° Sur, Chile. M Sc thesis, Univ Austral de Chile, Valdivia
Freiberg H-M (1985) Vegetationskundliche Untersuchungen an südchilenischen Vulkanen. Bonner Geogr Abh 70.
Gajardo R (1994) La vegetación natural de Chile. Clasificación y districución geográfica. Edit Univers, Santiago
Godley EJ (1960) The botany of southern Chile in relation to New Zealand and the Subantarctic. Proc Royal Soc, London 152: 457-475
Grace J (1989) Tree lines. Phil Transact. Royal Soc, London 324: 233-245
Haemet-Ahti L (1963) Zonation of the mountain birch forest in northernmost Fennoscandia. Ann Soc Zool Bot Fenn 'Vanamo' 34
Haemet-Ahti L (1986) The zonal position of *Nothofagus* forests. Veröff Geobot Inst ETH, Stiftung Rübel 91: 217-227
Haemet-Ahti L (1987) Mountain birch and mountain birch woodland in NW Europe. Phytocoenologia 15: 449-453
Haemet-Ahti L, Ahti T (1969) The homologies of the Fennoscandian mountain and costal birch forests in Eurasia and North America. Vegetatio 19: 208-219

Hall CAS, Stanford JA, Hauer ER (1992) The distribution and abundance of organisms as a consequence of energy balances along multiple environmental gradients. Oikos 65: 377-390

Hildebrand R (in press) Flora, vegetación y sinecología de los bosques de *Nothofagus pumilio* en su área de distribución central (38°- 48° l.s.). Phytocoenologia

Hildebrand-Vogel R (2002) Structure and dynamics of southern Chilean natural forests with special reference to the relation of evergreen versus deciduous elements. Folia Geobot 36: 107-128

Hildebrand-Vogel R, Godoy R, Vogel A (1990) Subantarctic-Andean *Nothofagus pumilio* forests. Vegetatio 89: 55-68

Hildebrand-Vogel R, Flores L, Godoy R, Vogel A (1998) Synecology of *Nothofagetea pumilionis* - comparison between South Andean and European deciduous forests. Stud Plant Ecol 20, 34

Holtmeier F-K (1973) Geoecological aspects of timberlines in northern and Central Europe. Arctic Alpine Res 5 (3, Pt 2), A45-A54

Holtmeier F-K (1980) Influence of wind on tree physiognomy at the upper timberline in the Colorado Front Range. In: Benecke U, Davis MR (eds) Mountain environments and subalpine tree growth, Forest Service 70. Wellington, New Zealand, pp 247-261

Holtmeier F-K (1981) What does the term 'krummholz' really mean? Observations with special reference to the Alps and the Colorado Front Range. Mountain Res Development 1: 253-260

Holtmeier F-K (1985) Die klimatische Waldgrenze. Linie oder Übergangssaum (Ökoton)? Erdkunde 39: 271-285

Holtmeier F-K (1989) Ökologie und Geographie der oberen Waldgrenze. Ber Reinh Tüxen-Ges 1: 15-45

Holtmeier F-K (1994) Ecological aspects of climatically-caused timberline fluctuations: review and outlook. In: Beniston M (ed) Mountain environments in changing climates. Routledge, London, pp 220-233

Holtmeier F-K (1995) Waldgrenzen und Klimaschwankungen. Ökologische Aspekte eines vieldiskutierten Phänomens. Geoökodynamik 16: 1-24

Holtmeier F-K (2003) Mountain Timberlines – Ecology, patchiness, and dynamics. Advances in Global Change Research, 14. Kluwer Academic, Dordrecht

Holtmeier F-K, Broll G (1992) The influence of tree islands and microtopography on pedological conditions in the forest-alpine tundra ecotone on Niwot Ridge, Colorado Front Range, USA. Arctic Alpine Res 24: 216-228

Hustich I (1983) Tree-line and tree growth studies during 50 years: some subjective observations. In: Morisset P, Payette S (eds) Tree-line ecology. Université Laval, Quebec, pp 181-188

Ives JD, Hansen-Bristow K J (1983) Stability and instability of natural and modified upper timberline landscapes in the Colorado Rocky Mountains, U S A. Mountain Res Develop 3: 149-155

Jackson WD (1973) Vegetation of the Central Plateau. In: Banks MR (ed) The lake country of Tasmania. Hobart, Royal Soc of Tasmania, pp 61-85

Jobbágy EG, Jackson RB (2000) Global controls of forest line elevation in the northern and southern hemispheres. Global Ecol Biogeog 9: 253-268
Kalela EK (1941a) Über die Holzarten und die durch die klimatischen Verhältnisse verursachten Holzartenwechsel in den Wäldern Ostpatagoniens. Ann Acad Sci Fenn Ser 5 A, IV Biol 2, pp 1-151
Kalela EK (1941b) Über die Entwicklung der herrschenden Bäume in den Beständen verschiedener Waldtypen Ostpatagoniens. Ann Acad Sci Fenn Ser 5 A, IV Biol 3, pp 1-65
Körner C (1998) Re-assessment of high elevation treeline positions and their explanation. Oecologia 115: 445-459
Kozdon P (1958) Die autochthonen Baumarten und die forstlichen Verhältnisse der südlichen (subarktischen) Andenkordillere. Schweiz Z Forstw 109: 325-347
Landrum LR (1988) The myrtle family (Myrtaceae) in Chile. Proc California Acad Sci 45: 277-317
Landrum LR (1999) Revision of *Berberis* (Berberidaceae) in Chile and adjacent southern Argentina. Ann Missouri Bot Gard 86: 793-834
Lara A, Aravena JC, Villalba R, Wolodarsky-Franke A, Luckman B, Wilson R (2001) Dendroclimatology of high-elevation *Nothofagus pumilio* forests at their northern distribution limit in the central Andes of Chile. Can J For Res 31: 925-936
Lavoie C, Payette S (1994) Recent fluctuation of the lichen-spruce forest limit in subarctic Québec. J Ecol 82: 725-734
Lechowicz MJ (1984) Why do temperate deciduous trees leaf out at different times? Adaptation and ecology of forest communities. Am Nat 124: 821-842
Lloyd AH, Graumlich LJ (1997) Holocene dynamics of treeline forests in the Sierra Nevada. Ecol 78: 1199-1210
Marticorena C, Quezada M (1985) Catálogo de la flora vascular de Chile. Gayana Bot 42: 1-157
Marticorena C, Rodríguez R (1995) Flora de Chile Vol I Pteridophyta-Gymnospermae. Universidad de Concepción, Concepción
Marticorena C, Rodríguez R (2001) Flora de Chile Vol II Winteraceae-Ranunculaceae. Universidad de Concepción, Concepción
Matthei O (1997) Las especies del genero *Chusquea* Knuth (Poaceae: Bambusoideae), que crecen en la X región, Chile. Gayana Bot 54: 199-220
McQueen DR (1976) The ecology of *Nothofagus* and associated vegetation in South America. Tuatara 22: 38-68, 233-244
Miehe G, Miehe S (2000) Comparative high mountain research on the treeline ecotone under human impact. Erdkunde 54: 34-50
Miller A (1976) The climate of Chile. In: Schwerdtfeger W (ed) Climates of Central and South America. World Survey of Climatology 12, Elsevier, Amsterdam, pp 113-131
Mueller-Dombois D, Ellenberg H (1974) Aims and methods in vegetation ecology. Wiley and Sons, New York
Nakhutsrishvili G (1999) The vegetation of Georgia (Caucasus). Braun-Blanquetia 15: 1-74

Nikolovski T (1970) Waldgesellschaften und Waldbäume an der oberen Grenze der Verbreitung in verschiedenen Gebirgssystemen der SR-Mazedonien. Mitt Ost-alpin-Dinarische Ges Vegetationskde 11: 151-160

Noble IR (1993) A model of the responses of ecotones to climatic change. Ecol Applications 3: 396-403

Norton DA, Schönenberger W (1984) The growth forms and ecology of *Nothofagus solandri* at the timberline, Craigieburn Range, New Zealand. Arctic Alpine Res 16: 361-370

Oberdorfer E (1960) Pflanzensoziologische Studien in Chile. Flora et Vegetatio Mundi 2: 1-208

Ogden J, Stewart GH, Allen RB (1996) Ecology of New Zealand *Nothofagus* forests. In: Veblen TT, Hill RS, Read J (eds) The ecology and biogeography of *Nothofagus* forests. Yale Univ Press, New Haven, pp 25-82

Pollmann W (2001) Caracterización florística y posición sintaxonómica de los bosques caducifolios de *Nothofagus alpina* (Poepp. et Endl.) Oerst. en el centro-sur de Chile. Phytocoenologia 31: 353-400

Pollmann W, Veblen TT (in press) *Nothofagus* regeneration dynamics in south-central Chile: A test of a general model. Ecology

Prohaska F (1976) The climate of Argentina, Paraguay and Uruguay. In: Schwerdtfeger W (ed) Climates of Central and South America World Survey of Climatology 12. Elsevier, Amsterdam, pp 13-112

Ramírez C, San Martín C, Ojeda P (1997) Muestreo y tabulación aplicados al estudio de los bosques nativos. Bosque 18: 19-27

Read J, Brown MJ (1996) Ecology of Australian *Nothofagus* forests. In: Veblen TT, Hill RS, Read J (eds) The ecology and biogeography of *Nothofagus* forests. Yale Univ Press, New Haven, pp 131-181

Rebertus AJ, Veblen TT (1993) Structure and tree-fall gap dynamics of old-growth *Nothofagus* forests in Tierra del Fuego, Argentina. J Veg Sci 4: 641-654

Rebertus AJ, Veblen TT, Kitzberger T (1993) Gap formation and dieback in Fuego-Patagonian *Nothofagus* forests. Phytocoenologia 23: 581-599

Reichelt G, Wilmanns O (1973) Vegetationsgeographie. Westermann, Braunschweig

Rikli M (1946) Das Pflanzenkleid der Mittelmeerländer. Bern

Rusch V (1993) Altitudinal variation in the phenology of *Nothofagus pumilio* in Argentina. Rev Chil Hist Nat 66: 131-141

Sakai A, Paton DM, Wardle P (1981) Freezing resistance of trees of the South temperate zone, especially subalpine species of Australasia. Ecol 62: 563-570

San Martín J, Troncoso A, Mesa A, Bravo T, Ramírez C (1991) Estudio fitosociológico del bosque caducifolio magellánico en el límite norte de su área de distribución. Bosque 12: 29-41

Schreiber HJ (1998) Waldgrenznahe Buchenwälder und Grasländer des Falakron und Pangäon in Nordostgriechenland. Arb Inst Landschaftsökol 4: 1-170

Seibert P (1996) Farbatlas Südamerika. Landschaften und Vegetation. Ulmer, Stuttgart

Skottsberg C (1910) Botanische Ergebnisse der schwedischen Expedition nach Patagonien und dem Feuerlande 1907-1909. I. Übersicht über die wichtigsten

Pflanzenformationen Südamerikas von 41° südl Br., ihre geographische Verbreitung und Beziehung zum Klima. Kungliga Svenska Ventensksakpsademiens Handlingar 46 (3): 1-28

Skottsberg C (1916) Botanische Ergebnisse der schwedischen Expedition nach Patagonien und dem Feuerlande 1907-1909. V. Die Vegetationsverhältnisse längs der Cordillera de los Andes von 41° südl Breite. Ein Beitrag zur Kenntnis der Vegetation in Chiloé, Westpatagonien, dem andinen Patagonien und Feuerland. Kungliga Svenska Vetenskapsakademiens Handlingar 56 (5): 1-366

Slatyer RO, Noble IR (1992) Dynamics of montane treelines. In: Hansen AJ, di Castri F (eds) Landscape boundaries: consequences for biotic diversity and ecological flows. Springer, New York, pp 346-359

Taljaard TJ (1972) Synoptic climatology of the southern hemisphere. Meteorol Monogr 13: 139-213

Tranquillini W (1979) Physiological ecology of the alpine timberline; tree existence at high altitudes with special reference to the European Alps. Springer, New York

Troll C (1948) Der asymmetrische Aufbau der Vegetationszonen und Vegetationsstufen auf der Nord- und Südhalbkugel. Jahresber Geobot Inst Rübel, Zürich 1947, pp 46-83

Troll C (1960) The relationship between the climates, ecology and plant geography of the southern cool temperate zone and of the tropical high mountains. Proc Royal Soc London 152: 529-532

Troll C (1973) The upper timberlines in different climatic zones. Arctic Alpine Res 5 (3, Pt 2), A3-A18

Vázquez FM, Rodríguez RA (1999) A new subspecies and two new combinations of *Nothofagus* Blume (Nothofagaceae) from Chile. Bot J Linnean Soc 129: 75-83

Veblen TT (1979) Structure and dynamics of *Nothofagus* forests near timberline in south-central Chile. Ecol 60: 937-945

Veblen TT (1982) Regeneration patterns in *Araucaria araucana* forests in Chile. J Biogeog 8: 211-247

Veblen, TT (1989a) *Nothofagus* regeneration in treefall gaps in northern Patagonia. Can J For Res 19: 365-371

Veblen TT (1989b) Tree regeneration responses to gaps along a transandean gradient. Ecol 70: 543-545

Veblen TT (1992) Regeneration dynamics. In: Glenn-Lewin DG, Peet RK, Veblen TT (eds) Plant succession. Chapman and Hall, London, pp 152-187

Veblen TT, Ashton DH (1978) Catastrophic influences on the vegetation of the Valdivian Andes, Chile. Vegetatio 36: 149-167

Veblen TT, Ashton DH (1979) Successional pattern above timberline in south-central Chile. Vegetatio 40: 39-47

Veblen TT, Ashton DH, Schlegel FM, Veblen AT (1977) Plant succession in a timberline depressed by volcanism in south-central Chile. J Biogeog 4: 275-294

Veblen TT, Donoso C, Schlegel FM, Escobar B (1981) Forest dynamics in south-central Chile. J Biogeog 8: 211-247

Veblen TT, Burns BR, Kitzberger T, Lara, A, Villalba R (1995) The ecology of the conifers of southern South America. In: Enright NJ, Hill RS (eds) Ecology of the southern conifers. Melbourne Univ Press, Melbourne, pp 120-155

Veblen TT, Donoso D, Kitzberger T, Rebertus AJ (1996) Ecology of southern Chilean and Argentinean *Nothofagus* forests. In: Veblen TT, Hill RS, Read J (eds) The ecology and biogeography of *Nothofagus* forests. Yale Univ Press, New Haven, pp 293-353

Veblen TT, Kitzberger T, Villalba R, Donnegan J (1999) Fire history in northern Patagonia: the roles of humans and climatic variation. Ecol Monog 69: 47-67

Villalba R, Veblen TT (1997a) Regional patterns of tree population age structures in northern Patagonia: climatic and disturbance influences. J Ecol 85: 113-124

Villalba R, Veblen TT (1997b) Spatial and temporal variation in tree growth along the forest-steppe ecotone in northern Patagonia. Can J For Res 27: 580-597

Villalba R, Veblen TT (1998) Influences of large-scale climatic variability on episodic tree mortality at the forest-steppe ecotone in northern Patagonia. Ecol 79: 2624-2640

Wardle JA (1970) The ecology of *Nothofagus solandri*: Parts 1 to 4. N Z J Bot 8: 494-646

Wardle JA (1974) Life history of mountain beech (*Nothofagus solandri* var. *cliffortioides*). Proc N Z Ecol Soc 21: 21-26

Wardle P (1963) Growth habits of New Zealand subalpine shrubs and trees N Z J Bot 1: 18-47

Wardle P (1965) A comparison of alpine timberlines in New Zealand and North America. N Z J Bot 3: 113-135

Wardle P (1971) An explanation for alpine timberline. N Z J Bot 9: 371-402

Wardle P (1973) New Zealand timberlines Arctic Alpine Res 5 (3, Pt 2), A127-A135

Wardle P (1974) Alpine timberlines. In: Ives JD, Barry R (eds) Arctic and alpine environments. Methuen Publ, London, pp 371-402

Wardle P (1978) Ecological and geographical significance of some New Zealand growth forms. Erdw Forschungen 11: 531-536

Wardle P (1981) Is the alpine timberline set by physiological tolerance, reproductive capacity, or biological interactions? Proc Ecol Soc Aust 11: 53-66

Wardle P (1998) Comparison of alpine timberlines in New Zealand and the Southern Andes. Roy Soc N Z Misc Ser 48: 69-90

Westhoff V, van der Maarel E (1973) The Braun-Blanquet approach. In: Whittaker RH (ed) Ordination and classification of communities. Handbook Veg Sci 5: 619-726

Whittaker RH (1975) Communities and Ecosystems. 2nd ed, MacMillan, New York

Whiters MA, Meentemeyer V (1999) Concepts of scale in landscape ecology. In: Kopatek JM, Gardner RH (eds) Landscape ecology analysis. Springer, New York, pp 205-252

Pocket Gopher – Actor under the Stage. Studies on Niwot Ridge, Colorado Front Range, U.S.A.

Hans-Uwe Schütz

Abstract

The Northern Pocket Gopher (*Thomomys talpoides* Richardson) is a small herbivory subterranean rodent. It is known as an eager digger and in areas with intense activity it is suspected of damaging young tree growth. Results from soil analysis and snow measurements are presented to describe the pocket gophers' summer and winter environment within a ribbon forest in the forest-alpine tundra-ecotone on Niwot Ridge, Colorado Front Range. The distribution of snow and the structure of snow cover are influencing the pocket gopher's life under cover. Soil analysis showed that pocket gopher activity modified the soil environment. The amount of coarse soil particles in the upper topsoil layers was increased compared to topsoil with no gopher disturbance. The distribution of organic matter within the upper 12 cm was more homogeneous in areas with gopher excavation. None of the damages found on young trees could be attributed to pocket gopher activities without question. The most obvious damages were reduced growth at the windward side (83 %) followed by crippled growth (60 %) and damages caused by snow fungi (32 %). In summary, the pocket gophers in the ribbon forests on Niwot ridge are actively modifying the soil environment by exposing mineral soil and increasing the likelihood of soil erosion by wind and water. There is, however, no evidence that pocket gophers are responsible for the visible prevalent damage on juvenile conifers.

1 Introduction

As a being with an anthropocentric view of 'our' world it is easy to define the stage as what we see around us while walking along. So the actors on this stage are you and I at first glance, and a lot more beings when we are willing to open up our eyes. Many of them create landscape patterns by their individual habits or as part of a greater choreography of a whole assembly in a certain landscape. The image, and not at least the beauty, of a landscape is not all a work of what we know already and which is visible to us as actors on the stage, but certain input to it is due as well to actors under the stage. The Northern Pocket Gopher is an example.

This little (average adult length:16-25 cm; Miller 1964) subterranean rodent (photo 1), a very reclusive character, eager for digging in almost every kind of soil (Miller 1964), helps to prepare the stage for actors that are more colorful and eye-catching (Fleischer and Emerick 1984; Andersen 1987; Davies 1994; Holtmeier 2002). Most obvious are the gophers' digging activities which result in mound systems during summertime and systems of soil eskers (sediment-filled snow tunnels) as a result of their digging activities during wintertime in the soil and the snow. This actvitiy forms the ground for later revegetation processes. An example is its role in revegetating ash covered areas following the Mount St. Helens volcanic eruption (Andersen and MacMahon 1985). Soil dug out by the pocket gophers and variations in the availability of nutrients form the stage we see in many alpine and subalpine landscapes in northern America. The pocket gophers influence the actors on the stage, making the daily repeated play more diverse for our eyes, every time we open them again. The investigations partly presented in this paper are aimed at a better understanding of the role of pocket gophers in natural systems.

Photo 1 Northern Pocket Gopher (*Thomomys talpoides* Richardson) (Schütz, 1990-09-08)

The focus of this work is the role of the northern pocket gopher (*Thomomys talpoides* Richardson) (photo 1). The study site (photo 2) is located in a 'ribbon forest' environment on Niwot Ridge, Colorado Front Range (USA). Ribbon forests (Billings 1969; Buckner 1977; Holtmeier 1982) are forest structures that develop perpendicular to prevailing winds in the subalpine belt and the forest-alpine tundra-ecotone. The closer one gets to upper treeline, the more crippled and discontinuous these structures become.

2 Study Site

The study area is located on Niwot Ridge, on the east slope of the Colorado Front Range (USGS Ward 7.5-minute quadrangle). At an altitude of 3290 to 3370 m, a ribbon forest developed on turf banked terraces (Benedict 1970) in the forest-alpine tundra-ecotone. The ribbons are 3 to 60 m wide while the glades measure 29 to 125 m (photo 2). The 13 ha study site is located on the northeast-facing slope of a glacial interfluve. Holocene and late Pleistocene (Bull Lake and Pinedale) solifluction deposits overlie the igneous rocks. These are Precambrian sillimanite- and biotite-gneiss with local quartz monzonite intrusions (Gable and Madole 1976).

Photo 2 Aerial overview of the ribbon forest on Niwot Ridge, Colorado. Exposure northeast (Schütz, 1990-07-11)

The climate is continental, with maximum of precipitation during the winter and spring months November to April (mostly snow) (table 1). The amount of winter precipitation varies from year to year. This is especially true in early winter. February and March are generally the months with greatest snowfall. The summers are dry but due to thunderstorms they are still more humid than in the eastern flatlands. Wind is a permanent companion in the forest-alpine tundra-ecotone. Peak wind gusts are normally above 30 m s^{-1} and can reach more than 80 m s^{-1}. The redistribution of snow by strong winds and its accumulation in depressions and leeward of barriers (rocks, trees and tree ribbons) are predominant factors forming the vegetation pattern in the forest-alpine tundra-ecotone (Holtmeier 1982; 1987a; 2003).

The heterogeneous patterns of vegetation are a mirror of soil conditions. Soil thickness, soil moisture, soil texture and the abundance and size of stones vary between topographic positions as can be seen in erosion cuts (photo 3). Soil accumulates to the lee of the turf-banked terrace fronts, while the terraces themselves have thin, stony soils.

Table 1 Monthly precipitation and air temperature, 'Niwot Ridge – Saddle' (1989-1991)

Month	**1**	**2**	**3**	**4**	**5**	**6**	**7**	**8**	**9**	**10**	**11**	**12**
Mean												
Temperature [°C]												
1989	-11.8	-12.2	-5.8	-3.9	1.1	4.4	9.4	6.2	3.5	-4.0	-7.3	-13.2
1990	-11.1	-10.6	-7.1	-3.4	-0.5	8.2	8.5	8.4	6.0	-2.0	-6.6	-13.6
1991	-13.5	-9.1	-11.0	-6.6	0.3	5.0	7.4	7.5	3.8	0.2	-9.0	/
Precipitation [mm]												
1989	107	184	136	150	70	43	66	35	28	72	129	323
1990	205	210	232	206	56	66	106	78	64	82	205	318
1991	211	190	277	77	>17	64	105	40	73	16	362	/

/ = data not challenged
Source: Losleben 1990 and monthly summary of climatological data of the station 'Niwot Ridge – Saddle', Institute of Arctic and Alpine Research, University of Colorado

Photo 3 Erosion cut at a forest ribbon with crippled-tree growth on Niwot Ridge (Schütz, 1989-07-06)

3 Material and Methods

Snow distribution and duration were mapped using aerial photographs taken by the author in June and July 1989, 1990 and 1991. On May 1, 1990 snow thickness was measured in 2 meter intervals along four transects using a 4 meter pole. Two snow pits were dug on May 2 and 3, 1990, and 100 cm^3 snow samples were collected in 10 cm depth increments to measure snow densities.

Within the ribbon forests the biotope of pocket gophers are the snow glades between the ribbons of trees. To document the temperature in this habitat ABB Goerz thermoscript recorders were positioned at soil depths of 5-10 cm and 15-20 cm in well-vegetated and sparsely vegetated locations. Soil temperature was continuously recorded from 07/14/1990 to 09/07/1990. For digitalization and statistical analysis of the soil-temperature data the author used the computer software 'Thermo', 'Thermo-Plot' and 'Thermo-Ebcdic' developed by Wolf (1991).

To verify the pocket gopher's influence on soil, soil texture, organic matter, and cation exchange capacity (CEC) were analyzed of four sites with gopher activities and four nearby areas without digging activities within the ribbon forest on Niwot Ridge. Soil samples were collected in September 1989 at standardized depths of 0-4 cm, 4-8 cm, and 8-12 cm, covering most of the living space of pocket gophers in the soil. The samples were analyzed with the following standard methods:

(1) Texture analysis of the fine earth fractions (< 2 mm) after organic matter removed by 30 % H_2O_2 with sieving and pipette method (Page et al. 1982) and estimation of the percentage of skeletal material (> 2 mm) after sieving,

(2) Organic carbon and total nitrogen content by an elemental analyzer (Carlo Erba NA 1500),

(3) Cation exchange capacity (CEC) by extraction with 1 M ammonium chloride (Trüby and Aldinger 1989).

Damage to 92 young trees ranging from saplings to trees as tall as 1.6 m were listed and described.

4 Results

4.1 Snow

The deepest snow patches in the study area survive until the end of July (figures 1-5). Most of the snow glades in the ribbon forests retain a snow cover until the second half of June (figure 1). The time of snow melt varied by two weeks between 1989 and 1991 (Schütz 1998). The locations of the latest snow fields in the ribbon forests were the same in the years 1989, 1990 and 1991 (Schütz 1998).

The snow depth on the leeward side of the ribbons generally reaches three meters and exceeds four meters at several locations in the ribbon forest on Niwot Ridge (figures 2-5). Greater snow depths can be found at distances of 6 m (figures 3 and 5) to 22 m (figures 2, 3 and 5) leeward of the ribbons. Maximum depths are observed 14 m to the lee of the tree ribbons (figures 2 and 3) . The relative shallow snow depth within the first two to three meters leeward of the tree ribbons (figures 2-5) is caused by the turbulence of the snow-carrying winds directly behind a barrier and by the black body effect of the evergreen trees during snow melting periods. The sudden increase of snow depths within the following meters is supported by the slope contour with a steeper angle behind the shoulders of the solifluction terraces.

The snow depths in figures 2-5 may exceed the actual snow accumulation, which can be proven exactly only with measurements perpendicular to the ground surface due to the vertical measurements and the varying relief of the

Figure 1 Duration of snow cover and locations of snow transects and snow profiles in the ribbon forest on Niwot Ridge

ground below the snow cover. In some places, the thickness of the snow cover exceeded four meters and could not be accurately measured with available equipment (figure 5). The distribution of snow depths in figure 4 (transect E-F) is presumed to be caused by older icy surfaces of the snow pack which could not be penetrated by the pole used for the snow measurements along 30 meters of the transect. Such icy layers were also documented in the two snow profiles (figures 6 and 7).

No correlation was found between areas of pocket gopher activity (1989-1991) and measured snow depths or duration of the snow cover (figures 3 and 5) (Schütz 1998).

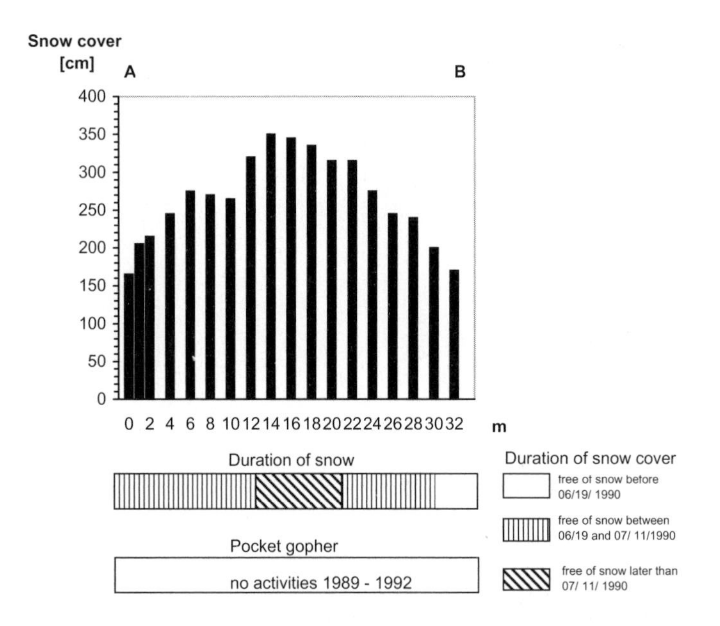

Figure 2 Snow thickness and gopher activities at transect A - B

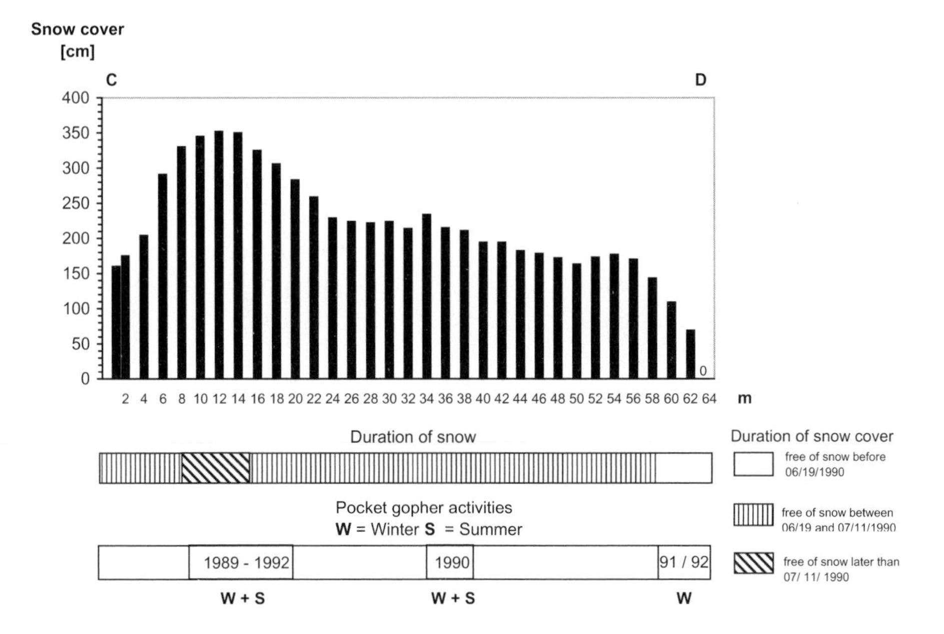

Figure 3 Snow thickness and gopher activities at transect C - D

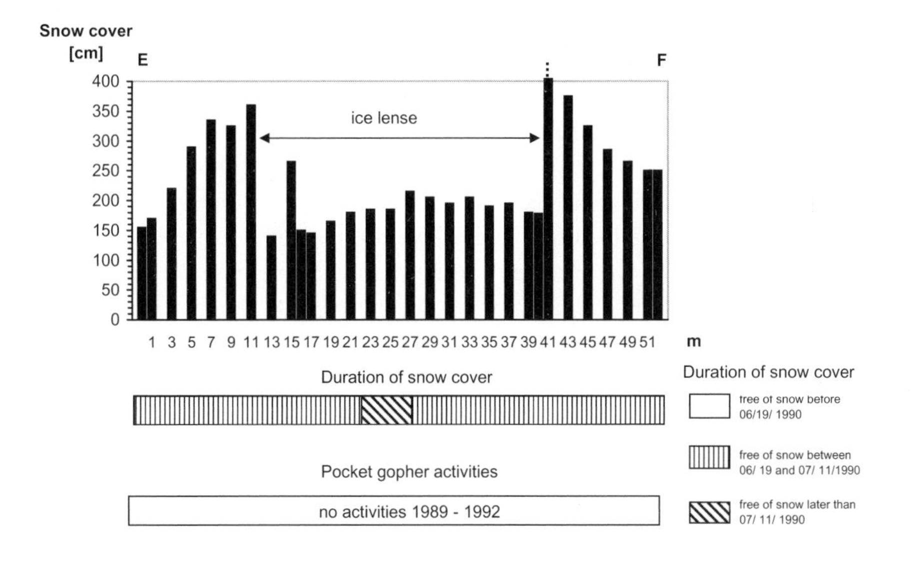

Figure 4 Snow thickness and gopher activities at transect E - F (⋮ => 400 cm)

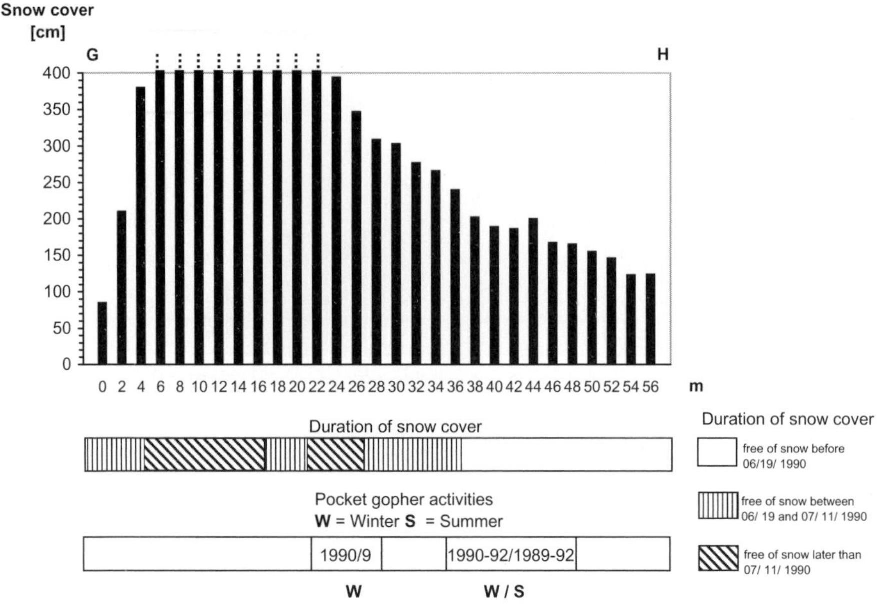

Figure 5 Snow thickness and gopher activities at transect G - H (⋮ => 400 cm)

Both snow profiles show icy or crusted layers as a result of older buried snow surfaces (wind crusts or wind slabs) and / or a result of a period with a high frequency of freeze and thaw cycles (Colbeck 1991). The icy or crusted layers are not represented strictly by the densities of the snow samples in figures 6 and 7. This is due to the thinness of crusted layers (often less than 4 cm), whose densities are not accurately recorded by snow samples taken with a 4 cm high metal frame. Both profiles are good examples for the heterogeneous composition of the snow cover, both within a single profile and two profiles at the same location. Every layer indicated by a direction change in the density data of the profiles can be assumed to represent a different event of precipitation or accumulation of transported snow. High-density layers reflect not only springtime melting, but also the origin of the snow as a result of transportation by the wind (rounded crystals and greater compaction). A depth hoar layer with relatively low snow densities (360-380 kg m^{-3}) and a granular structure developed at the base of the snow pack in profile 2 (figure 7). This layer is relatively unstable and can cause the overlying snow pack to swim. Snow creep along the slope of the ribbon forest results in the curved stem form of many trees especially at the edges of the forest ribbons.

The presence of frozen ground at the base of the snow profiles indicates that only little or no snow cover was present when temperatures dropped below freezing in the early winter of 1989.

Like large stones in the soil, icy layers and frozen ground restrict pocket gopher mobility during winter. A snow cover like that in profile 2 is more suitable for winter activities within the snow than conditions in profile 1 (figures 6 and 7). A beneficial consequence of icy layers to pocket gophers can be the protection they provide against penetrating melt water, which flows off on the icy or crusted layers before these layers collapse during melting (figure 6).

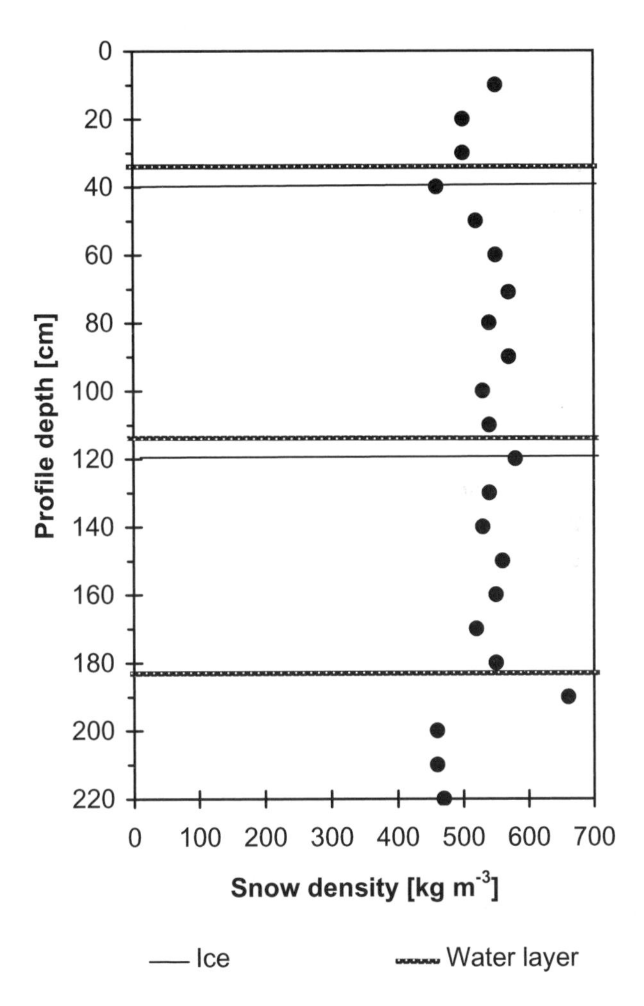

Figure 6 Snow densities of snow profile 1 (1990-02-05)

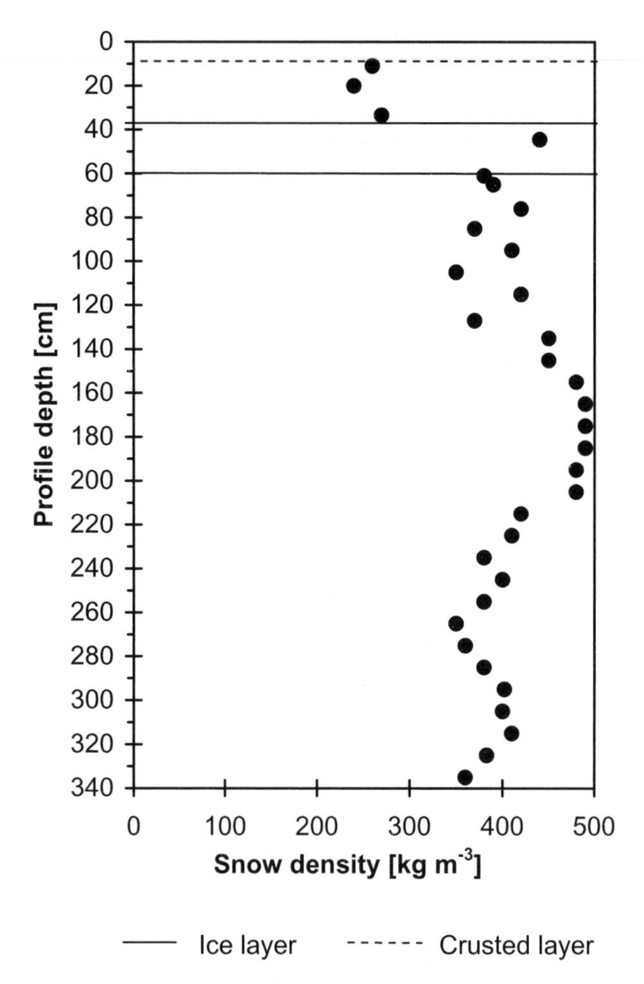

Figure 7 Snow densities of snow profile 2 (1990-03-05)

4.2 Soil

Soil temperatures were measured at two locations during the growing season. Bad waterproofing of the instruments caused a few data gaps. The temperature at a soil depth of 5-10 cm at site 1 (snow glade area without vegetation cover) was only recorded until 8/17/1990, when the instrument was destroyed by trampling of deer or elk.

Temperature measuring strips were analyzed by digitizing one data point every half hour on the continuous measuring curve. The arithmetic average of theses data points is assumed to sufficiently represent the actual mean value of soil temperature. Slight variation of the mean value for the 5-10 cm depth at site 1 may be due to the shorter period of soil temperature recording.

Soil temperature varied more at site 1 (without vegetation cover) than at site 2 (with vegetation cover, *Deschampsia caespitosa*-stand), both diurnally and over the entire measurement period. This variation is most pronounced at 5-10 cm soil depth (table 2).

Minimum and maximum values document the temperature range in the living environment of pocket gophers. The lowest temperatures and the most extreme temperature gradient during the summer occurred at a depth of 5-10 cm at site 1 (data only until 8/17/1990) (table 2, figure 8). Soil temperatures at 15-20 cm depth at the site without vegetation are similar to temperatures at 5-10 cm soil depth at the site with vegetation (table 2, figures 9 and 10).

Table 2 Summer soil temperatures [°C] in the ribbon forest on Niwot Ridge (07/14 - 09/07/1990)

Site	**1**			**2**		
	Snow glade without vegetation cover			Snow glade with vegetation cover		
Soil depth	mean	min.	max.	mean	min.	max.
5 - 10 cm	9.2[1]	1.5[1]	21.9[1]	9.9	5.5	16.3
15 - 20 cm	9.8	5.0	16.3	9.4	5.2	13.0

[1] recording stopped at 08/17/1990

The soil temperature data from 5-10 cm depth may approximate the temperature regime for seed beds of tree seedlings. Patten (1963) and Safford (1970) mentioned a minimal temperature of 7°C for a successful germination of seeds from *Picea*. Temperatures at site 1 commonly remained below this threshold (figure 8). At site 2, the germination threshold was exceeded on 23 of 46 days (figure 10). Only on two days during the preferred germination period in July the soil temperature remained below 7°C. The temperature regime at 15-20 cm soil depth at site 2 (with vegetation cover) is characterized by a relatively narrow temperature range (between 5 and 13°C). The apportionment of temperature values around the mean value of 9.4°C is uniform (figure 11).

Pocket gophers are mostly active in areas with vegetation cover. In areas of sparse vegetation cover digging activities must be more intense to reach plant roots they can feed on. Soil temperature data under the snow cover during the winter period could not be collected with the kind of instrumentation used in this experiment.

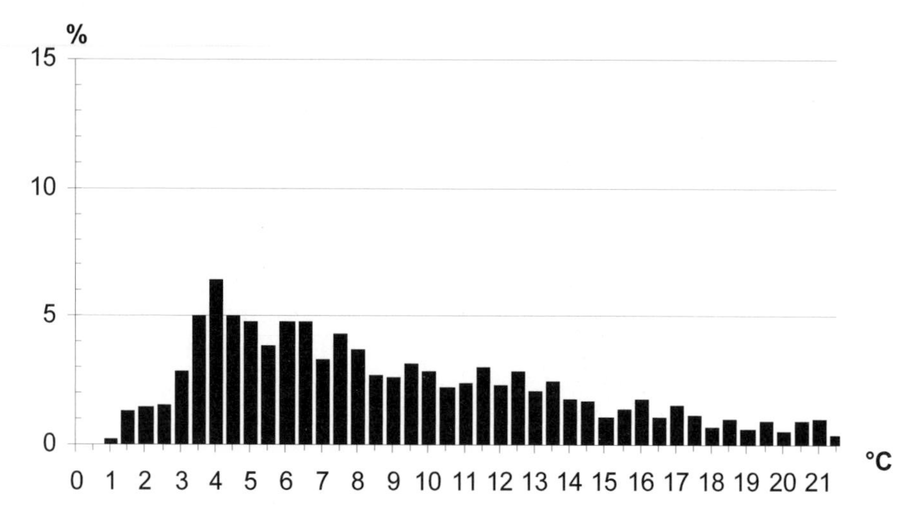

Figure 8 Apportionment of half an hour soil temperature values at 5-10 cm soil depth, Site 1, without vegetation cover. Temperature ranges: 0 = 0.0-0.4°C, 0.5 = 0.5-0.9°C, 1 = 1.0-1.4°C, 1.5 = 1.5-1.9°C, 2...

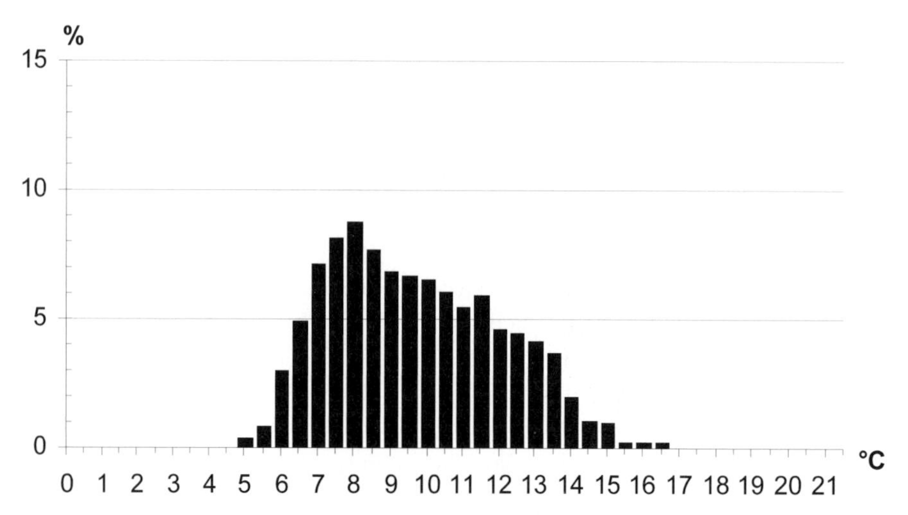

Figure 9 Apportionment of half an hour soil temperature values at 5-10 cm soil depth, Site 2, with cover of *Deschampsia caespitosa* - complex. Temperature ranges: 0 = 0.0-0.4°C, 0.5 = 0.5-0.9°C, 1 = 1.0-1.4°C, 1.5 = 1.5-1.9°C, 2...

The results of the soil analyses are displayed in figure 12 and tables 3-5. The skeletal fraction was estimated in volume percentage. The soil is generally coarser in areas with gopher activities than without (figure 12), with the exception of site 1 at 4-8 and 8-12 cm depth. In sites without gopher activities the coarse soil fraction only once exceeds 25 % in 8-12 cm soil depth. At sites

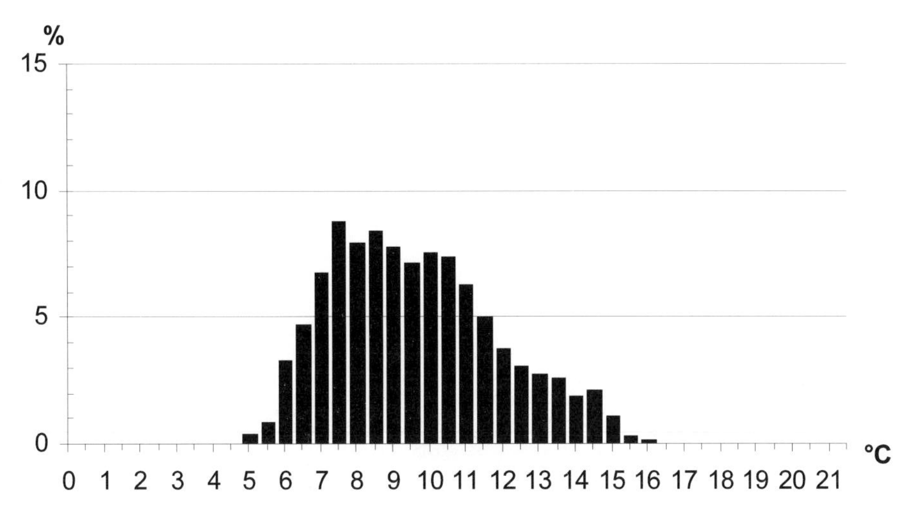

Figure 10 Apportionment of half an hour soil temperature values at 15-20 cm soil depth, Site 1, without vegetation cover. Temperature ranges: 0 = 0.0-0.4°C, 0.5 = 0.5-0.9°C, 1 = 1.0-1.4°C, 1.5 = 1.5-1.9°C, 2...

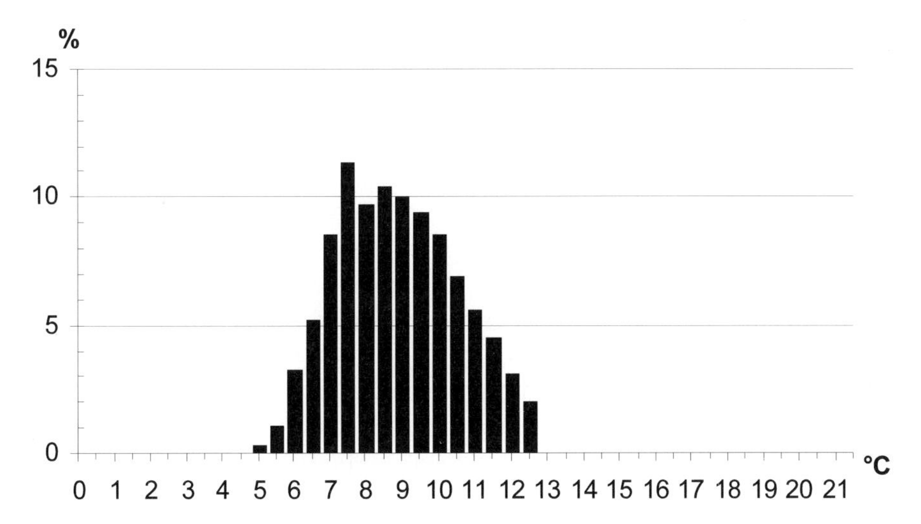

Figure 11 Apportionment of half an hour soil temperature values at 15-20 cm soil depth, Site 2, with cover of *Deschampsia caespitosa* - complex. Temperature ranges: 0 = 0.0-0.4°C, 0.5 = 0.5-0.9°C, 1 = 1.0-1.4°C, 1.5 = 1.5-1.9°C, 2...

with obvious digging activities the volume of the soil fraction larger than 2 mm exceeds 30 %, with only one exception. The same is true for the soil samples of the structures formed by pocket gophers (figure 12) (mounds and soil eskers).

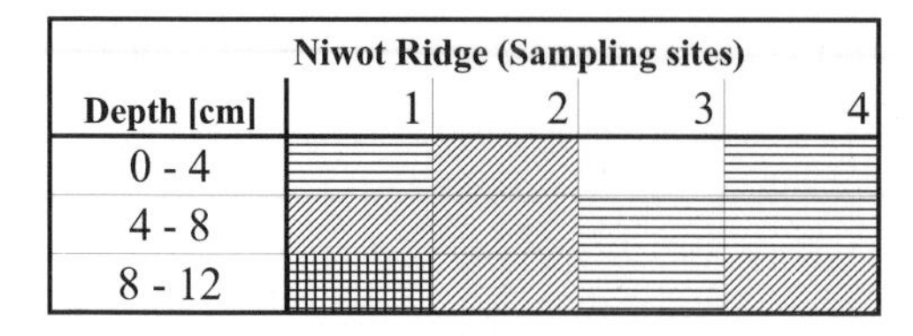

A) without pocket gopher activities

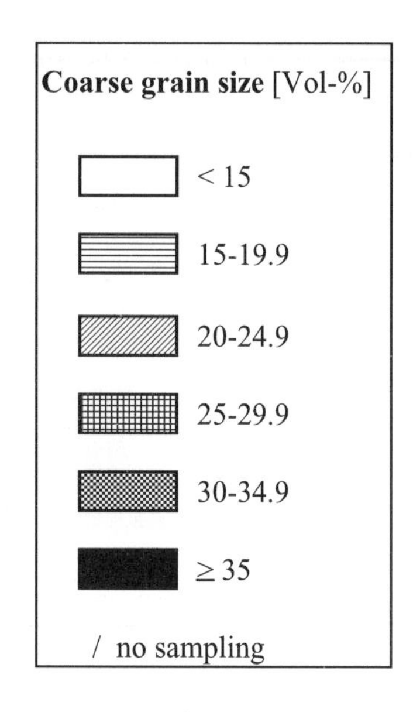

Depth [cm]	Niwot Ridge (Sampling sites) 1a	2a	3a	4a
0 - 4				
4 - 8				
8 - 12				
mounds			/	
eskers		/	/	

B) with pocket gopher activities

Figure 12 Coarse grain size distribution with soil depth (soil sampling 1989)

Grain sizes smaller 2 mm were determined only at five sites due to greater analytical expenditure. At three sites no actual pocket gopher activities occur. The results in table 3 show only slight variation of the soil texture classes with soil depth and as well at different sites. Even the absolute values don't verify a pocket gopher's influence on fine soil texture like an uniform distribution of sand, silt, and clay. The dominant grain size is sand with a moderate to high content of silt (27.7-45.8 %) and clay (9-14.9 %) (table 3). Only in a few sampling depths the soil has a loamy texture: a silt rich loam with 50.4 % silt in 4-8 cm at site 1a (with gopher activities) and loam with a weak portion of sand (site 4, at 8-12 cm depth) and a medium portion of sand (site 2, at 4-8 cm depth) but relative high amounts of clay (at about 19 %), both at sites without pocket gopher activities. Site 4 located in the lower part of the ribbon forest at Niwot Ridge has in general higher amounts of silt (> 40 %) within all three sampling depths.

The sites with pocket gopher activities (1a, 2a and 3a) show greater homogeneity in the distribution of organic material throughout the profiles than the comparable sites without digging activities (table 4). Generally the organic-matter content is higher at the upper soil depth at undisturbed sites. At site 4 higher amounts of organic material characterize as well the second sampling depth (4-8 cm).

Table 3 Grain size (% sand, silt, clay) of undisturbed and disturbed sites

Site	**1**			**1a[1]**			**2**		
soil depth \ grain size[2]	sand	silt	clay	sand	silt	clay	sand	silt	clay
0 - 4 cm	52.2	36.6	11.3	49.9	38.5	12.3	46.4	34.5	19.2
4 - 8 cm	63.4	27.7	9.0	32.3	50.4	11.8	58.1	28.9	13.1
8 - 12 cm	56.2	33.1	10.8	56.6	29.6	13.8	52.7	33.9	13.5
earth mounds	–	–	–	59.8	27.8	12.5	–	–	–
soil eskers	–	–	–	61.4	28.0	10.7	–	–	–

Site	**2a[1]**			**4**		
soil depth \ grain size[2]	sand	silt	clay	sand	silt	clay
0 - 4 cm	58.2	28.1	13.9	45.2	40.3	14.6
4 - 8 cm	58.3	29.0	12.8	39.3	45.8	14.9
8 - 12 cm	56.6	31.5	12.0	37.8	43.4	18.8
earth mounds	64.0	26.0	10.0	–	–	–
soil eskers	–	–	–	–	–	–

[1] disturbed site with pocket gopher activities

[2] sand: 0.063 - 2.0 mm; silt: 0.002 - 0.063 mm; clay: < 0.002 mm

The soil eskers may have as much organic material, total nitrogen, and the same C/N ratio as the upper horizons of undisturbed sites. The soil mounds consist of material similar to soils at the depth of 8-12 cm.

The results of the analysis of cation exchange capacity (CEC) are presented in table 5. It is obvious that calcium contents are inversely related to the amount of protons and Al cations. Furthermore, high values of total CEC are always determined by high concentrations of calcium.

The amount of calcium is highest in the first soil sampling depth (0-4 cm) in every site. It always exceeds 3 $cmol_c\ kg^{-1}$ in the first sampling depth of

Table 4 Organic matter content (C_{org} x 1.724), total nitrogen and C/N ratio of undisturbed and disturbed sites

	organic matter [mass %]							
soil depth \ site	1	1a[1]	2	2a[1]	3	3a[1]	4	4a[1]
0 - 4 cm	10.0	4.2	14.8	7.8	17.9	7.5	13.7	10.8
4 - 8 cm	4.9	4.3	8.1	6.8	9.1	6.5	9.9	7.5
8 - 12 cm	4.0	1.8	7.4	6.1	5.3	5.4	7.6	4.4
earth mounds	–	3.4	–	3.8	–	–	–	8.0
soil eskers	–	6.0	–	–	–	–	–	8.6

Table 4 continued

soil depth \ site	1	1a[1]	2	2a[1]	3	3a[1]	4	4a[1]
	N_t [mass %]							
0 - 4 cm	0.5	0.2	0.5	0.4	0.8	0.4	0.6	0.5
4 - 8 cm	0.2	0.2	0.4	0.4	0.4	0.3	0.5	0.3
8 - 12 cm	0.2	0.1	0.4	0.2	0.3	0.3	0.3	0.2
earth mounds	–	0.2	–	0.3	–	–	–	0.4
soil eskers	–	0.3	–	–	–	–	–	0.4

soil depth \ site	1	1a[1]	2	2a[1]	3	3a[1]	4	4a[1]
	C / N							
0 - 4 cm	12.9	11.4	11.8	11.1	13.3	11.8	13.8	13.3
4 - 8 cm	12.6	11.5	11.1	11.4	12.9	11.4	13.0	12.9
8 - 12 cm	12.0	7.5	11.0	11.7	11.1	11.0	12.9	12.8
earth mounds	–	11.7	–	10.9	–	–	–	13.0
soil eskers	–	13.0	–	–	–	–	–	13.0

[1] disturbed site with pocket gopher activities

Table 5 Cation exchange capacity of undisturbed and disturbed sites

soil depth \ site	1	1a[1]	2	2a[1]	3	3a[1]	4	4a[1]
	Ca [$cmol_c$ kg^{-1}]							
0 - 4 cm	3.6	2.1	5.8	1.8	4.6	1.9	4.7	4.7
4 - 8 cm	0.8	1.5	2.0	0.4	0.4	1.0	2.6	1.7
8 - 12 cm	0.5	0.9	1.3	0.2	0.2	0.7	1.8	0.8
earth mounds	–	0.6	–	0.1	–	–	–	1.9
soil eskers	–	3.7	–	–	–	–	–	2.2

soil depth \ site	1	1a[1]	2	2a[1]	3	3a[1]	4	4a[1]
	H + Al [$cmol_c$ kg^{-1}]							
0 - 4 cm	3.2	2.6	1.6	2.3	2.4	2.0	2.4	1.9
4 - 8 cm	4.0	3.1	3.0	4.2	3.9	3.4	3.7	4.5
8 - 12 cm	4.3	2.3	3.3	4.2	4.9	3.5	4.9	4.0
earth mounds	–	3.1	–	4.0	–	–	–	3.4
soil eskers	–	1.5	–	–	–	–	–	3.3

Table 5 Continued

	CEC [$cmol_c$ kg^{-1}]							
soil depth \ site	1	1a[1]	2	2a[1]	3	3a[1]	4	4a[1]
0 - 4 cm	8.2	5.7	9.4	5.2	8.5	5.1	9.4	8.6
4 - 8 cm	5.3	5.2	6.1	5.3	4.9	5.1	8.1	7.3
8 - 12 cm	5.0	3.4	5.5	4.8	5.4	4.6	7.7	5.5
earth mounds	–	4.3	–	4.3	–	–	–	6.4
soil eskers	–	6.4	–	–	–	–	–	6.4

[1] disturbed site with pocket gopher activities

undisturbed sites while it measures at about 2 $cmol_c$ kg^{-1} in three out of four sites with pocket gopher activities. At all sites the amount of calcium and the CEC are decreasing with greater soil depth.

4.3 Damage to young trees

Ninety-two young trees in the ribbon forest on Niwot Ridge, ranging from saplings to 1.60 m in height, were listed and their above ground damages described (table 6). None of the above ground damage could be directly or obviously attributed to pocket gophers. Damage to the root system were not analyzed due to the destructiveness to the young trees. Only 5 % of all listed

Table 6 Most frequent damage to young trees in the ribbon forest on Niwot Ridge, July 1991

Types of damage	***Abies lasiocarpa* (25)[1]**	***Picea engelmannii* (67)**	**total (92)**
reduced growth at windward side of the trees	19 [76][2]	57 [85]	76 [83]
crippled growth / multiple tree tops	18 [72]	37 [55]	55 [60]
snow fungi	3 [12]	26 [39]	29 [32]
damage by trampling	5 [20]	20 [30]	25 [27]
deformation by heavy snow loads	6 [24]	11 [16]	17 [19]
spruce galls	/	9 [13]	9 [10]
without any damage	1 [4]	4 [6]	5 [5]

[1] number of listed individuals
[2] [%] percent, in relation to total individuals of one species; the sum of damages exceeds 100 % because most trees show multiple damages

trees were without damage. These trees grew within *Salix*-shrubs or close to the ribbons in leeward positions. One of these trees was surrounded by gopher mounds for two years but did not show any sort of aboveground damage. Damage was either climatically induced (e.g. reduced growth at windward side of 83 % of the trees) or caused by other animals than pocket gophers (e.g. trampling by mule deer or elk [27 %]) (table 6). Deep snow cover is also critical to the young trees. The pressure of creeping snow masses leads directly to deformation of young trees. Long duration of wet snow is a perfect environment for snow fungi to infect the coniferous trees. Both types of damage affiliated with the snow cover were noticed at more than 50 % of the young trees.

The different apportionment of types of damages at *Abies lasiocarpa* and *Picea engelmannii* are due to the preferences for different site conditions of these trees. Half of the young *Abies* trees were found within protective *Salix* shrubs. When these trees grow beyond the protection of these shrubs they will experience reduced growth at the windward side due to ice-particle abrasion. Because these sites do not have prolonged wet snow cover, damage to young *Abies* trees by snow fungi is rare. Another reason for the different patterns of damage is an infestation that occurred only in spruce trees: 13 % of all spruces were infested by spruce galls, causing minor growth reduction to the trees.

5 Discussion

Soil eskers were found every spring during the course of this study (1989-1991). These provide evidence that snow cover is a regular part of the northern pocket gopher's environment. The snow cover in the ribbon forests is dependent on the amount of snow fall and the amount of transported snow accumulating behind barriers like forest ribbons or relief features (e.g. the fronts of solifluction terraces). The depth of the redeposited snow is related to the heights of the barriers and the speed of winds. The origin of the transported snow may be not far away. Tabler and Schmidt (1973) show results from measurements in Wyoming with transport distances of 460 m to 900 m. In this study area the ridge top of Niwot Ridge is an appropriate deflation area. Here wind speed can exceed 50 m s^{-1} (Institute of Artic and Alpine Research 1989; 1990; 1991). This is the lower wind-speed limit for deflation snow storms (Dyunin and Kotlyakov 1980). But even winds with a speed greater 40 m s^{-1} will cause redistribution of snow on Niwot Ridge as was experienced by the author in March and April 1990. These events are related to Chinook wind situations, which are most common in January (Benedict 1991; Barry 1992). Therefore several deflation events during the winter and spring are responsible for the snow accumulating in the snow glades of the ribbon forest. In between these snow redistribution events, deep frost and periods of surface melting form snow profiles that contain ice lenses. These lenses, some of which extend more than 30 m (figure 4: snow transect E-F), may prevent some of the gopher's activity space from flooding during early snowmelt but may also hinder easy movement through the snow cover. The snow cover provides thermal insulation and a good protection, especially against predatory birds.

Climatic factors such as early deep frost without snow cover, number of deflation events, and periods of snowmelt or deep frost during the winter may vary from year to year, but late snow beds are found each year in the same areas in the snow glades. The location of deep snow cover will vary from year to year only when there are changes in the height and configuration of barriers (the forest ribbons), e.g. following windfall of trees or pest infestations.

The layer of depth hoar close to the ground in snow profile 2 (figure 7) is characterized as sufficient for small mammal movement beneath the snow (Auerbach and Halfpenny 1991; Marchand 1996). But the snow crystals of depth hoar have little cohesion and do not allow the existence of persistent tunnels in this layer. Movement might be possible in depth hoar layers but it is not as energy efficient as excavating tunnels that can be used several times at low energy cost. By moving through the snowpack the pocket gopher can easily bypass barriers such as rocks, streams and wet or frozen soils (Davis 1939 in Tryon 1947; Vaughan 1963; Stoecker 1976). Hansen and Reid (1973) found snow tunnels in the snow as much as 45 cm above ground surface. Noble and Alexander (1975) mentioned that *Thomomys talpoides* crossed a 70 cm high close-meshed fence within the snowpack. Within a snow glade of the ribbon forest on Niwot Ridge the author discovered a nest and two food chambers in melting snow, 85 cm above the ground surface. Benedict and Benedict (2001) describe very detailed subnivean nests and food chambers of a montane vole (*Microtus montanus nanus*) in the forest-alpine tundra-ecotone of Fourth of July valley, Colorado Front Range. In this case nests and food chambers were built in the snowpack above unfrozen ground shortly after a September snowstorm created a satisfactory snowpack. This leads to the hypothesis that foraging winter-active rodents do not just escape into the snowpack, but use it as a natural refrigerator and a suitable living area.

The time period that pocket gophers spend within the snow is uncertain. Hansen (1962) supposes water accumulation under the snow cover during snowmelt to be the reason for pocket gopher activities within the snowpack. In general, food chambers should be relocated into the soil with the beginning of snowmelt (Hansen and Morris 1968). Tryon (1947) assumes that snow tunnels are constructed and refilled during early winter, after the first snow accumulates above unfrozen ground. Tryon states that the ground would never freeze under the snow cover in many high mountain areas. Investigations in the ribbon forest on Niwot Ridge show that the ground can freeze even under a snow cover of more than 2 or 3 meters (figures 6 and 7). Soils are very shallow in the ribbon forest on Niwot Ridge. In most places the soil is too rocky for pocket gophers to burrowing at depths greater than 30 cm. This might be not deep enough to burrow beneath frozen ground (Hansen and Morris 1968). If the ground is frozen they must move in the snowpack to obtain isolation from cold winter temperatures while still foraging at or close to the ground surface.

The early summer snowmelt period is the most dangerous time in the life of *Thomomys talpoides* in the ribbon forest at Niwot Ridge. Melt water can break into the burrowing system of the pocket gopher. It can cause flooding of the tunnel systems (Hansen 1962; Armstrong 1975; Stoecker 1976), and where tunnels are leading downslope gully erosion can be initiated at a low scale (Thorn 1978). During the snowmelt period spring-like water emergences (sometimes artesian) are commonly seen where tunnels meet the ground surface on the lower slopes and terrace platforms in the snow glades. This might be a reason that gopher activity was noticed where snow cover exceeded even 3 m height (figures 3 and 5) in upper slopes within the snow glades.

Burrowing by pocket gophers causes turbation of soil. Many times it results in soil surfaces that are bare or have sparse vegetation cover. This changes the temperature regime in the upper part of the soil. Where gopher activities continue for years, the soil itself and the vegetation cover are in a state of permanent succession (Cortinas and Seastedt 1996). In such places the vertical temperature gradient within the soil is steeper, and extreme temperatures are more common than in places without gopher activities and with complete vegetation cover.

The results of the soil analyses are also a mirror of burrowing activities. Most obvious with all parameters is a greater homogeneity of the substrate in all three soil depths. This results in a greater portion of coarse soil fraction in the upper soil (figure 12) and a relative reduction of organic matter and thus calcium for example (tables 4 and 5) in the upper soil. In contrast to investigations of Chapin et al. (1986) none of the 24 soil samples taken from soil profiles along a transect in 1990 showed an enrichment of the soil with nitrogen due to fecal excretions (Schütz 1998).The results of the soil analysis lead to the hypothesis that lateral transport of nutrients by melt water exceeds the influence of enrichment by gopher excrement (Schütz 1998). Profiles in the lower part of the transect show higher concentrations of C, N, and Ca than comparable horizons in profiles in the upper part of the transect.

Bovis (1978) noted that alpine and subalpine belts of the Colorado Front Range experience little soil loss due to erosion. But soil erosion is triggered wherever vegetation is removed or covered with soil. Material transported onto the soil surface by the pocket gophers is exposed to possible redistribution, especially by melt water runoff in early summer. In contrast to Sherrod and Seastedt (2001), erosion by wind in the snow glades between tree ribbons is of minor influence due to low wind velocities. With their intense digging and tunneling, pocket gophers influence the downslope transport of sediments. As a result, small-gully erosion pathways form parallel to the direction of the slope. These erosion pathways may cause gaps in tree ribbons of ribbon forests. Erosion is most intense where trampling by deer or elk leads to additional removal of vegetation and cutting of gopher burrows (see development of terracettes in Thorn 1978).

To estimate the pocket gopher's importance to tree regeneration within the ribbon forest the damage to young trees caused by gopher was compared with damages by other causes. Within three ribbon forests in the Colorado Front Range, only one young tree had aboveground damage that might be due to peeling by pocket gophers (Schütz 1998). In the ribbon forest on Niwot Ridge, no above ground damage could be found. Underground damage was not investigated. The author's opinion is that pocket gophers do not harm tree roots directly as long as there is enough diverse herbaceous food within the gopher's territory. Most damage might be caused by digging along the root systems.

In general, living conditions within the ecotone are hard for young trees (table 6). A high rate of mortality (60-92 % in the first year) in spruce and fir saplings is also documented in Nobel and Alexander (1977), Heidmann (1976), and Cui and Smith (1991). Reasons were drought, picking birds, and frost heaving. Fiedler et al. (1985) mention a high survival rate (41 %) of *Abies lasiocarpa* and *Picea engelmannii* during the first ten years of growth in small protected sites in the upper subalpine zone of Dixie National Forest (Utah). In the author's opinion, such high survival rates will require :

- Early winter snow cover that continues through the wintertime.
- Snow depths and exposure to solar insolation that allow complete snowmelt by the second half of June.
- Protection from continuous cold melt water runoff.
- Lack of a game trail at the site of tree establishment.
- Absence of prolonged pocket-gopher activities.

6 Conclusions

The data presented in this paper show that the northern pocket gopher influences the soil and vegetation (e.g. tree) pattern by its burrowing activities. But *Thomomys talpoides* cannot be considered a key species in the ribbon forest environment. In particular the amount of erosion cannot be attributed solely to pocket gophers. The cumulative effects of gophers, voles, deer, and elk are responsible for the amount and pattern of erosion.

Population data for pocket gophers in ribbon forests are still missing. As well investigations about the occupation of ribbon forests by pocket gophers along the Colorado Front Range could be useful to evaluate the importance of *Thomomys talpoides* to the further development of ribbon forests (and generally to the forest-alpine tundra ecotone) more precisely. Parallel analysis of the deer and elk populations and their grazing trails will help to understand the influence of different species to their environment.

Long snow duration, infection by the brown snowmelt fungus (*Herpotrichia juniperi*), and damage by deer and elk contribute decisively to the mortality of young trees in the snow glades. Melt water, responsible for a lateral redistribution of nutrients, is a key factor for rich vegetation pattern.

The actual vegetation pattern (the scenery of a never-ending play) formed by ribbon forests along solifluction terraces (the multidimensional stage) and their changes are due to the overlapping influence of snowpack duration, pocket gophers, voles, deer, elk, and other poorly understood factors (the local players). As spectators who are dramaturgically involved, men will have an increasing influence on the quality of the play (disturbances caused by men) and its actors (changing behavior and population growth), by increasing population pressure and non-sustainable recreational behavior in mountain regions (Holtmeier 1987b; 2003).

Acknowledgements

The author wants to thank the University of Münster for funding the dissertation work in 1989 and 1990, and the Gottlieb Daimler- and Karl Benz-Foundation for financing a field trip in 1991. The German Academic Exchange Service (DAAD) paid some of the traveling costs. Logistical support was provided by the Mountain Research Station of the University of Colorado at Boulder and by the Landscape Ecology Laboratory of the University of Münster. Prof. Dr. Broll (Vechta, Germany) and above all Dr. James B. Benedict

(Colorado, USA) and Dr. Sabine Mellmann-Brown (Montana, USA) made helpful suggestions improving the manuscript.

Many of the author's thanks go to Prof. Dr. Holtmeier (Münster) who motivated the author to visit the Colorado Front Range and to study the fascinating development of its landscape pattern. As my work proceeded, I began to feel more than just a spectator in the auditorium of the stage.

References

Andersen DC (1987) Below-ground herbivory in natural communities: a review emphasing fossorial animals. The Quarterly Review of Biology 62 (3): 261-286

Andersen DC, MacMahon JA (1985) Plant succession following the Mount St. Helens volcanic eruption: faciliation by a burrowing rodent, *Thomomys talpoides*. The American Midland Naturalist 114 (1): 62-69

Armstrong DM (1975) Rocky Mountain mammals. A Handbook of mammals of Rocky Mountain National Park and Shadow Mountain National Recreation Area, Colorado. Rocky Mountain Nature Association

Auerbach NA, Halfpenny JC (1991) Snowpack and the subnivean environment for different aspects of an open meadow in Jackson Hole, Wyoming, USA. Arctic Alpine Res 23 (1): 41-44

Barry RG (1992) Mountain weather and climate. 2nd ed, London New York

Benedict AD (1991) The Southern Rockies. The Rocky Mountain regions of southern Wyoming, Colorado, and northern New Mexico. A Sierra Club Naturalist's Guide, San Francisco

Benedict JB (1970) Downslope soil movement in a Colorado alpine region: rates, processes, and climatic significance. Arctic Alpine Res 2 (3): 165-226

Benedict JB, Benedict AD (2001) Subnivean root caching by a montane vole (*Microtus montanus nanus*), Colorado Front Range, USA. Western North American Naturalist 61 (1): 241-244

Billings WD (1969) Vegetational pattern near alpine timberline as affected by fire-snowdrift interactions. Vegetatio 19: 192-207

Bovis MJ (1978) Soil loss in the Colorado Front Range: Sampling design and areal variation. Z Geomorph NF, Suppl 29: 10-21

Buckner DL (1977) Ribbon-forest development and maintenance in the central Rocky Mountains of Colorado, Dissertation; Department of Environmental, Population, and Organismic Biology. University of Colorado, Boulder

Chapin III FS, Vitousek PM, VanCleve K (1986) The nature of nutrient limitation in plant communities. The American Naturalist 127 (1): 48-58

Colbeck SC (1991) The layered character of snow covers. Reviews of Geophysics 29 (1): 81-96

Cortinas MR, Seastedt TR (1996) Short- and long-term effects of gophers (*Thomomys talpoides*) on soil organic matter dynamics in alpine tundra. Pedobiologia 40: 162-170

Cui M, Smith WK (1991) Photosynthesis, water relations and mortality in *Abies lasiocarpa* seedlings during natural establishment. Tree Physiology 8: 37-46

Davies EF (1994) Disturbance in alpine tundra ecosystems: the effects of digging by northern pocket gophers (*Thomomys talpoides*). Masters thesis, Utah State University, Logan, Utah

Davis WB (1939) The recent mammals of Idaho. Caldwell, Idaho

Dyunin AK, Kotlyakov VM (1980) Redistribution of snow in the mountains under the effect of heavy snow-storms. Cold Reg Sci Technol 3: 287-294

Fiedler CE, McCaughey WW, Schmidt WC (1985) Natural regeneration in intermountain spruce-fir forests. A gradual process. Research Paper INT 343, Intermountain Research Station Ogden, UT 84401, United States Department of Agriculture, Forest Service

Fleischer MC, Emerick JC (1984) From grassland to glacier. The natural history of Colorado. Boulder (Johnson Books)

Gable DJ, Madole RF (1976) Geologic map of the Ward quadrangle, Boulder County, Colorado [scale 1: 24.000]. Map GQ-1277, Department of the Interior, United States Geological Survey, Reston

Hansen RM (1962) Movements and survival of *Thomomys talpoides* in a mima-mound habitat. Ecology 43 (1): 151-154

Hansen RM, Morris MJ (1968) Movement of rocks by northern pocket gophers. J Mammal 49 (3): 391-399

Hansen RM, Reid VH (1973) Distribution and adaptations of pocket gophers. In: Turner GT, Hansen RM, Reid VH, Tietjen HP, Ward AL (eds) Pocket gopher and Colorado Mountain rangeland. Bulletin, Fort Collins, Colorado State University, Agriculture Experiment Station, pp 1-19

Heidmann LJ (1976) Frost heaving of tree seedlings: a literature review of causes and possible control. USDA Forest Service General Technical Report RM-21, Rocky Mountain Forest and Range Experiment Station, Fort Collins, Colorado

Holtmeier F-K (1982) ‚Ribbon forest' und ‚Hecken'; streifenartige Verbreitungsmuster des Baumwuchses an der oberen Waldgrenze in den Rocky Mountains. Erdkunde 36: 142-153

Holtmeier F-K (1987a) Beobachtungen und Untersuchungen über den Ausaperungsverlauf und einige Folgewirkungen in ‚ribbon forests' an der oberen Waldgrenze in der Front Range, Colorado. Phytocoenologia 15 (3): 373-396

Holtmeier F-K (1987b) Human impacts on high altitude forests and upper timberline with special reference to middle latitudes. In: Fujimori T, Kimura M (eds) Human impacts and management on mountain forests. Ibaraki/Japan (Forestry and Forest Products Research Institute), pp 9-22

Holtmeier F-K (2002) Tiere in der Landschaft. Einfluss und ökologische Bedeutung. 2nd ed, Stuttgart

Holtmeier F-K (2003) Mountain Timberlines – Ecology, patchiness, and dynamics. Advances in Global Change Research, 14. Kluwer Academic, Dordrecht

Institute of Arctic and Alpine Research (1989; 1990; 1991) Climatological data summary. Station: Saddle. Unpublished

Losleben M (1990) Climatological data from Niwot Ridge, East Slope, Front Range, Colorado, 1989. Long-Term Ecological Research Data Report 90/1. Institute of Artic and Alpine Research. University of Colorado, Boulder

Marchand PJ (1996) Life in the cold. 3rd ed, Hannover and London (University Press of New England)

Miller RS (1964) Ecology and Distribution of Pocket Gophers (Geomyidae) in Colorado. Ecology 45 (2): 256-272

Noble DL, Alexander RR (1975) Rodent exclosures for the subalpine zone in the central Colorado Rockies. Rocky Mountain Forest and Range Experiment Station. USDA Forest Service Research Note RM - 280

Page AL, Miller RH, Keeney DR (eds) (1982) Methods of soil analysis. Part 2: Chemical and microbiological properties. 2nd ed, Madison

Patten DT (1963) Light and temperature influence on Engelmann spruce seed germination and subalpine forest advance. Ecology 44: 817-818

Safford SO (1970) Picea A. Dietr. Spruce. Data filed. USDA Northeastern Forest Experiment station. Upper Darby, Pennsylvania

Schütz H-U (1998) Untersuchungen zur Ökologie von 'ribbon forests' der 'Colorado Front Range' (Rocky Mountains, USA) unter besonderer Berücksichtigung von *Thomomys talpoides* (Geomyidae). Dissertation, University of Münster, Germany

Sherrod SK, Seastedt TR (2001) Effects of the northern pocket gopher (*Thomomys talpoides*) on alpine soil characteristics, Niwot Ridge, Co. Biogeochemistry 55 (2): 195-218

Stoecker R (1976) Pocket gopher distribution in relation to snow in the alpine tundra. In: Steinhoff HW, Ives JD (eds) Ecological impacts of snowbank augmentation in the San Juan Mountains of Colorado. Final Report of the San Juan Ecology Project. Colorado State University Publ, Fort Collins, pp 281-287

Tabler RD, Schmidt RA (1973) Weather conditions that determine snow transport distances at a site in Wyoming. In: The role of snow and ice in hydrology. Proceedings of the Banff Symposia, vol 1, Geneva Budapest Paris (UNESCO-WMO-IAHS), pp 118-127

Thorn CE (1978) A preliminary assessment of the geomorphic role of pocket gophers in the alpine zone of the Colorado Front Range. Geografiska Annaler 60A (3-4): 181-187

Trüby P, Aldinger E (1989) Eine Methode zur Bestimmung austauschbarer Kationen in Waldböden. Z f Pflanzenernährung u Bodenkunde 152: 301-306

Tryon CA Jr (1947) The biology of the pocket gopher (*Thomomys talpoides*) in Montana. Technical Bulletin 448, Montana State College, Agricultural Experiment Station, Bozeman, Montana

Vaughan TA (1963) Movements made by two species of pocket gophers. American Midland Naturalist 69 (2): 367-372

Wolf P (1991) Untersuchungen zur Ökologie epigäischer Arthropoden auf dem Niwot Ridge (Colorado Front Range, USA) unter besonderer Berücksichtigung der Laufkäfer (Coleoptera, Carabidae). Master's thesis, Inst f Geography, University of Münster, Germany

The Impact of Seed Dispersal by Clark's Nutcracker on Whitebark Pine: Multi-scale Perspective on a High Mountain Mutualism

Diana F. Tomback

Abstract

The holarctic mutualisms between *Cembrae* Pines (*Pinus*, Pinaceae) and nutcrackers (*Nucifraga*, Corvidae) comprise important textbook examples of interaction and coevolution. However, only recently we have learned to what extent nutcracker seed caching influences the ecology and biology of these pines. The North American mutualism between whitebark pine (*Pinus albicaulis*) and Clark's nutcracker (*Nucifraga columbiana*) demonstrates the greatest array of consequences to the pine, particularly in fire-dependent communities in the northern Rocky Mountains. The impacts of Clark's nutcracker seed-caching on whitebark pine are evident at multiple spatial and temporal scales, from tree growth form and fine-scale genetic structure and the timing of regeneration and community development to regional and range-wide genetic structure, and post-Pleistocene range expansion.

1 Introduction

Within the past quarter century, ecologists have elucidated the details of fascinating high elevation, holarctic mutualistic relationships between the birds known as 'nutcrackers' (genus *Nucifraga*, family Corvidae) and the 'stone pines' - five-needled white pines of subsection *Cembrae* (genus *Pinus*, subgenus *Strobus*, family Pinaceae). Although naturalists in Europe and Asia were well aware for centuries of the close ties between the different populations and subspecies of the Eurasian or spotted nutcracker (*N. caryocatactes*) and several stone pines, for example in Russia the 'kedrovka' and the 'kedr' (Siberian stone pine, *Pinus sibirica*) and the 'kedrovy stlannik' (Japanese stone pine, *Pinus pumila*), in the Swiss Alps the 'Tannenhäher' and 'die Arve' (Swiss stone pine, *Pinus cembra*), and in the French Alps 'Cassenoix' and 'l'Arolle' (Swiss stone pine), it was not until recently that the strength of the interaction and the ecological details of these seed dispersal mutualisms were understood (e.g., Crocq 1978; Mattes 1978; 1982; Hayashida 1994; Hutchins et al. 1996; Kajimoto et al. 1998). In western North America, the fact that Clark's nutcrackers (*N. columbiana*) disperse whitebark pine (*Pinus albicaulis*) seeds (e.g. Tomback 1978; 1982; Hutchins and Lanner 1982) was not accepted in the forestry community until recently, and even now erroneous information is perpetuated.

For all the nutcracker-*Cembrae* pine interactions, the nutcrackers act as dispersal agents by harvesting pine seeds from cones, transporting seeds in their unique sublingual pouch (Bock et al. 1973) and burying clusters of seeds in tens of thousands of shallow caches throughout the forest terrain (e.g. Crocq 1978; Mattes 1978; Tomback 1978; Hutchins et al. 1996). The birds retrieve their caches when other foods are scarce. Unretrieved seeds typically germinate during snowmelt or after summer rains (e.g. Vander Wall and Hutchins 1983; Kajimoto et al. 1998; Tomback et al. 2001a). The *Cembrae* pines are distinguished from other white pines (subgenus *Strobus)* by having large, wingless seeds and cones that remain closed even after seeds ripen (Price et al. 1998). These characteristics ensure that seeds are retained in cones until removed by nutcrackers, and preclude seed dispersal by wind, the predominant means of seed dissemination for the majority of Pinaceae species. The *Cembrae* pine-nutcracker interaction is a coevolved mutualism, i.e., each species has influenced the other's evolutionary history (Lanner 1980; Tomback 1983; Tomback and Linhart 1990).

The consequences of seed dispersal by nutcrackers go beyond forest regeneration, and are apparent at multiple spatial and temporal scales, affecting the ecology, timing of regeneration, distribution, and population genetic structure of the *Cembrae* pines. These effects are most apparent in the interaction between Clark's nutcracker and the only North American *Cembrae* pine, the whitebark pine.

In this paper, I provide background information on the taxonomy and evolution of nutcrackers and *Cembrae* pines. Then, I describe the interaction between Clark's nutcracker and whitebark pine, focusing on how seed dispersal by nutcrackers impacts regeneration biology, fine- and larger-scale population genetic structure, successional status, community development, and the distribution of whitebark pine.

2 Distinguishing Traits, Taxonomic Relationships, and Evolutionary Origins

Understanding the taxonomic relationships of both the nutcrackers and the stone pines provides insight as to what extent these taxa were pre-adapted for mutualistic interaction based on phylogenetic affinities and the evolutionary origin of these mutualisms.

Nutcrackers possess behavioral traits that are commonly associated with species of Corvidae, as outlined by Amadon (1944; cited and quoted in Hope 1989): 'The following traits are usually present: long-continued courtship feeding; nest building and feeding of young by both male and female; incubation and brooding by female only; burying or hiding of food; breaking of food with the bill while the food in held in the feet; loud and usually harsh notes; omnivorous and more or less predatory feeding habits; bold and inquisitive nature.' General corvid morphological traits include larger size than most passerines (songbirds), a sturdy, relatively long bill, and an expandable esophagus or buccal cavity, which is used to transport food for storage or feeding young (e.g. Amadon 1944 cited in Hope 1989; Turcek and Kelso 1968). Based on both osteological and DNA hybridization studies, genus *Nucifraga* is closely related to genus *Corvus*, the crows and ravens, and is placed within a clade of holartic and Old World corvids (Sibley and Ahlquist 1985; Hope 1989).

Nutcracker adaptations for an annual cycle based on fresh and stored conifer seeds and hazelnuts (*Corylus avellena*) are enhancements of basic corvid features. These specialized traits include a sturdy, long, sharp bill that is used to dig into the fibrous and resinous scales of closed pine cones, peck or crack open seeds, and prepare and retrieve seed caches (e.g. Crocq 1990; Tomback 1998 and references therein); a well-developed spatial memory that enables these birds to re-locate thousands of seed caches within a year of storing them (e.g. Balda 1980; Vander Wall 1982; Kamil and Balda 1985); a well-developed incubation patch in males as well as in females, facilitating male incubation and brooding while females retrieve seed caches (Mewaldt 1952; Swanberg 1956); an early nesting strategy ensuring that young are independent in time to cache their own seeds in late summer (Vander Wall and Balda 1977; Tomback 1978); and the sublingual pouch, an anatomical feature unique to genus *Nucifraga* (Bock et al. 1973; Crocq 1990) that can hold 100 or more pine seeds and 20 or more hazelnuts (e.g., Mattes 1978; Tomback 1978). The sublingual pouch, which is a sac-like diverticulum of the floor of the mouth, with an opening under the tongue (Bock et al. 1973), enables nutcrackers to transport seeds to caching sites or to the nest (e.g. Mewaldt 1956; Vander Wall and Balda 1977; Mattes 1978; Tomback 1978).

Thus, *Nucifraga* was generally pre-adapted for an annual cycle based on the use of fresh and stored pine seeds, but achieved greater specialization with the evolution of the sublingual pouch and a male incubation patch, which is very rare among passerines (Swanberg 1956).

The 'stone pines' are classified in subsection *Cembrae,* in section *Strobus* under subgenus *Strobus* (i.e. the soft or haploxylon pines) (Price et al. 1998) (table 1). The seeds of subgenus *Strobus* pines in general are larger than those of subgenus *Pinus* pines (i.e. the hard or diploxylon pines) in general (table 3 in Tomback and Linhart 1990). All pines in section *Strobus* have needles in fascicles of five, with the umbo (point) of the cone scale terminal rather than dorsal in position. *Cembrae* pines share the derived character traits of large, wingless seeds and indehiscent cones – that is the cones remain closed after seeds are ripe (Price et al. 1998). In addition, the *Cembrae* pines (except for the creeping, matlike Japanese stone pine, *Pinus pumila*) assume a distinctive growth form with profuse branching, resulting in a flat-topped, shrubby canopy. The upper branches are vertically directed with horizontally-oriented cones at the tips, which makes cones more visible and accessible to nutcrackers (Tomback 1978; Lanner 1980; Lanner 1982). A similar growth form also occurs in the distantly related piñon pines (*Pinus edulis* and *Pinus monophylla*) and in the more closely related limber pine (*Pinus flexilis*); these are also corvid-dispersed (Tomback and Linhart 1990). These derived *Cembrae* traits presumably evolved to facilitate seed dispersal by *Nucifraga* (Lanner 1980; Bruederle et al. 2001), which has been well documented for all *Cembrae* pines (e.g. Reimers 1959; Crocq 1978; Mattes 1978; Tomback 1978; Hayashida 1994; Hutchins et al. 1996) (table 1).

Also within section *Strobus* is subsection *Strobi*, a taxon closely related to *Cembrae* that includes species with a mixture of seed dispersal strategies –

Table 1 Pines of Section *Strobus* (subgenus *Strobus*, genus *Pinus*, family Pinaceae). Subsection *Cembrae* probably evolved from an ancestral pine in subsection *Strobi* (based on Price et al. 1998, RA Price, personal communication)

Subsection *Strobi*		Subsection *Cembrae*
P. armandii	*P. lambertiana*	*P. albicaulis*
P. ayacahuite	*P. monticola*	*P. cembra*
P. bhutanica	*P. morrisonicola*	*P. koraiensis*
P. chiapensis	*P. parviflora*	*P. pumila*
P. dabeshanensis	*P. peuce*	*P. sibirica*
P. dalatensis	*P. strobus*	
P. fenzeliana	*P. wallichiana*	
P. flexilis	*P. wangii*	

small, winged seeds; large, winged seeds; and large, wingless seeds (Bruederle et al. 2001) (table 1). However, the cones open when seeds are ripe for all *Strobi* pines. Recent phylogenetic studies based on DNA sequencing indicate that *Cembrae* pines are closely related to *Strobi* pines, confirm that *Cembrae* pines are monophyletic but include *Pinus armandii*, and suggest a recent origin of *Cembrae* pines from subsection *Strobi* (Liston et al. 1999; Wang et al. 1999; R Price, personal communication). In fact, the fossil record indicates that *Cembrae* was the most recent pine subsection to evolve (Millar 1998). The evolution of *Cembrae* pines thus resulted in a clade with specialized morphology for seed dispersal by nutcrackers (R Price, personal communication; Bruederle et al. 2001; McCaughey and Schmidt 2001).

Circumstantial evidence suggests that both *Nucifraga* and the *Cembrae* pines originated in Eurasia: There are four *Cembrae* pines in Eurasia, not including *P. armandii*, and only whitebark pine in North America. There is only one form of Clark's nutcracker in North America, but at least 8 to 10 described subspecies of the Eurasian nutcracker (Roselaar 1994). Lanner (1980) suggested that the greater number of nutcracker and stone pine forms in Eurasia reflects a longer presence on that continent. He further suggested that both *Nucifraga* and an ancestral *Cembrae* pine together dispersed across a Beringian land bridge into the New World, with subsequent isolation and speciation resulting in the modern forms of Clark's nutcracker and whitebark pine.

Based on genetic distances from allozyme studies, Krutovskii et al. (1994) estimated the time of separation between whitebark pine and Eurasian stone pines at 0.6 million to 1.3 million years ago, corresponding to a Bering Strait land bridge that first appeared in the Pliocene but endured through the Pleistocene. Espinoza de Los Monteros and Cracraft (1997) date the arrival of a *Perisoreus* (Siberian jay) ancestral form from Eurasia to the early Miocene and the subsequent evolution of New World jays continuing through the Pliocene, making the Clark's nutcracker the most recently arrived seed dispersing corvid. After reaching North America, the nutcracker then expanded its range beyond that of whitebark pine, encountering other western pines with large, wingless seeds, including limber and southwestern white pine (*Pinus strobiformis* same as *P. ayacahuite*), and the piñon pines (Tomback 1983; Tomback 2001). The origin of the piñon pines (subsection *Cembroides*) predated the evolution of New World jays and arrival of the nutcracker (Espinoza de Los Monteros and Cracraft 1997; Millar 1998), and was probably associated with seed dispersal by rodents and later also jays and Clark's nutcracker.

3 Clark's Nutcrackers as Dispersers of Whitebark Pine Seeds

Within the range of whitebark pine, the phenology of cone ripening and availability of fresh and stored whitebark pine seeds are major determinants of the annual cycle of Clark's nutcrackers (Tomback 1978; 1994a). Nutcrackers break into whitebark pine cones as early as mid-July, extracting pieces of unripe seeds, which they eat or feed to dependent juveniles (Tomback 1978). They retrieve whitebark pine seed caches made the previous fall throughout the summer (Vander Wall and Hutchins 1983; Tomback unpublished observations). If new cones are available, by mid to late August they harvest whole seeds with mature, brown seed coats, and initiate seed caching (Tomback 1978; Hutchins and Lanner 1982). Seed caching continues until cones are depleted – as late as October or early November if cone production is good (Hutchins and Lanner 1982; Tomback 1982). During fall, many nutcrackers descend to lower elevations in search of other large-seeded conifers with cones; most birds overwinter and breed below the subalpine zone, retrieving caches at these elevations (Mewald 1956; Tomback 1978; Tomback 1998). In May or June, depending on snowpack depth, nutcrackers return to subalpine elevations, many in family groups, where adult birds forage primarily by retrieving whitebark pine seeds from caches (Tomback 1978; Vander Wall and Hutchins 1983). Unretrieved seed caches from previous years germinate as snow melts or after rain, producing clusters of seedlings (Tomback 1982; Vander Wall and Hutchins 1983; Tomback et al. 2001a).

The following summarizes nutcracker caching behavior with respect to whitebark pine seeds (see Tomback 1998; 2001 for overview): Nutcrackers typically store from 1 to 15 whitebark pine seeds per cache, with averages of 3 to 5 seeds per cache (Tomback 1978; 1982; Hutchins and Lanner 1982). Caches with two or more seeds accounted for 65 % of all caching observations by Hutchins and Lanner (1982). The seeds are buried 1 to 3 cm deep under forest litter, mineral soil, pumice, or gravelly substrate. With mineral soil or forest litter, a nutcracker uses side-swiping bill motions to dig a shallow trench. The seeds are placed in the trench one at a time, and again the bird uses side-swiping motions to cover the seeds and fill in the trench. For looser substrate i.e. gravel or pumice, the seeds are pushed, one at a time, into the soil. In both cases, the finished cache shows no signs of substrate disturbance (Tomback 1978). Nutcrackers generally make several caches within an area, spacing them from 10 to 300 cm apart, and then fly to another location nearby to make another group of caches, and so on until the sublingual pouch is empty (Tomback 1978). Nutcrackers sometimes retrieve seeds from caches shortly after making them and then recache them elsewhere (Tomback 1978; Hutchins and Lanner 1982).

The microsites selected by nutcrackers for caching are reasonably diverse; the sites themselves may be chosen for purposes of spatial orientation (see Vander Wall 1982; Kamil and Balda 1985). Nutcrackers place many caches next to trees, either near the trunk or exposed roots; these tend to experience faster snowmelt in spring than adjacent terrain (Tomback 1978). They also store caches near rocks, next to fallen trees, in open terrain, among plants, in rock fissures, on rocky ledges; in trees themselves, in cracks, holes, and under the bark; at the edges of meadows among rocks and vegetation, and at treeline among krummholz (matlike) patches of whitebark pine (figure 10 in Tomback 1978; Tomback and Kramer 1980; Tomback 1986). Recently burned terrain or clearcuts also attract nutcrackers for seed caching; nutcrackers readily cache seeds in charred soil (Tomback 1986; Tomback et al. 2001a; WW McCaughey, personal communication).

After harvesting seeds, nutcrackers either cache them within a few to several hundred meters from source trees or fly to steep and often south-facing communal storage areas, usually frequented by the local population of nutcrackers (Tomback 1978; Hutchins and Lanner 1982). The subalpine communal storage areas are usually within 2 to 4 km of seed source stands (Tomback 1978; Hutchins and Lanner 1982). Nutcrackers also store whitebark pine seeds in lower elevation communal slopes, as far as 12.5 km from source trees and 800 m elevation downslope, where whitebark pine does not grow (Tomback 1978). Maximum seed dispersal distances by nutcrackers are reported as 22 km for Colorado piñon pine seed caching in northern Arizona (Vander Wall and Balda 1977).

Nutcrackers have also been observed caching bristlecone pine (*Pinus aristata*) seeds in alpine tundra (Baud 1993) and no doubt do this for whitebark pine seeds.

The numbers of whitebark pine seeds stored per individual nutcracker from August through fall have been estimated separately by Tomback (1982) and Hutchins and Lanner (1982), using different assumptions. These numbers provide a sense of the magnitude of seed dispersal accomplished by nutcrackers caching collectively within a geographical area. Working in Squaw Basin in the Absaroka Mountains, Wyoming, Hutchins and Lanner (1982) observed nutcrackers store whitebark pine seeds for about 80 days, and assumed that pouchloads were all near capacity (93 seeds) and that nutcrackers cached for 9 hours per day. They estimated that one nutcracker cached as many as 98000 whitebark pine seeds in a season, and at a mean of 3.2 seeds per cache; this represented more than 30600 caches per bird. Given these estimates, a population of 25 nutcrackers would store more than 750000 caches.

At Mammoth Mountain on the eastern slope of the Sierra Nevada, California, Tomback (1978; 1982) observed nutcrackers store whitebark pine seeds for 38 days in 1974 and 47 days in 1975. Because of high daytime temperatures, nutcrackers stored seeds only during the early morning and late afternoon — about 4.5 hrs per day. Tomback (1982) used an average (77) for numbers of seeds per pouchload, and assumed that each nutcracker made one trip per day to lower elevations to store seeds. An estimated 35000 whitebark pine seeds were stored per bird in about 9500 caches at an average of 3.7 seeds per cache, with about 32000 stored at subalpine elevations alone. However, after storing whitebark pine seeds, nutcrackers traveled to lower elevations, where they stored Jeffrey (*Pinus jeffreyi*) and single-leaf piñon pine (*P. monophylla*) seeds for another six weeks. Tomback (1982) further calculated that a population of nutcrackers required only about 55 % of their stored whitebark pine seeds to feed themselves and their young from April through July, the time that whitebark pine seed caches were retrieved in the Mammoth Lakes area. If the local population consisted of 25 nutcrackers, the excess seeds represented about 95000 caches. Rodents probably retrieve a portion of these caches, and many caches are placed in sites unsuitable for germination or seedling survival, but recruitment would still be considerable if only a tiny fraction germinated and produced mature trees.

4 Consequences of Seed Dispersal by Nutcrackers at Multiple Scales

Seed dispersal by Clark's nutcrackers has complex consequences for whitebark pine, and other bird-dispersed pines. These consequences go beyond simple coevolutionary interaction, which is most commonly portrayed as the dynamics of reciprocal selection pressures (Thompson 1994). Some tolerances of pines for seed dispersal by nutcrackers, both their caching mode and site selection, may be phylogenetically-based – both intrinsic to subsection *Strobus* and refined during the subsequent evolution of subsection *Cembrae* pines. The unique effects of seed dispersal by Clark's nutcracker on whitebark pine may be the product of the climatic conditions and fire regimes of western North America, particularly over the last 10000 years.

4.1 Fine-scale Spatial Effects: Growth Form and Population Genetic Structure

As previously described, nutcrackers disperse whitebark pine seeds in caches typically containing 1 to15 seeds or more, with average cache sizes varying with terrain type (e.g. Tomback 1978; Tomback et al. 2001a). This mode of seed dispersal has major consequences for the fine-scale genetic structure of whitebark pine populations (figure 1).

Seeds within caches usually germinate synchronously or, more rarely, over several years, resulting in a cluster of seedlings roughly the same age (Tomback et al. 1993b; Tomback et al. 2001a). Several to many seedlings may grow together to produce a 'tree cluster' growth form, which appears to be a single tree with multiple stems (Linhart and Tomback 1985; Tomback and Linhart 1990). The stems are either fused at the base or higher up the stem, contiguous at the base, or in any combination of these morphologies

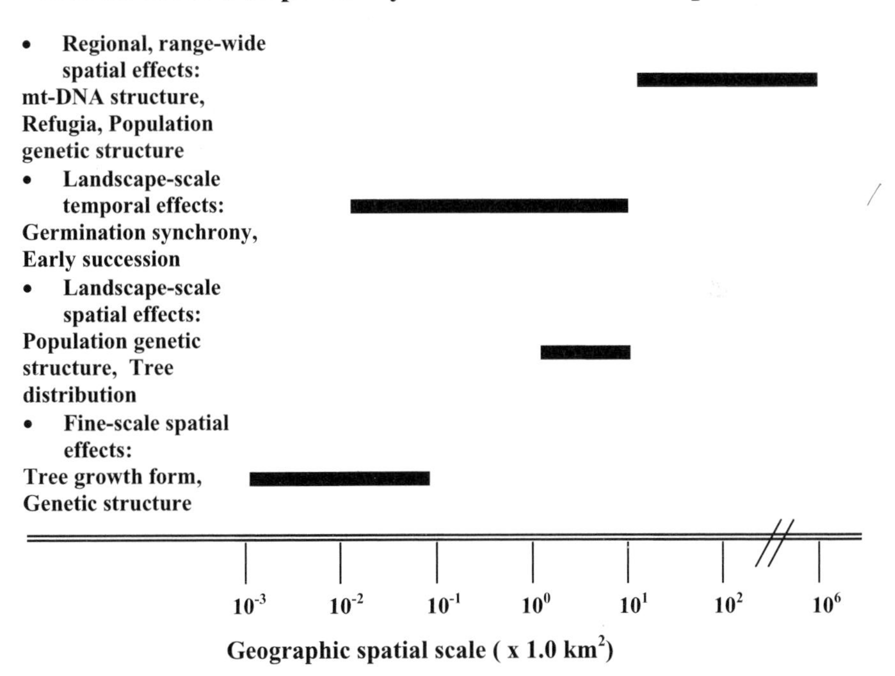

Figure 1 The effects of nutcracker seed caching on whitebark pine at different spatial and temporal scales, from local to range-wide, across western North America. The unit of spatial reference here is square kilometers. The temporal unit indicates synchrony in seed germination over a single growing season and in early succession over several growing seasons following fire or other disturbance

(e.g. Tomback and Linhart 1990; Tomback and Schuster 1994). For example, on Mammoth Mountain in the eastern Sierra Nevada, and on Palmer Mountain in the Absaroka Mountains, 90 % and 58 %, respectively, of the whitebark pine tree sites supported two to five or more trunks or stem groups (table 4 in Tomback and Linhart 1990).

Several genetic studies using allozyme analysis have confirmed that most whitebark pine stem groups consist of two or more distinct individuals, arising from the same nutcracker cache (Linhart and Tomback 1985; Furnier et al. 1987; Rogers et al. 1999). However, these same analyses also reveal that some older stem groups are actually multiple trunks of the same genotype, and probably result from a loss of apical dominance through mechanical damage to the trunk (Schuster and Mitton 1991). Distinguishing these two forms requires genetic analysis. Nonetheless, the tree cluster growth form represents an extreme example of a clumped population dispersion pattern.

This clumped distribution is associated with a fine-scale population genetic structure unique to bird-dispersed pines. Because nutcrackers harvest several seeds to a full pouchload from the same tree, odds are that the seeds within a cache are genetically related, particularly caches made early in the season when cones are not depleted (Tomback 1988). Furthermore, within a given area nutcrackers cache at random with respect to the caches already present, bringing in seeds from different parent trees; and thus neighboring caches may not end up from the same pouchload of seeds. Several studies have demonstrated specifically for whitebark pine that stems from the same cache are more closely related on average to one another than to stems in neighboring caches, reflecting the randomness of this caching behavior (Furnier et al. 1987; Rogers et al. 1999). Rogers et al. (1999) showed that the stems in some groups were related on average as closely as half-siblings to full-siblings. In other words, neighboring stem groups are generally less related to one another than are the stems within a group, and the population is structured to some extent in kin clusters. This unusual fine-scale genetic structure has been found in two other nutcracker-disseminated pines – limber pine and Swiss stone pine, and will no doubt be found in others (Bruederle et al. 2001 for review).

In limber pine, competition among stems in tree clusters apparently has constrained canopy development, pollen cone production, and radial growth (Feldman et al. 1999). Yet to be explored are whether crowding tolerance and reduced competition in limber pine and other nutcracker-dispersed pines have evolved by kin selection, are a by-product of slow growth rates, or are compensated by the benefits of seed dispersal by nutcrackers (Tomback and Linhart 1990; Feldman et al. 1999).

4.2 Landscape-scale Spatial Effects: Population Genetic Structure and Tree Distribution

A number of larger spatial scale effects are the consequence of nutcracker seed dispersal (figure 1). First of all, genetic sampling within three watersheds in the eastern Sierra Nevada indicated that the treeline krummholz thickets of whitebark pine and erect tree stem groups from the upper subalpine have similar gene frequencies (Rogers et al. 1999). Furthermore, Rogers et al. (1999) found minimal genetic differences among the whitebark pine populations in three adjacent watersheds. They attributed this lack of differentiation among populations to the 'homogenizing effect of seed dispersal by birds', given that nutcrackers can transport seeds over distances – as far as 12 to 22 km – adequate to reach all three drainages on a single caching trip. In contrast, conifers with wind-dispersed seeds may show genetic differentiation associated with different slope exposures over short distances (Mitton et al. 1977). Thus, there was little genetic differentiation among whitebark pine populations at a landscape scale, although the populations themselves were genetically diverse.

At the landscape scale, the cache-site preferences of nutcrackers largely determine where whitebark pine will grow (Tomback 2001). Whether whitebark pine establishes on a given site or in a given area depends on its tolerances for aspect, substrate, water availability, and climatic conditions (Weaver 2001). Although nutcrackers will generally store seeds throughout the subalpine zone, they prefer steep slopes, open forests, open terrain, recent burns, clearcuts, and even small forest openings for caching whitebark pine seeds (e.g. Tomback 1978; Tomback et al. 2001a; WW McCaughey, personal communication; RE Keane, personal communication). Because nutcrackers often cache seeds at treeline and above (Tomback 1986; Baud 1993), and cache below the subalpine zone (Tomback 1978; SF Arno, personal communication), the elevational distribution of whitebark pine can change fairly rapidly with respect to climatic warming or cooling trends.

4.3 Landscape-scale Temporal Effects: Delayed Seed Germination and Early Succession

Seed caching by nutcrackers following landscape-level disturbance, particular fire, influences the timing of whitebark pine regeneration and subsequent subalpine community development (figure 1). These processes are of particular importance in the northern Rocky Mountains, including Wyoming, Idaho, Montana, and the southern Canadian Rockies, where whitebark pine occurs in successional, fire-prone communities in the upper subalpine zone, as well as in climax communities and at treeline (Arno 2001). Fire regimes, whether typically stand-replacing or mixed severity, and fire return intervals vary geographically and with forest types, as short as 13 to 46 years in Wyoming to 300 to 400 years within areas in Yellowstone National Park (table 4-5 in Arno 2001 and references therein). In the absence of fire, whitebark pine is replaced over time by shade-tolerant conifers. The extent of fire-dependence of whitebark pine in its center of abundance, namely the northern Rocky Mountains, may be unusual among the *Cembrae* pines, although *P. sibirica* forests periodically experience stand-replacing fire as well (Smolonogov 1994).

Seed dispersal by nutcrackers has led to the phenomenon of delayed seed germination in whitebark pine seeds, which in turn results in a general temporal synchrony in the appearance of new seedlings (McCaughey and Tomback 2001; Tomback et al. 2001a). Controlled experiments based on simulating nutcracker caches (McCaughey 1990; 1993) and natural experiments following the 1988 Yellowstone fires (Tomback 1994b; Tomback et al. 2001a) demonstrate that whitebark pine seeds usually require two or more winter dormancy cycles before they germinate. Thus, most whitebark pine seeds do not germinate in the growing season following cone production, but one or more years beyond. Good cone crops in whitebark pine occur every 3 to 5 years (Krugman and Jenkinson 1974), so delayed seed germination provides continuity in recruitment over time and may be a bet-hedging strategy for seedling survival, especially where moisture availability varies greatly from year to year (Tomback et al. 2001a).

The underlying causes of delayed germination in whitebark pine seeds include underdeveloped embryos, physiological resistance to germination, and/or seed coats that are semi-impervious to water uptake or restrain embryo expansion (Tomback et al. 2001a and references therein). Underdeveloped embryos may be the result of seed caching by nutcrackers as early as mid-August, preventing complete seed development in the short growing season of the subalpine zone (Tomback et al. 2001). The other delaying factors may function as checks on germination until soil moisture achieves a certain threshold; this may promote seedling survival (see discussion in Tomback et al. 2001a). Support for a connection between caching and delayed seed germination comes from the fact that three other *Cembrae* pines show this tendency (Tomback et al. 2001a and references therein); and, Japanese stone pine germinates in greater densities following high levels of moisture (e.g., Kajimoto et al. 1998), which has also been shown for whitebark pine seedlings (Tomback et al. 1993b; Tomback et al. 2001a).

Because nutcrackers cache whitebark pine seeds after disturbances, whitebark pine is classified as a pioneering species (Lanner 1980; 1996 and references therein; Tomback and Linhart 1990). Although seedlings of whitebark pine and other conifers may appear early after fire, the hardy whitebark pine seedlings tolerate harsh post-fire conditions on nearly all slope aspects and conditions (Arno and Hoff 1990; Tomback et al. 1990; 1993a; Tomback et al. 2001a). On exposed, stressful sites, whitebark pines may facilitate community development, producing favorable microsites with moisture, shade, and shelter from wind (Lanner 1980; 1996), and even serving as 'nurse' trees to shade-tolerant competitors (Calaway 1998).

4.4 Regional to Range-wide Spatial Effects: Genetic structure, Glacial Refugia, and Range Expansion

Seed dispersal by nutcrackers can occur over far greater distances than seed dispersal by wind and even move seeds against prevailing winds (McCaughey et al. 1986; Tomback et al. 1990). Genetic structure in whitebark pine at the regional and range-wide scale comes from two mechanisms of gene-flow: seed dispersal by nutcrackers and pollen dispersal by wind. Pollen dispersal distances exceed those of seed dispersal, potentially homogenizing regional conifer populations with respect to the nuclear and chloroplast genomes, in the absence of strong, local selection (e.g. Latta and Mitton 1997; Schuster and Mitton 2000 and references therein; Richardson et al. 2002a).

Allozyme studies, both regional and range-wide, indicate that whitebark pine has lower levels of genetic diversity (i.e. both percentage of polymorphic loci and numbers of alleles per locus) and lower levels of genetic differentiation among populations than do most other pines (Yandell 1992; Jorgensen and Hamrick 1997; Bruederle et al. 1998; Bruederle et al. 2001 for overview). In other words, most allozyme genetic variation is attributed to individual genetic variation within populations rather than genetic differences among populations, although some genetic differentiation occurs in relation to geographic distance (Jorgensen and Hamrick 1997). Several explanations have been proposed for this low genetic diversity and differentiation, including recent evolution of whitebark pine, and the homogenizing effects of seed dispersal by nutcrackers (Bruederle et al. 2001).

More apparent is the influence of seed dispersal distance by nutcrackers on the post-glaciation mitochondrial DNA (mtDNA) genetic structure of whitebark pine populations. In the Pinaceae, mtDNA is maternally inherited through seed tissue, so nutcrackers are the main vector for mtDNA. Recall that nutcrackers disperse seeds a maximum distance of 22 km (Vander Wall and Balda 1977), with 12 km observed for whitebark pine (Tomback 1978).

Sampling throughout the northern and western range of whitebark pine, Richardson et al. (2002a) found three mitochondrial haplotypes, each possibly representing an historical Pleistocene glacial refugium. Richardson et al. (2002a) speculated that after glacial retreat, whitebark pine expanded its range, but the historical genetic substructure remains: mtDNA haplotype 1 characterizes the northernmost populations and several in central Idaho; haplotype 2 characterizes populations in the greater Yellowstone and northwestern Montana; haplotype 3 characterizes populations in the central and southern Cascades and the Sierra Nevada. Only in a few populations did more than one haplotype occur. The boundaries between haplotypes may well indicate areas beyond the seed dispersal distances flown by nutcrackers. A case in point is an abrupt transition between two haplotypes in the Washington Cascades, where Snoqualmie Pass provided a 30 km barrier (Richardson et al. 2002b).

Using similar techniques for limber pine, Mitton et al. (2000) identified 8 mtDNA haplotypes and several potential glacial refugia in the southwest. For both pines, nutcrackers transporting seeds are responsible for the pattern of migration of these haplotypes out of refugia, and distance limits to seed dispersal result in population substructuring. These studies underscore both the historical importance and limitations of nutcracker seed dispersal in expanding the ranges of these pines, when conditions became more favorable.

5 Discussion

Seed dispersal by nutcrackers has had important and similar effects on the evolution, ecology, distribution, and population structure of *Cembrae* pines in North America, Asia, and Europe. However, it appears that more aspects of the life history of whitebark pine than any other *Cembrae* pine have been impacted by nutcrackers, possibly the result of genetics, climate, and other ecological factors. For example, Weaver (2001) reports that average July through September precipitation is less for whitebark pine than for any other *Cembrae* pine, which could influence the biology of regeneration. The aridity, fuel loads, and prevalence of summer lightning in many western forests also makes them fire-prone. In the northern Rocky Mountains, whitebark pine forms successional, fire-dependent communities, which are an important forest type (Arno 1986; 2001). There, nutcrackers initiate whitebark pine regeneration soon after disturbance, which provides some advantage to whitebark pine over its competitors (e.g., Tomback et al. 2001a). Whitebark pine seedlings are reasonably hardy, tolerating exposed sites and poor seedbeds, and facilitating the successional process (Arno and Hoff 1990; Tomback et al. 2001a).

Swiss stone pine frequently occurs in stem groups like whitebark pine (Tomback et al. 1993b), and the seeds delay germination (Krugman et al. 1974); but communities are generally self-replacing, and fire and droughty conditions are infrequent occurrences. Whereas Siberian stone pine mixed forest communities are widely fire-prone, the stone pines themselves are not successionally replaced over time by shade-tolerant spruce and fir (Smolonogov 1994). Tree clusters are uncommon in Siberian stone pine, where typically one individual per cache survives, and thus the fine-scale kin structure of stem groups is lacking (Krutovskii and Politov, unpublished data, reported in Bruederle et al. 2001). Japanese stone pine grows in stem clusters (Kajimoto et al. 1998; Kajimoto 2002), but the pine forms mat-like growth forms on windy, subalpine zone sites in mixed shrub communities, and is not dependent on fire for community renewal (Kajimoto 2002). Korean pine grows taller than the other stone pines and has faster growth rates, which may explain why seedling and sapling clusters thin over time, producing trees with single trunks (Hutchins et al. 1996). Thus, the kin structure is also lacking in this pine. There is some evidence for intermittent disturbances in old growth Korean pine forests, such as fire and windthrow, but the Korean pine is not successional (Hutchins et al. 1996).

In the absence of Clark's nutcracker, recruitment of whitebark pine would be significantly reduced, particularly in large, burned areas and clearcuts (e.g. Tomback et al. 1990); and, its distribution might be greatly restricted, particularly for steep, exposed sites and at treeline. Other potential dispersers are less efficient and far more restricted in their movements (Hutchins and Lanner 1982; Tomback 1982). The absence of seed dispersal by nutcrackers would greatly slow post-fire regeneration and the early successional process.

In contrast, nutcracker populations would weather the loss of whitebark pine, but at a much reduced carrying capacity. Nutcrackers use other pines, such as limber pine, the piñons, and large-seeded wind-dispersed pines as additional food sources (Tomback and Kendall 2001).

Unfortunately, this latter experiment is currently taking place: whitebark pine is declining rapidly in its northern range from the introduced fungal disease, white pine blister rust (*Cronartium ribicola*) as well as advancing succession from previous fire suppression practices. Blister rust is spreading throughout the range of whitebark pine, and infection rates are intensifying through time (Kendall and Keane 2001; Tomback et al. 2001b). Forest managers are just beginning to realize the extent of the problem and the precarious future for whitebark pine.

Acknowledgements

I wrote this paper during my 2002-03 sabbatical leave, supported by a Faculty Fellowship from the University of Colorado at Denver. I thank the many undergraduate and graduate students and colleagues who have collaborated with me through the years on studies of Clark's nutcrackers and whitebark pine. My sincere congratulations to Professor Dr. Holtmeier on the occasion of his retirement, and I gratefully acknowledge his early contributions to the nutcracker-pine literature. Many thanks also for his interest in the Clark's nutcracker-whitebark pine studies, and for good discussions on both the nutcracker-pine interaction and treeline dynamics.

References

Arno SF (1986) Whitebark pine cone crops – a diminishing source of wildlife food? Western Journal of Applied Forestry 1: 92-94

Arno SF (2001) Community types and natural disturbance processes. In: Tomback DF, Arno SF, Keane RE (eds) Whitebark pine communities: ecology and restoration. Island Press, Washington, DC, USA, pp 74-88

Arno SF, Hoff RJ (1990) *Pinus albicaulis* Engelm. Whitebark pine. In: Burns RM, Honkala BH (technical coordinators) Silvics of North America. USDA Forest Service, Agriculture Handbook 654, Washington, DC, USA, pp 268-279

Balda RP (1980) Recovery of cached seeds by a captive *Nucifraga caryocatactes*. Zeitschrift für Tierpsychologie 52: 331-346

Baud KS (1993) Simulating Clark's Nutcracker caching behavior: germination and predation of seed caches. MA thesis. University of Colorado at Denver, USA

Bock WJ, Balda RP, Vander Wall SB (1973) Morphology of the sublingual pouch and tongue musculature in Clark's Nutcracker. Auk 90: 491-519

Bruederle LF, Tomback DF, Kelly KK, Hardwick RC (1998) Population genetic structure in a bird-dispersed pine, *Pinus albicaulis* (Pinaceae). Canadian Journal of Botany 7: 83-90

Bruederle LP, Rogers DL, Krutovskii KV, Politov DV (2001) Population genetics and evolutionary implications. In: Tomback DF, Arno SF, Keane RE (eds) Whitebark pine communities: ecology and restoration. Island Press, Washington, DC, USA, pp 137-153

Calaway RM (1998) Competition and facilitation on elevation gradients in subalpine forests of the northern Rocky Mountains, USA. Oikos 82: 561-573

Crocq C (1978) Écologie du Casse-noix (*Nucifraga caryocatactes* L.) dans les Alpes françaises du sud: ses relations avec l'Arolle (*Pinus cembra* L.). Dissertation, L'Université de Droit D'Économie et de Sciences D'Aix-Marseille, France

Crocq C (1990) Le Casse-noix moucheté. Lechevalier~R. Chabaud, France

Espinosa de los Monteros A, Cracraft J (1997) Intergeneric relationships of the New World jays inferred from cytochrome b gene sequences. The Condor 99: 490-502
Feldman R, Tomback DF, Koehler J (1999) Cost of mutualism: competition, tree morphology, and pollen production in limber pine clusters. Ecology 80: 324-329
Furnier GR, Knowles P, Clyde MA, Dancik BP (1987) Effects of avian seed dispersal on the genetic structure of whitebark pine populations. Evolution 41: 607-612
Hayashida M (1994) Role of nutcrackers on seed dispersal and establishment of *Pinus pumila* and *P. pentaphylla*. In: Schmidt WC, Holtmeier F-K (compilers) Proceedings – International workshop on subalpine stone pines and their environment: the status of our knowledge. USDA Forest Service Intermountain Research Station, General Technical Report INT-GTR-309, Ogden, Utah, USA, pp 159-162
Hope S (1989) Phylogeny of the avian family Corvidae. Dissertation, City University of New York, New York, USA
Hutchins HE, Lanner RM (1982) The central role of Clark's Nutcracker in the dispersal and establishment of whitebark pine. Oecologia 55: 192-201
Hutchins HE, Hutchins SA, Liu B-W (1996) The role of birds and mammals in Korean pine (*Pinus koraiensis*) regeneration dynamics. Oecologia 107: 120-130
Jorgensen SM, Hamrick JL (1997) Biogeography and population genetics of whitebark pine, *Pinus albicaulis*. Canadian Journal of Forest Research 27: 1574-1585
Kajimoto T (2002) Factors affecting seedling recruitment and survivorship of the Japanese subalpine stone pine, *Pinus pumila*, after seed dispersal by nutcrackers. Ecological Research 17: 481-491
Kajimoto T, Onodera H, Ikeda S, Daimaru H, Seki T (1998) Seedling establishment of subalpine stone pine (*Pinus pumila*) by nutcracker (*Nucifraga*) seed dispersal on Mt. Yumori, Northern Japan. Arctic and Alpine Research 30: 408-417
Kamil AC, Balda RP (1985) Cache recovery and spatial memory in Clark's Nutcracker (*Nucifraga columbiana*). Journal of Experimental Psychology: Animal Behavior Processes 11: 95-111
Kendall K, Keane RE (2001) Whitebark pine decline: infection, mortality, and population trends. In: Tomback DF, Arno SF, Keane RE (eds) Whitebark pine communities: ecology and restoration. Island Press, Washington, DC, USA, pp 221-242
Krugman SL, Jenkinson JL (1974) *Pinus* L. Pine. In: Schopmeyer CS (technical coordinator) Seeds of woody plants in the United States. USDA Forest Service, Agriculture Handbook No 450, Washington, DC, USA, pp 598-638
Krugman SL, Stein WI, Schmitt DM (1974) Seed biology. In Schopmeyer CS (technical coordinator) Seeds of wood plants in the United States. USDA Forest Service, Agriculture Handbook No 450, Washington, DC, USA, pp 5-40
Krutovskii KV, Politov DV, Altukov YP (1994) Genetic differentiation and phylogeny of stone pine species based on isozyme loci. In: Schmidt WC, Holtmeier F-K (comps) Proceedings – International workshop on subalpine stone pines and their environment: the status of our knowledge. USDA Forest Service Intermountain Research Station, General Technical Report INT-GTR-309, Ogden, Utah, USA, pp 19-30

Lanner RM (1980) Avian seed dispersal as a factor in the ecology and evolution of limber and whitebark pines. In: Dancik BP, Higginbotham KO (eds) Proceedings of Sixth North American Forest Biology Workshop. University of Alberta, Edmonton, Alberta, Canada, pp 15-48

Lanner RM (1982) Adaptations of whitebark pine for seed dispersal by Clark's Nutcracker. Canadian Journal of Forest Research 12: 391-402

Lanner RM (1996) Made for each other: A symbiosis of birds and pines. Oxford University Press, New York, New York, USA

Latta RG, Mitton JB (1997) A comparison of population differentiation across four classes of gene marker in limber pine (*Pinus flexilis* James). Genetics 146: 1153-1163

Linhart YB, Tomback DF (1985) Seed dispersal by Clark's nutcracker causes multi-trunk growth form in pines. Oecologia 67: 107-110

Liston A, Robinson WA, Piñero D, Alvarez-Buylla ER (1999) Phylogenetics of *Pinus* (Pinaceae) based on nuclear ribosomal DNA internally transcribed spacer region sequences. Molecular Phylogenetics and Evolution 11: 95-109

Mattes H (1978) Der Tannenhäher (*Nucifraga caryocatactes* L) im Engadin: Studien zu seiner Ökologie und Funktion im Arvenwald. Münsterische Geographische Arbeiten 2

Mattes H (1982) Die Lebensgemeinschaft von Tannenhäher, *Nucifraga caryocatactes* (L.) und Arve, *Pinus cembra* L., und ihre forstliche Bedeutung in der oberen Gebirgswaldstufe. Eidgenössische Anstalt für das forstliche Versuchswesen, Berichte Nr. 241,

McCaughey WW (1990) Biotic and microsite factors affecting *Pinus albicaulis* establishment and survival. Dissertation. Montana State University, Bozeman, Montana, USA

McCaughey WW (1993) Delayed germination and seedling emergence of *Pinus albicaulis* in a high elevation clearcut in Montana, USA. In: Edwards DGW (comp, ed) Dormancy and barriers to germination. Proceedings International Symposium IUFRO Project Group P2.04-00 (Seed Problems). Victoria, British Columbia, Canada. Forestry Canada, Pacific Forestry Centre, Victoria, BC

McCaughey WW, Schmidt WC (2001) Taxonomy, distribution, and history. In: Tomback DF, Arno SF, Keane RE (eds) Whitebark pine communities: ecology and restoration. Island Press, Washington, DC, USA, pp 29-40

McCaughey WW, Tomback DF (2001) The natural regeneration process. In: Tomback DF, Arno SF, Keane RE (eds) Whitebark pine communities: ecology and restoration. Island Press, Washington, DC, USA, pp 105-120

McCaughey WW, Schmidt WC, Shearer RC (1986) Seed-dispersal characteristics of conifers in the Inland Mountain West. In: Proceedings – Conifer tree seed in the Inland Mountain West symposium. General Technical Report, Intermountain Research Station, INT-203, Ogden, Utah, USA, pp 50-62

Mewaldt LR (1952) The incubation patch of the Clark Nutcracker. Condor 54: 361

Mewaldt LR (1956) Nesting behavior of the Clark Nutcracker. Condor 58: 3-23

Millar CI (1998) Early evolution of pines. In: Richardson DM (ed) Ecology and biogeography of *Pinus*. Cambridge University Press, Cambridge, United Kingdom, pp 69-91

Mitton JB, Linhart YB, Hamrick JL, Beckman JS (1977) Observations on the genetic structure and mating system of ponderosa pine in the Colorado Front Range. Theoretical & Applied Genetics 51: 5-13

Mitton JB, Kreiser BR, Latta RG (2000) Glacial refugia of limber pine (*Pinus flexilis* James) inferred from the population structure of mitochondrial DNA. Molecular Ecology 9: 91-97

Price RA, Liston A, Strauss SH (1998) Phylogeny and systematics of *Pinus*. In: Richardson DM (ed) Ecology and biogeography of *Pinus*. Cambridge University Press, Cambridge, United Kingdom, pp 49-68

Richardson BA, Brunsfeld SJ, Klopfenstein NB (2002a) DNA from bird-dispersed seed and wind-disseminated pollen provides insights into postglacial colonization and population genetic structure of whitebark pine (*Pinus albicaulis*). Molecular Ecology 11: 215-227

Richardson BA, Klopfenstein NB, Brunsfeld SJ (2002b) Assessing Clark's nutcracker seed-caching flights using maternally inherited mitochondrial DNA of whitebark pine. Canadian Journal of Forest Research 32: 1103-1107

Reimers NF (1959) Birds of the cedar-pine forests of south-central Siberia and their role in the life of the cedar-pine. Trudy Biologischeskogo Instituta, Sibirskogo Otdelenie Akad. Nauk, USSR 5, pp 121-166 (translated from Russian by L Kelso)

Rogers DL, Millar CI, Westfall RD (1999) Fine-scale genetic structure of whitebark pine (*Pinus albicaulis*): associations with watershed and growth form. Evolution 53: 74-90

Roselaar CS (1994) *Nucifraga caryocatactes* Nutcracker. In: Cramp S, Perrins CM, Brooks DJ (eds) Handbook of the Birds of Europe, the Middle East, and North Africa, vol VIII. Oxford University Press, Oxford, England, pp 76-95

Schuster WSF, Mitton JB (1991) Relatedness within clusters of a bird-dispersed pine and the potential for kin interactions. Heredity 67: 41-48

Schuster WSF, Mitton JB (2000) Paternity and gene dispersal in limber pine (*Pinus flexilis* James). Heredity 84: 348-361

Sibley CG, Ahlquist JE (1985) The phylogeny and classification of the Australo-Papuan passerine birds. Emu 85: 1-14

Smolonogov EP (1994) Geographical differentiation and dynamics of Siberian stone pine forests in Eurasia. In: Schmidt WC, Holtmeier F-K (comp.) Proceedings – International workshop on subalpine stone pines and their environment: the status of our knowledge. USDA Forest Service Intermountain Research Station, General Technical Report INT-GTR-309, Ogden, Utah, USA, pp 275-279

Swanberg PO (1956) Incubation in the Thick-billed Nutcracker, *Nucifraga c. caryocatactes* (L.). In: Wingstrand KG (ed) Bertil Hanström zoological papers in honour of his sixty-fifth birthday November 20th, 1956. Zoological Institute, Lund, Sweden, pp 279-297

Thompson JN (1994) The coevolutionary process. University of Chicago Press, Chicago, Illinois, USA

Tomback DF (1978) Foraging strategies of Clark's Nutcracker. Living Bird 16; pp 123-161

Tomback DF (1982) Dispersal of whitebark pine seeds by Clark's Nutcracker: a mutualism hypothesis. Journal of Animal Ecology 51: 451-467

Tomback, DF (1983) Nutcrackers and pines: coevolution or coadaption? In: Nitecki MH (ed) Coevolution. University of Chicago Press, Chicago, Illinois, USA, pp 179-223

Tomback DF (1986) Post-fire regeneration of krummholz whitebark pine: a consequence of nutcracker seed caching. Madroño 3: 100-110

Tomback DF (1988) Nutcracker-pine mutualisms: multi-trunk trees and seed size. In: Ouellet H (ed) Acta XIX Congressus Internationalis Ornithologici, vol 1. University of Ottawa Press, Ottawa, Ontario, Canada, pp 518-527

Tomback DF (1994a) Ecological relationship between Clark's nutcracker and four wingless-seed *Strobus* pines of western North America. In: Schmidt WC, Holtmeier F-K (comps) Proceedings – International workshop on subalpine stone pines and their environment: the status of our knowledge. USDA Forest Service Intermountain Research Station, General Technical Report INT-GTR-309, Ogden, Utah, USA, pp 221-224

Tomback DF (1994b) Effects of seed dispersal by Clark's nutcracker on early postfire regeneration of whitebark pine. In: Schmidt WC, Holtmeier F-K (comps) Proceedings – International workshop on subalpine stone pines and their environment: the status of our knowledge. USDA Forest Service Intermountain Research Station, General Technical Report INT-GTR-309, Ogden, Utah, USA, pp 193-198

Tomback DF (1998) Clark's Nutcracker (*Nucifraga columbiana*), No 331. In: Poole A, Gill F (eds) The birds of North America. The Birds of North America, Inc, Philadelphia, Pennsylvania, USA

Tomback DF (2001) Clark's nutcracker: Agent of regeneration. In: Tomback DF, Arno SF, Keane RE (eds) Whitebark pine communities: ecology and restoration. Island Press, Washington, DC, USA, pp 89-104

Tomback DF, Kendall KC (2001) Biodiversity losses: the downward spiral. In: Tomback DF, Arno SF, Keane RE (eds) Whitebark pine communities: ecology and restoration. Island Press, Washington, DC, USA, pp 243-262

Tomback DF, Kramer KA (1980) Limber pine seed harvest by Clark's Nutcracker in the Sierra Nevada: timing and foraging behavior. The Condor 82: 467-468

Tomback DF, Linhart YB (1990) The evolution of bird-dispersed pines. Evolutionary Ecology 4: 185-219

Tomback DF, Schuster WSF (1994) Genetic population structure and growth form distribution in bird-dispersed pines. In: Schmidt WC, Holtmeier F-K (comps) Proceedings – International workshop on subalpine stone pines and their environment: the status of our knowledge. USDA Forest Service Intermountain Research Station, General Technical Report INT-GTR-309, Ogden, Utah, USA, pp 43-50

Tomback DF, Hoffmann LA, Sund SK (1990) Coevolution of whitebark pine and nutcrackers: implications for forest regeneration. In: Schmidt WC, McDonald KJ (comps) Proceedings – Symposium on whitebark pine ecosystems: ecology and management of a high-mountain resource. USDA Forest Service Intermountain

Research Station, General Technical Report INT-270, Ogden, Utah, USA, pp 118-129

Tomback DF, Sund SK, Hoffmann LA (1993a) Post-fire regeneration of *Pinus albicaulis*: height-age relationships, age structure, and microsite characteristics. Canadian Journal of Forest Research 23: 113-119

Tomback DF, Holtmeier F-K, Mattes H, Carsey KS, Powell M (1993b) Tree clusters and growth form distribution in *Pinus cembra*, a bird-dispersed pine. Arctic and Alpine Research 25: 374-381

Tomback DF, Anderies AJ, Carsey KS, Powell ML, Mellmann-Brown S (2001a) Delayed seed germination in whitebark pine and regeneration patterns following the Yellowstone fires. Ecology 82: 2587-2600

Tomback DF, Arno SF, Keane RE (2001b) The compelling case for management intervention. In: Tomback DF, Arno SF, Keane RE (eds) Whitebark pine communities: ecology and restoration. Island Press, Washington, DC, USA, pp 3-25

Turcek FJ, Kelso L (1968) Ecological aspects of food transportation and storage in the Corvidae. Communications in Behavioral Biology, Part A, I, pp 277-297

Vander Wall SB (1982) An experimental analysis of cache recovery in Clark's Nutcracker. Animal Behavior 30: 84-94

Vander Wall SB, Balda RP (1977) Coadaptations of the Clark's Nutcracker and piñon pine for efficient seed harvest and dispersal. Ecological Monographs 47: 89-111

Vander Wall SB, Hutchins HE (1983) Dependence of Clark's Nutcracker, *Nucifraga columbiana*, on conifer seeds during postfledging period. Canadian Field-Naturalist 97: 208-214

Wang X-R, Tsumura Y, Yoshimaru H, Nagasaka K, Szmidt AE (1999) Phylogenetic relationships of Eurasian pines (*Pinus,* Pinaceae) based on chloroplast *rbcl*, *MAT*K, *RPL20-RPS18* spacer, and *TRNV* intron sequences. American Journal of Botany 88: 1742-1753

Weaver T (2001) Whitebark pine and its environment. In: Tomback DF, Arno SF, Keane RE (eds) Whitebark pine communities: ecology and restoration. Island Press, Washington, DC, USA, pp 41-73

Yandell UG (1992) An allozyme analysis of whitebark pine (*Pinus albicaulis* Engelm.). Master's thesis. University of Nevada, Reno, USA

Mountain Ecosystems

Studies in Treeline Ecology

Regional Treeline Studies in Europe

Humus Forms and Reforestation of an Abandoned Pasture at the Alpine Timberline (Upper Engadine, Central Alps, Switzerland)

Bettina Hiller and Andreas Müterthies

Abstract

Humus forms as well as the natural regeneration of European larch (*Larix decidua* Mill.) and Swiss stone pine (*Pinus cembra* L.) were mapped along an altitudinal gradient from the subalpine forest to the alpine zone in the Upper Engadine (Switzerland). The establishment and distribution patterns of larch and stone pine as well as the humus forms are controlled by microtopography which influences other site factors. In the subalpine forest, Mor humus forms are very common, while in the timberline ecotone the humus forms can be described as Mor and Moder humus forms, and in the alpine zone Moder humus forms dominate. The thickness of the organic layers decreases from the subalpine forest to the alpine zone. The density of naturally regenerated European larch and Swiss stone pine has increased since the pastures have been abandoned. The analysis of the density and growth of both tree species lead to the conclusion that conditions above 2300 m a.s.l. are unsuitable for a successful reforestation of the study site.

1 Introduction and Objectives

The alpine timberline, one of the most important altitudinal boundaries of the vegetation in high mountain areas marks the transition from the subalpine forest to the alpine zone. It is an ecotone made up of different microsites in which elements of both ecosystems, the subalpine forest and the alpine zone are nested (Holtmeier and Broll 1992; Bütler and Domergue 1997). Within the timberline ecotone soil types, humus forms and plant species shift from those typical for the subalpine forest to the characteristic ones of the alpine zone. The distribution pattern and development of European larch (*Larix decidua* Mill.) and Swiss stone pine (*Pinus cembra* L.) as well as the occurrence of different humus forms are controlled by microtopography influencing other site factors such as the duration of snow cover, soil temperature and soil moisture (Hiller et al. 2002). The humus forms, which are influenced by vegetation and soil organisms, as well as abiotic factors like relief and climate are suitable indicators for site conditions. Both the site conditions and the general modes of seed dispersal cause a heterogeneous regeneration of the main tree species which is reflected in the distribution of seedlings and saplings in the ecotone. In the European Alps, the present structure of most alpine timberlines is the result of anthropogenic influences and site conditions. Natural regeneration of trees had been almost completely suppressed by cattle grazing and other anthropogenic disturbances for centuries.

The aim of this study was, firstly to analyse the changes of humus forms and regeneration patterns of European larch (*Larix decidua*) and Swiss stone pine (*Pinus cembra*) along an altitudinal gradient from the forest to the alpine zone. Secondly, the study focused on the reforestation and the humus forms developed since alpine pasturage ended in the 1950ies. Thirdly, the future development of humus forms and the natural regeneration of the main tree species were of interest.

2 Study Area

The study area is the north-west facing slope of the Piz da Staz (2847 m a.s.l.) in the Upper Engadine, Central Alps, Switzerland (figure 1). The climate of the study area is slightly continental, with relatively low precipitation and high solar radiation. The mean annual air temperature in Samedan (1706 m a.s.l.) is 1.3°C; the total annual precipitation is 657 mm. The prevailing microclimatic conditions are strongly influenced by the locally varying microtopography. Talus deposits of gabbrodiorite, diorite and essexite make up the parent material (Staub 1946). The topography is characterized by rocky outcrops, knolls and small ridges alternating with depressions. Leptosols, Cambisols and Podzols are common in the subalpine Larch-Stone pine forest as well as in the timberline ecotone (Müller 1983). Above the timberline, shallow Leptosols and Cambisols are widespread (Müller 1983; Neuwinger 1987; Burns 1990). Due to alpine pasturage, which was ended in the 1950ies, the timberline of the study area had retreated to 2200 m a.s.l. Swiss stone pine and European larch form the timberline, which also dominate the subalpine Larch-Stone pine forest (*Larici-Pinetum cembrae*) (Keller et al. 1998). Due to microtopography, the locally varying site conditions cause a mosaic-like vegetation pattern. Dwarf shrubs, like *Vaccinium gaultherioides* and *Vaccinium myrtillus*, are common. Wind-exposed locations are characterized by *Loiseleuria procumbens* and various lichen species. In depressions where

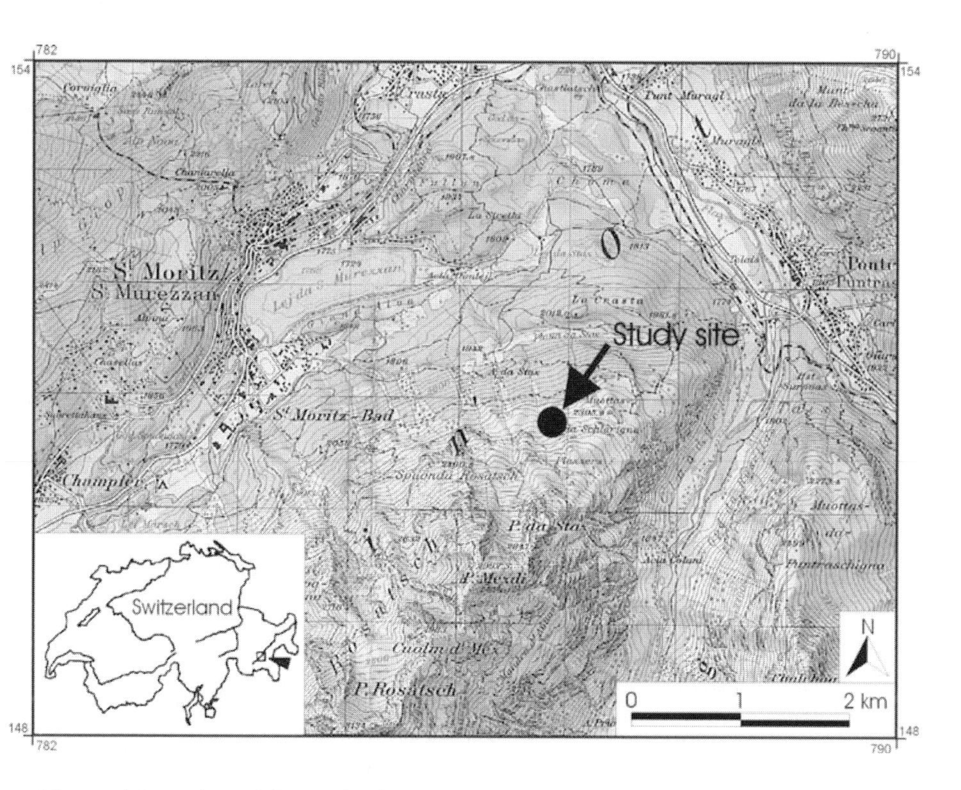

Figure 1 Location of the study site (Upper Engadine, Central Alps, Switzerland) Reproduced with permission of swisstopo (BA046336)

the snow cover persists longer than in the surrounding area and where moist to wet conditions during the growing season prevail, the vegetation is mostly dominated by sedges, grasses and mosses. In the alpine zone the microclimatic conditions cause a vegetation mosaic of *Carex curvula* swards alternating with other communities (Ellenberg 1996).

3 Material and Methods

The study site extends over the whole timberline ecotone between 2200 m and 2400 m a.s.l. In 1997 and 1998, microtopography, snow melt-out, vegetation, soil conditions and humus forms were mapped using a 10 m grid. In addition, these factors were also recorded at study sites in the subalpine Larch-Stone pine forest and in the alpine zone. The humus forms of the study site were described and classified according to Green et al. (1993) as well as the German classification system (AK Standortskartierung 1996).

The distribution of Swiss stone pine and European larch larger than 2 m were mapped. For smaller trees (0.4-2 m high), the stem density was mapped on the basis of a 10 m grid for the whole study site. Spatial interpolation was calculated by using the IDW – Inverse Distance Weighted method (Philip and Watson 1982; Watson and Philip 1985). The calculation and visualization of the results was done using a Geographic Information System (ArcView).

4 Results and Discussion

Microtopography controls the microclimatic conditions as well as the distribution of plant communities, soils and humus forms. It also influences the snow cover thickness, snow cover duration and snow melt-out. The duration of the snow cover on convex sites is usually shorter than on concave topography. Depending on microtopography pedogenesis shows a high heterogeneity. Shallow Podzols, Cambisols and Leptosols are very common, while in depressions pedogenesis is often influenced by accumulation of organic and mineral material. The shallow and stony soils are well-drained and characterized by high organic matter contents (cf. Gracanin 1972; Legros 1975, 1993; Robert et al. 1980; Bütler and Domergues 1997; Prichard et al. 2000). Furthermore, charcoal and ash remains are widespread in soils of the study site indicating former fire clearing activities in the timberline ecotone. Ash can built up a shallow horizon between the organic layer and the eluvial horizon or can be distributed within the organic layer and the A horizon.

4.1 Humus Forms

The humus forms of the study site in the timberline ecotone can be described as Mor and Moder humus forms (figure 2b). Moder and Mor humus forms are very common on silicate rocks in the timberline ecotone of the Austrian or Swiss Central Alps (cf. Blaser 1980; Mosimann 1985; Lüscher 1991; Bednorz 2000). On a steep and stony slope between 2240 and 2250 m a.s.l., Mor humus forms are very widespread. Sometimes L or F horizons are lacking or are very poorly developed. In particular, the translocation of litter and organic fine material on these eroded sites by wind and water can explain such thin or lacking organic layers. On a plateau with a bog-like vegetation at 2260 m a.s.l., the humus forms can be described as Moder. On these poorly to very poorly drained sites Hydromoder or Saprimoder were observed. Due to the microtopography the humus forms show a high heterogeneity on the slope between 2270 m and up to 2400 m a.s.l. At wind-exposed mounds

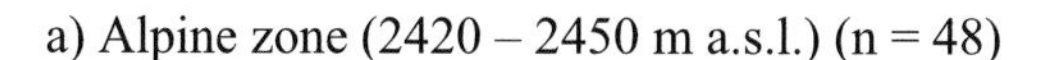

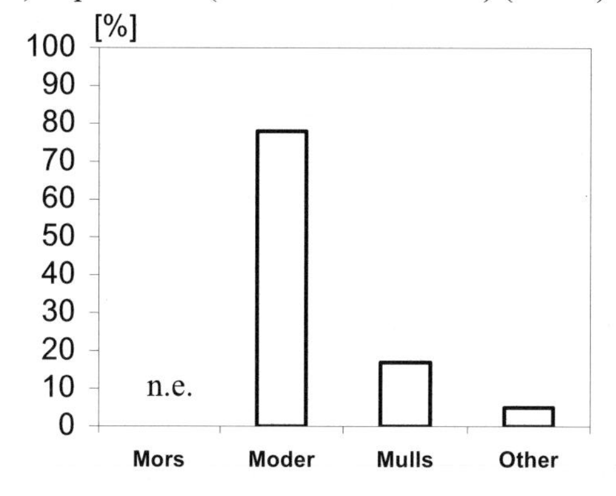

b) Timberline ecotone (2240 – 2400 m a.s.l.) (n = 132)

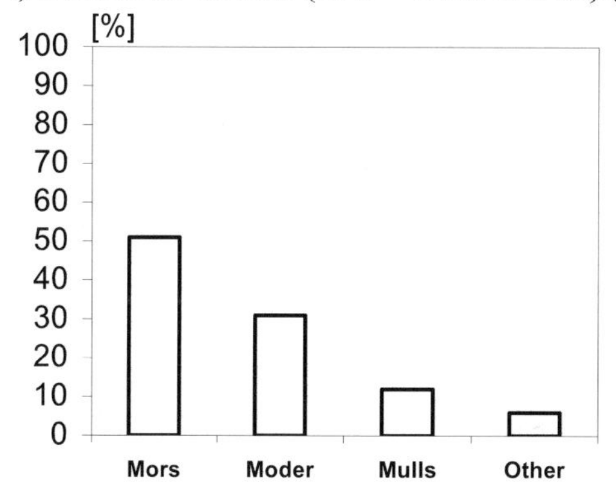

c) Subalpine forest (2180 m a.s.l.) (n = 14)

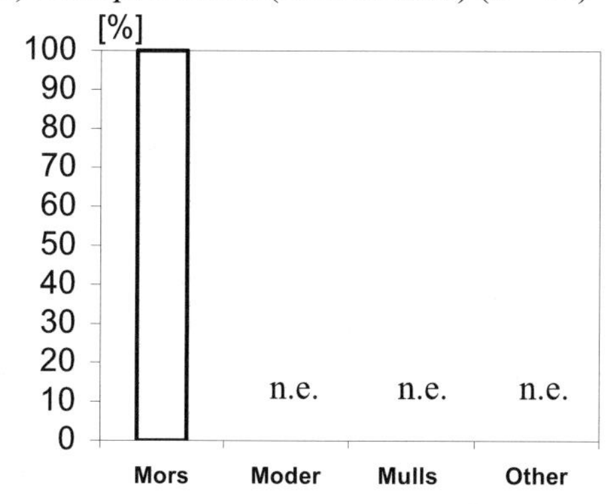

Figure 2 Humus forms in the subalpine forest, in the timberline ecotone, and in the alpine zone (n.e.: not existing)

under scattered vegetation of dwarf shrubs and lichens the humus forms are Mors, mainly Humimors or Tenuic Humimors (Green et al. 1993). Furthermore, at steep slopes or in depressions under dwarf shrub stands the humus forms can also be described as Humimors or Tenuic Humimors. The organic layers of the Mor humus forms are characterized by shallow L and F transitional horizons and H horizons (figure 3). The thickness of the H horizons exceeds the thickness of the L and F horizons. In Humimors the thickness of the H horizon is up to 7 or 8 cm, while in the shallower Tenuic Humimors the

Humusform: Tenuic Humimor (Green et al. 1993)
Feinhumusreicher rohhumusartiger Moder (AG Boden 1994)
Vegetation: dwarf shrubs (*Loiseleuria procumbens, Vaccinium gaultherioides*) and lichens
Elevation: 2300 m

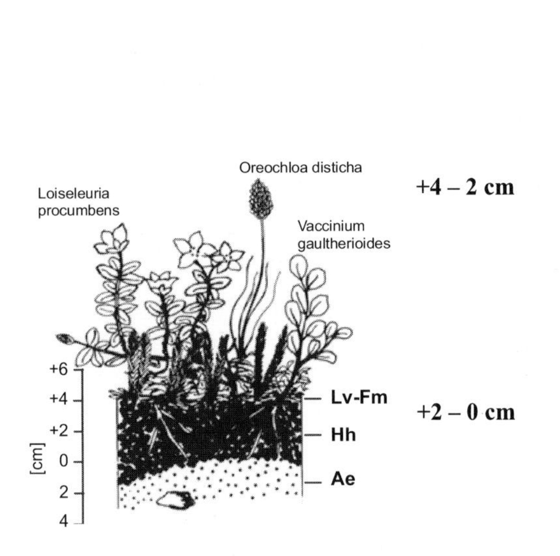

+4.5 – 4 cm **Lv-Fm**
dry; brown-black to black (discoloration); Vaccinium and Loiseleuria leaves and stems; single particle, seldom compact; loose; very few fungal mycelia;
boundary: gradual, wavy

+4 – 2 cm **Hh1**
moist; 5YR 2/1; moderate, granular; friable; greasy;
roots ∅ < 2 mm: plentiful,
roots ∅ > 2 mm: common;
no visible faunal droppings;
boundary: clear, wavy

+2 – 0 cm **Hh2**
moist; 7.5YR 3/1; moderate, granular; friable; greasy, gritty;
roots ∅ < 2 mm: common,
roots ∅ > 2 mm: few;
no visible faunal droppings;
boundary: clear, wavy

0 – 4 cm **Ae**
moist; 7.5YR 6/2; moderate, granular; friable; gritty;
roots ∅ < 2 mm: common;
roots ∅ > 2 mm: very few

Figure 3 Humus forms in the timberline ecotone – Tenuic Humimor

H horizons are only 3 to 5 cm thick. The boundary with the underlying A-horizon, which is usually affected by podzolization, is clear.

In depressions under dense grass heath vegetation the humus forms are classified as Rhizic Leptomoder, Rhizic Mullmoder or Rhizomulls (Green et al. 1993). The Moder humus forms are characterized by very abundant roots in a mat-like structure and very shallow L and F horizons. Being 6 to 12 cm deep, the mean thickness of the organic layers of the Moder humus forms in depressions exceeds those of the Mor humus forms described above. The L and F transitional horizons contain mainly leaves and stems of grasses and herbs mixed with erect dead material of living plants in a loose matted structure. The boundary to the underlying A horizon is gradual (figure 4).

The site conditions at the timberline with its short growing season hamper the complete decomposition and favor the accumulation of organic matter. However, during the short growing season the site conditions like soil temperature and soil water conditions support activity of soil organisms and decomposition. Due to microtopography and the duration snow cover persists, the vegetation and also the humus forms of exposed sites and depressions differ (cf. Lüscher 1991; Bednorz et al. 2000; Hiller et al. 2002). The vegetation, but especially the quality and quantity of litter influence decomposition (cf. Klinka et al. 1990, Beyer 1996). Furthermore, the vegetation determines the number and kind of soil organisms and also decomposition processes (cf. Bal 1970, Steltzer and Bowmann 1998). Under vegetation dominated by dwarf shrubs and lichen, the litter amount is markedly higher than under grass heath vegetation. However, under grass dominated vegetation the upper centimeters of the soil are characterized by a dense net of fine roots (cf. Hiller 2001). These provide additional decomposable organic matter which remains in the soil (cf. Rehder and Schaefer 1978; Titlyanova et al. 1999). Furthermore, the low litter accumulation is partly the result of the low litter production of grasses compared to dwarf shrubs. Dead leaves often remain on the living plants. During the growing season the leaves start withering at the tips while the remaining parts of the leaves are still assimilating. However, not only litter amount but also the quality of plant residues determines decomposition. Higher nutrient contents and lower C/N values of the grass-dominated sites compared to the dwarf shrub sites are caused by the plants' composition (Hiller 2001). Furthermore, the former alpine pasturage supports nutrient enrichment in depressions. Wind and water deposition of organic material from adjacent slopes could increase the accumulation of nutrients. The higher nutrient contents, especially higher N-contents of the organic layers, favor decomposition processes. However, it was expected that under the climatic and edaphic conditions in the timberline ecotone L and F horizons should be

Humusform: Rhizic Leptomoder (Green et al. 1993)
Feinhumusreicher Moder (AG Boden 1994)

Vegetation: dominated by sedges, grasses and herbs (*Carex spec., Avena versicolor, Poa alpina, Bartsia alpina*)

Elevation: 2300 m

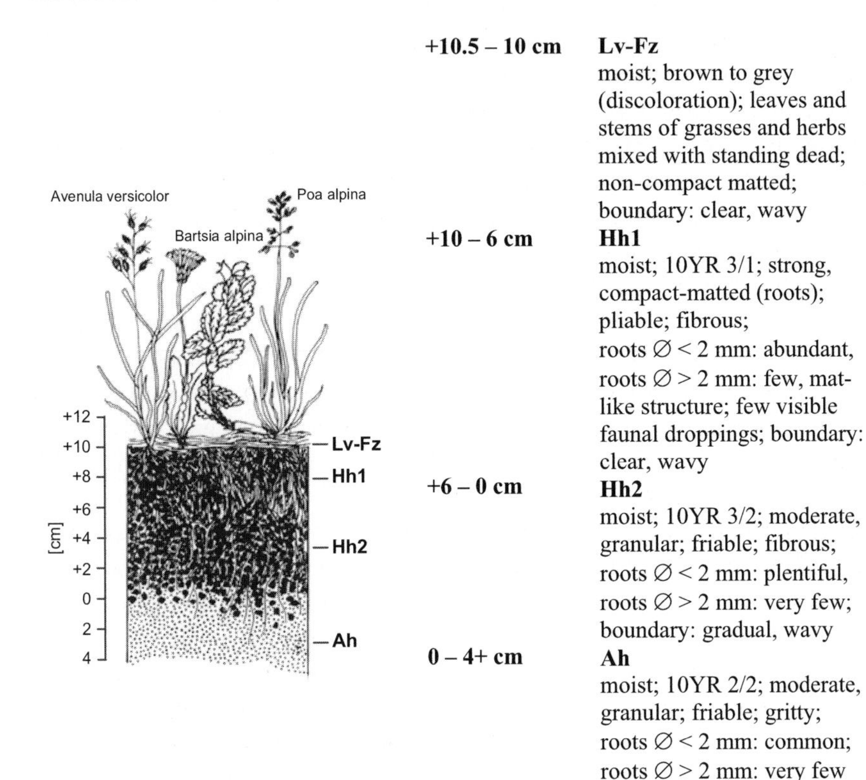

Figure 4 Humus forms in the timberline ecotone – Rhizic Leptomoder

markedly thicker. A low litter input and the translocation of litter and organic fine material with wind and water may hamper the development of thicker L and F horizons. The tradition of alpine pasturage over centuries is thought to be the main reason for the humus form morphology. After the alpine pasturing was ended in the 1950ies, the regeneration of the vegetation and reforestation started. Dwarf shrubs which had formerly retreated spread out, while favored species like *Nardus stricta* or *Gentiana punctata* were forced back. It can be assumed that the time period was too short since the alpine pastures had been abandoned to allow for the development of thicker organic layers (see Hiller 2001 for more literature).

Above the timberline the microclimatic conditions cause a vegetation pattern of *Carex curvula* swards. The humus forms are defined as Leptomoder, Mullmoder or Rhizomull (figure 2a). The morphology of these humus forms is very similar to the organic layers of grass heath sites in the timberline ecotone. As a result, the organic layers above the timberline are also characterized by L and F transitional horizons and H horizons built up mainly by finely dispersed organic matter. Compared to humus forms in the timberline ecotone, the thickness of the organic layer is slightly lower. In the alpine zone, humus form development is also influenced by microclimatic conditions and the vegetation. Low temperatures, wet conditions during snow melt and occasional drought during summer months may restrict biological activity and decomposition. However, in spite of the short growing season and the unfavorable climatic conditions above the timberline no Mor humus forms develop. In contrast to sites under dwarf shrubs, the litter produced by the alpine swards is highly decomposable. Although biological activity is high during the snow-free period, summer is too short for complete decomposition of litter. Thus, Moder humus forms are very typical under *Carex curvula* swards above the timberline (cf. Neuwinger and Czell 1959; Posch 1980; Bochter 1981; Müller 1983; Hiller et al. subm.).

In the subalpine Larch-Stone pine forest Mor humus forms are very common (figure 2c). The humus forms are characterized by L, F and H horizons. In contrast to humus forms in the timberline ecotone, it is possible to differentiate between L and F horizons. Furthermore, the thickness of the organic layers and also the thickness of L and F horizons is higher in the subalpine forest than in the timberline ecotone. Climatic conditions and litter amount as well as litter quality favor the development of Mor humus forms. At the forest study site the soil temperatures during the growing season are generally lower than in the timberline ecotone. Furthermore, the litter amount is markedly higher than in the alpine timberline ecotone and has a higher C/N value (cf. Hiller 2001).

4.2 Reforestation

As can be seen from the aerial photographs taken in 1979 when the alpine pastures were abandoned (figure 5), the tree density of larch and stone pine declined from closed forest to the alpine timberline. Almost no changes can be detected between 1979 and the actual situation in adult trees (figures 5 and 6). Nevertheless, the density of natural regeneration of larch and stone pine increased since the abandonment of the pastures (Müterthies 2002). Between timberline (2200 m a.s.l.) and tree line (2300 m a.s.l.) a dense

Figure 5 Aerial photo of the study site taken in 1979
© Aufnahme des Bundesamtes für Landestopographie

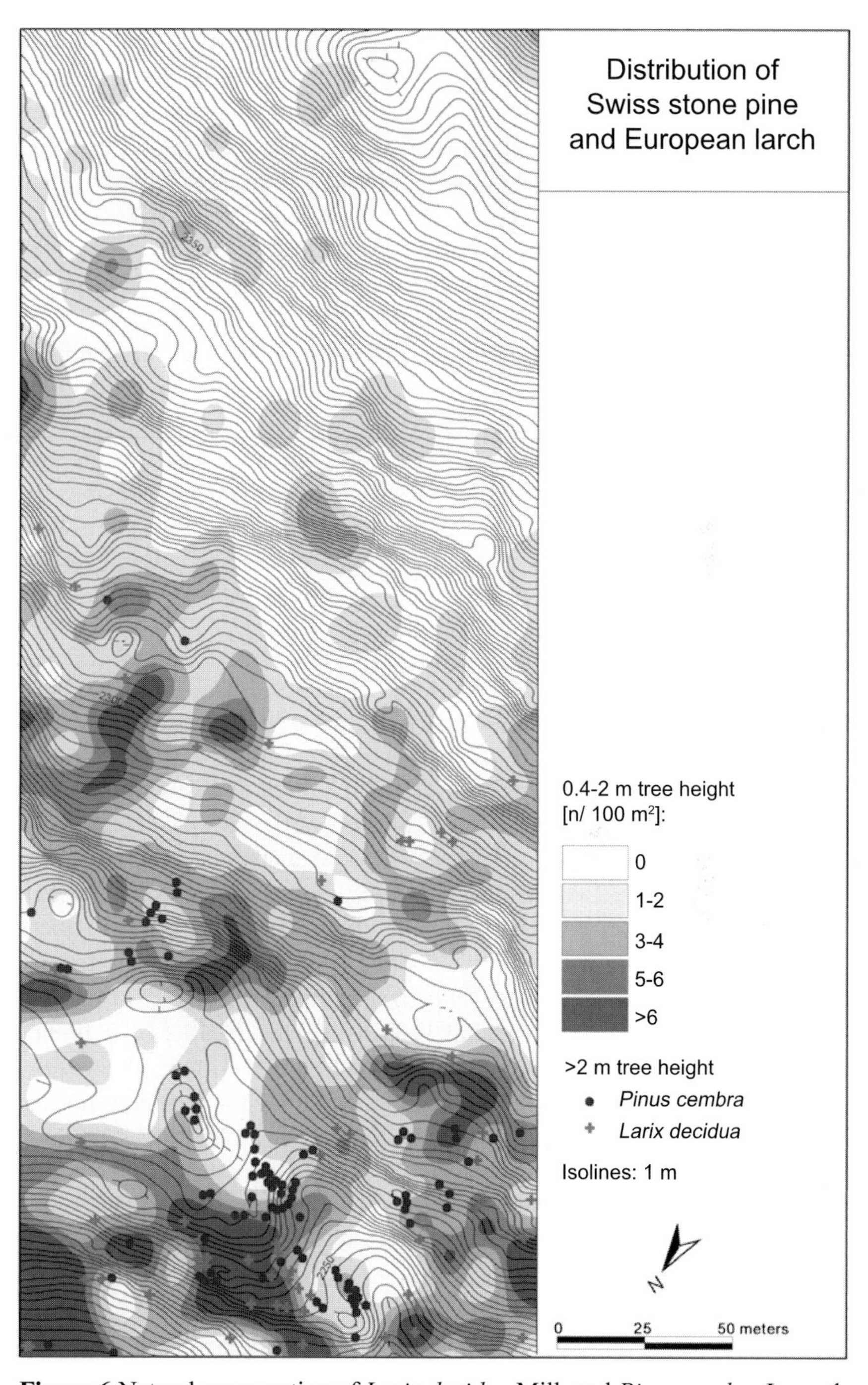

Figure 6 Natural regeneration of *Larix decidua* Mill. and *Pinus cembra* L. on the study site

regeneration primarily of stone pine occurs, whereas at higher elevations regeneration is minimal even today. The main reason why the natural regeneration of larch and stone pine at higher altitudes is restricted is the combined effect of a shorter vegetation period due to the snow cover persisting for a longer time period with an increase in frost damage (for references see Tranquillini 1979; Sakai and Larcher 1987; Holtmeier 2000, 2003) to these trees (Müterthies 2002; 2003). Within the next decades, a successful reforestation of the study site might therefore only take place up to about 2300 m a.s.l. (Müterthies 2002).

Beside the altitudinal gradient of the distribution of larch and stone pine, a strong horizontal differentiation of the natural regeneration and adult trees which is caused by the microrelief can be observed (Holtmeier 2000, 2003; Hiller et al. 2002; Müterthies 2002). On ridges and rocky outcrops the tree density of larch and stone pine is significantly higher than in depressions. In depressions both tree species suffer from the accumulation and longer presence of the snow cover. At these sites particularly Swiss stone pine is attacked by *Phacidium infestans*, a fungi which infects needles and shoots of pines that are covered by snow for too long (Roll-Hansen 1989).

4.3 Future Development of Humus Forms and Reforestation

With the progressive reforestation the timberline type is changing from an anthropogenic to a climatic timberline. The shift of the altitude of the upper timberline is therefore at first a consequence of the abandonment of the pastures, whereas the new altitudinal limit of the forest is restricted by the climatic conditions and thus approaches the potential timberline. Based on the numbers of stems per area and the growth of regenerated cembran pines and larches, the altitude of the potential timberline can be determined to be at 2300 m a.s.l. at present. Numbers of stems per area and the growth of the trees at sites at higher elevations are insufficient for a successful reforestation. Humus forms also reflect the succession of vegetation since the 1950ies. Mor humus forms are widespread at the study site and it is assumed that the percentage of Mor humus forms will increase. Due to natural regeneration of larch and pine forest at the study site, a larger amount of Mor humus forms as well as an increased thickness of the organic layer and the L and F horizons can be expected in the future.

However, an essential factor for the future timberline dynamics could be human impact. Economic aspects like agriculture and tourism as well as aesthetic demands directed at the landscape could lead to more or less intensive impacts on the natural dynamics of the upper timberline of the larch-cembran pine forest in the Upper Engadine.

5 Conclusions

After the alpine pasturing had ended in the 1950ies, the regeneration of vegetation and reforestation began. The density of naturally regenerated European larch and Swiss stone pine in the timberline ecotone shows an altitudinal differentiation, which is superimposed by the influence of the microrelief. The change of the humus forms from timberline to the alpine zone is overlaid by the influence of the microrelief in a similar way (Hiller 2001; Hiller et al. 2002). Consequently, the distribution patterns of the humus forms and of naturally regenerated larch and stone pines are similar in the timberline ecotone (Hiller et al. 2002; Müterthies 2002); e.g. at sites with intensive regeneration the humus form Tenuic Humimor prevails, whereas at sites with sparse or no regeneration at all Leptomoder or Mullmoder are often found (Hiller et al. 2002). Nevertheless, this does not mean that the similarity between the distribution of humus forms and the natural regeneration of both tree species can be attributed to a direct interference.

The altitudinal shift of the upper timberline is at first a consequence of the abandonment of the pastures, whereas the new altitudinal limit of the forest is set by the climatic conditions and thus approaches the potential timberline. Humus forms also reflect the abandonment of alpine pastures. Due to the increasing natural regeneration of larch and pine at the study site a larger amount of Mor humus forms as well as an increased thickness of the organic layer and the L and F horizons can be expected in the future.

References

AG Boden (1994) Bodenkundliche Kartieranleitung. 4. Auflage, Hannover

AK Standortskartierung (1996) Forstliche Standortsaufnahme: Begriffe, Definitionen, Einteilungen, Kennzeichnungen, Erläuterungen. 5. Auflage, München

Bal L (1970) Morphological investigation in two moder-humus profiles and the role of the soil fauna in their genesis. Geoderma 4: 5-36

Bednorz F (2000) Der Abbau der organischen Substanz im Waldgrenzökoton am Stillberg (Dischmatal/Schweiz). Arbeiten aus dem Institut für Landschaftsökologie 7. Münster

Bednorz F, Reichstein M, Broll G, Holtmeier F-K, Urfer W (2000) Humus forms in the forest-alpine tundra ecotone at Stillberg (Dischmatal, Switzerland): spatial heterogeneity and classification. Arctic, Antarctic and Alpine Research 32 (1): 21-29

Beyer L (1996) Humusformen und -typen. In: Blume HP, Felix-Henningsen P, Fischer W, Frede H-G, Horn R, Stahr K (eds) Handbuch der Bodenkunde. 1. Erg. Lieferung 12/96. Landsberg/Lech

Blaser P (1980) Der Boden als Standortfaktor bei Aufforstungen in der subalpinen Stufe (Stillberg, Davos). Mitteilungen der Eidgenössischen Anstalt für das forstliche Versuchswesen 56: 527-611

Bochter R (1981) Humus- und Bodenformen der montanen und subalpinen Stufe des Alpennationalparks Berchtesgaden. Mitteilungen der Deutschen Bodenkundlichen Gesellschaft 32: 593-598

Burns SF (1990) Alpine Spodosols: Cryaquods, Cryohumods, Cryorthods, and Placaquods above treeline. In: Kimble JM, Yeck RD (eds) Characterization, classification, and utilization of Spodozols: Proceedings of the 5th International Soil Correlation Meeting. USDA, Washington: 46-62

Bütler R, Domergue F-L (1997) Valeurs indicatrices de la végétation et des sols d'une moraine granitique de l'étage subalpin. Rev. Ecol. Alp. Grenoble 4: 1-12

Ellenberg H (1996) Vegetation Mitteleuropas mit den Alpen in ökologischer, dynamischer und historischer Sicht. Stuttgart

Gracanin Z (1972) Die Böden der Alpen. In: Ganssen R (ed) Bodengeographie. Stuttgart: 72-160

Green RN, Trowbridge RL, Klinka K (1993) Towards a taxonomic classification of humus forms. Forest Science Monograph 29: 1-49

Hiller B (2001) Humusformen im Waldgrenzökoton (Oberengadin, Schweiz). Arbeiten aus dem Institut für Landschaftsökologie 9, Münster

Hiller B, Müterthies A, Broll G, Holtmeier F-K (2002) Investigations on spatial heterogeneity of humus forms and natural regeneration of larch (*Larix decidua*) and Swiss stone pine (*Pinus cembra*) in an alpine timberline ecotone (Upper Engadine, Central Alps, Switzerland). Geographica Helvetica 57 (2): 81-90

Hiller B, Nübel A, Broll G, Holtmeier F-K Snowbeds on silicate rocks in the Upper Engadine (Central Alps, Switzerland) – Pedogenesis and interactions between soil, vegetation, and snow cover. submitted.

Holtmeier F-K (2000) Die Höhengrenze der Gebirgswälder. Arbeiten aus dem Institut für Landschaftsökologie 8, Münster

Holtmeier F-K (2003) Mountain Timberlines – Ecology, patchiness, and dynamics. Advances in Global Change Research, 14. Kluwer Academic, Dordrecht

Holtmeier F-K, Broll G (1992) The influence of tree islands on microtopography and pedoecological conditions in the forest-alpine tundra ecotone on Niwot Ridge, Colorado Front Range. Arctic and Alpine Research, 24 (3): 216-228

Keller W, Wohlgemuth T, Kuhn N, Schütz M, Wildi O (1998) Waldgesellschaften der Schweiz auf floristischer Grundlage. Mitteilungen der Eidgenössischen Forschungsanstalt für Wald, Schnee und Landschaft 73 (2), Birmensdorf: 93-357

Klinka K, Wang Q, Carter RE (1990) Relationships among humus forms, forest floor nutrient properties, and understory vegetation. Forest Science 36 (3): 564-581

Legros JP (1975) Occurrence des podzols dans l'Est du Massif Central français. Science du Sol 1: 37-49

Legros JP (1992) Soils of alpine mountains. In: Martini JP, Chesworth W (eds.) Weathering, soils and paleosols. Developments in Earth Surface Processes 2. Amsterdam

Lüscher UP (1991) Humusbildung und Humusumwandlung in Waldbeständen. Dissertation ETH Zürich Nr. 9572. Zürich

Mosimann T (1985) Untersuchungen zur Funktion subarktischer und alpiner Geoökosysteme (Finnmark (Norwegen) und Schweizer Alpen). Basler Beiträge zur Physiogeographie, Physiogeographica 7. Basel

Müller M (1983) Bodenbildung auf Silikatunterlage in der alpinen Stufe des Oberengadins. Dissertation ETH Zürich Nr. 7352. Zürich

Müterthies A (2002) Struktur und Dynamik der oberen Grenze des Lärchen-Arvenwaldes im Bereich aufgelassener Alpweiden im Oberengadin. Arbeiten aus dem Institut für Landschaftsökologie 11, Münster

Müterthies A (2003) The potential timberline: Determination with dendrochronological methods. In: Schleser G, Winiger M, Bräuning A, Gärtner H, Helle G, Jansma E, Neuwirth B, Treydte K (eds) TRACE - Tree Rings in Archeology, Climatology and Ecology, Vol. 1: 94-100

Neuwinger I, Czell A (1959) Böden in den Tiroler Zentralalpen. Mitteilungen der Forstlichen Bundesversuchsanstalt Mariabrunn 59. Wien: 371-410

Neuwinger I (1987) Bodenökologische Untersuchungen im Gebiet Obergurgler Zirbenwald Hohe Mut. In: Patzelt G (eds) MaB-Projekt Obergurgl. Veröffentlichungen des Österreichischen MaB-Programms 10. Innsbruck: 173-190

Philip GM, Watson, DF (1982) A precise method for determining contoured surfaces. Australian Petroleum Exploration Association Journal 22: 205-212

Posch A (1980) Bodenkundliche Untersuchungen im Bereich der Glocknerstraße in den Hohen Tauern. In: Franz H (ed) Untersuchungen an alpinen Böden in den Hohen Tauern 1974-1978. Stoffdynamik und Wasserhaushalt. Veröffentlichungen des Österreichischen MaB-Hochgebirgsprogramms 3. Innsbruck: 91-107

Prichard SJ, Peterson DL, Hammer RD (2000) Carbon distribution in subalpine forests and meadows of the Olympic Mountains, Washington. Soil Science Society of America, Journal 64: 1834-1845

Rehder H, Schaefer A (1978) Nutrient turnover studies in alpine ecosystems. IV. Communities of the Central Alps and comparative survey. Oecologia 34: 309-327

Robert M, Cabidoche Y-M, Berrier J (1980) Pédogenèse et minéralogie des sols de haute montagne cristalline (Etages Alpin et Subalpin) – Alpes-Pyrénées. Science du Sol 4: 313-334

Roll-Hansen F (1989) *Phacidium infestans* - a literature review. European Journal of Forest Pathology 19: 237-250

Sakai A, Larcher W (1987) Frost survival of plants - responses and adaptation to freezing stress. Ecological studies 62: 321

Staub R (1946) Geologische Karte der Berninagruppe und ihrer Umgebung im Oberengadin, Bergell, Val Malenco, Puschlav und Livigno, 1:50000, Nr. 118. Herausgegeben von der Geologischen Kommission der Schweizerischen Naturforschenden Gesellschaft. Zürich

Steltzer H, Bowman WD (1998) Differential influence of plant species on soil nitrogen transformations within moist meadow alpine tundra. Ecosystems 1: 464-474
Titlyanova AA, Romanova IP, Kosykh NP, Mironycheva-Tokareva NP (1999) Pattern and process in above-ground and below-ground components of grassland ecosystems. Journal of Vegetation Science 9: 307-320
Tranquillini W (1979) Physiological ecology of the alpine timberline, Springer, Berlin
Watson DF, Philip GM (1985) A refinement of inverse distance weighted interpolation. Geo-Processing, 2: 315-327

A Discontinuous Tree-ring Record AD 320-1994 From Dividalen, Norway: Inferences on Climate and Treeline History

Andreas Joachim Kirchhefer

Abstract

Tree-ring widths of living and dead Scots pines (*Pinus sylvestris* L.) from Dividalen, intra-alpine northern Norway, were combined to two partial chronologies, AD 320-1167 and 1220-1994. Abundant pine remains indicate good growing conditions for Scots pine during the Viking Age and early Medieval times, whereas few logs are preserved from the period 1000-1450. A cold period around 1130 probably triggered massive pine mortality at the forest line and the depression of the pine tree-line by ca. 50 m. The forest-line stands recovered first after a second cooling event around 1457. July temperatures were reconstructed for the time windows 587-980 and 1507-1993. Shorter spells of warm summers occurred around 607, 728, 1565, and 1762, and longer warm periods during the Viking Ages (ca. 819-957) and the 20th century since 1915. Cold summers prevailed ca. 749, 769-818, ca. 866, 1573-1624, ca. 1645, 1785-1826, ca. 1842 and 1864-1914.

1 Introduction

Previous dendrochronological investigations in northern Norway resulted in tree-ring chronologies of Scots pine, *Pinus sylvestris* L., with total lengths of 120 to 1123 years (Ording 1941; Schweingruber 1985; Ruden 1987; Thun and Vorren 1996; Kirchhefer 1999; 2000; 2001). The ring-width variability in the region is relatively homogeneous and principally reflects July-temperature variations. In the intra-alpine valleys, however, the decadal growth pattern during the mid-20th century climatic optimum deviated from the July-temperature trends (Kirchhefer 2001). As triggering factors, stress due to oceanic winters and summer drought were proposed. This particular growth pattern is developed most strongly in Dividalen. Coincidentally, this valley contains abundant pine remains preserved on dry forest soils. A radiocarbon date (G. Jacoby, pers. comm.) suggested that some of this material could be up to a thousand years old. The initial pine chronologies for northern Finland and Lake Torneträsk, northern Sweden, were also based on such pine remains (Sirén 1961; Bartholin 1983). Thus, sampling in Dividalen was extended to stumps and logs in order to explore the possibility for a millennial chronology for the valley. This article presents the preliminary tree-ring and palaeoclimate record, based on samples taken until 1998.

2 Study Sites

The valley of Dividalen belongs to the Målselv river catchment (figure 1). Running from south to north, it is sheltered by mountains of up to 1717 m in the west and 1503 m in the east. With an annual precipitation sum of 282 mm during 1961-1990, Dividalen is one of the driest localities in Norway (Moen et al. 1999). Climate is sub-continental with July temperatures of 12.7°C, January temperatures of -9.4°C and a July precipitation maximum (57 mm). For comparison, the nearest meteorological stations Øverbygd and Bardufoss (figure 1) are more oceanic, but differ mostly in the precipitation characteristic, with slightly above 650 mm yr^{-1} and a broad autumn maximum.

The study area (68°50'N, 19°38'E) stretches over 8 km along the west-facing shoulder between the main valley and a hanging valley below Lake Devdisvatn to the southwest-facing slope of Mt. Skrubben (848 m a.s.l.). Soils generally are Podzols developed on glacial tills, but outcrops of basic and acidic bedrock occur. Ground vegetation is characterised by lichens, *Arctostaphylos uva-ursi* and *Vaccinium vitis-idaea* on the driest sites and by *V. myrtillus*, *V. uliginosum* and herbs on mesic sites. Trees growing closer than 10 m from open water and mires were avoided in order to reduce diversity in site conditions and related climate responses. Also, on mesic sites, wood is more exposed to decay.

According to Moen et al. (1999) the pine forests of Dividalen belong to the middle boreal zone, whereas the sub-alpine birch forests belong to the northern boreal zone. Favoured by intra-alpine climate, the tree line is relatively high. The pine forest ascends to about 390 m, and scattered pine trees grow up to ca. 500 m a.s.l. The alpine tree line is formed by birch (*Betula pubescens* Ehrh.) and is situated around 600 m a.s.l. An indicator of the local climatic continentality is the occurrence of 50 cm tall, stunted individuals of Scots pine at up to 650 m a.s.l. on the south-facing slope of Mt. Skrubben. Macrofossils reveal that 9000 years ago pines and birch grew on top of Mt. Skrubben, and retreated to their present position around 4500-4000 cal. BP (Vorren et al. 2000; Jensen et al. 2002).

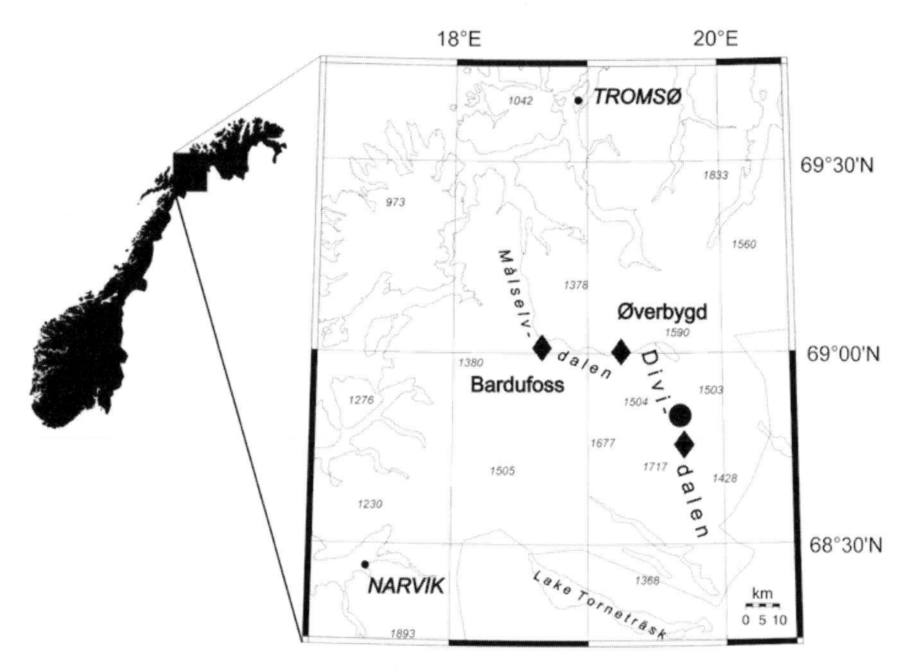

Figure 1 Map of Troms county with field site in Dividalen (dot) and relevant climate stations (diamonds). The elevations of selected mountains are indicated

3 Material and Methods

Samples were taken from 42 living Scots pines and 54 pine remains in dry, open-canopy habitats between 460 and 300 m a.s.l. Pine remains were mainly lying trunks, but also seven standing dead trees, a couple of stumps, and a few pieces of wood, here termed coarse woody debris. Cores were extracted from standing wood. From most other pine remains, discs were cut. Sampling height was 0.5-1.2 m above ground. Initially, 20 living trees were cored on glacial till at the south-west-facing slope of Mt. Skrubben. In the search for old trees, the study area was extended along the west-facing slope of Mt. Devdisfjellet southwards to the stream Kvernelva. Here, sampling finally concentrated on rock outcrops at 390 m a.s.l north of the stream Sleppelva, where abundant pine remains are preserved.

The surface of cores and stem discs were prepared with razor blades or by sanding, respectively. After discarding several samples due to low numbers of rings or irregular growth, ring widths were measured to the nearest 0.001 mm. Tree-ring series were cross-dated by COFECHA (Holmes et al. 1986) and visual compared with the samples, ring width series from other trees and with master chronologies. A mean ring-width chronology from the nearby Lake Torneträsk AD 441-1980 served as a reference series for cross-dating (Bartholin 1983; Grudd pers. comm. 1996). The dates of the inner- and outermost rings were recorded in order to gain a rough picture of past population dynamics. When the pith was lacking from the sample, the number of missing inner rings was estimated geometrically. A hundred rings were added when no realistic estimate could be given, for instance, in case of extensive heart rot. Analogously, a hundred rings were added to the date of the outermost ring when the entire sapwood was eroded from the surface of older specimens.

For detrending and standardisation of the ring-width series, negative exponential or non-ascending straight lines were fitted to the raw data of each radius (Holmes et al. 1986; Kirchhefer 2000). The climate-growth relationship was assessed by bootstrap multiple regression on tree-ring indices and climate data for 13 months from previous to present August (Till and Guiot 1990). This procedure involved principal component analysis (PCA) on the climate parameters, which were monthly mean temperatures and precipitation sums. Climate data were available from the nearby meteorological station Dividalen, located close to the valley bottom at 228 m a.s.l. and about 6 km (min. 2.0 to max. 10.5 km) south-southeast of the sampled trees. Missing climate data between June 1977 and January 1980 were interpolated by means of data from Øverbygd (precipitation only) and Bardufoss (figure 1). Based on the well-replicated parts of the tree-ring chronology, the Dividalen record of July temperatures was extended back in time using a bootstrapped transfer function (Till and Guiot 1990).

4 Results

The oldest living pines were more than 550 years old (figure 2). The pith of one individual 80 cm above ground was dated to 1453. The innermost ring of another pine at 1 m above ground was dated to 1494 (photo 1). Adding 67 rings to account for heart rot made this pine more than 575 years old. The piths of 21 of the total 93 trees were dated or estimated to between 1470 and 1535 (figure 2). Half of those pines were alive. The youngest individuals germinated around 1800. The chronology based on living trees extends back to 1454. Rings from eight pine remains prolonged the recent part of the Dividalen chronology back to 1220.

A total of 30 pine remains originate from the first millennium AD (photo 2). Their rings combined to form a chronology covering the years 320-1167. The innermost rings, 320-339, however, are not replicated and their

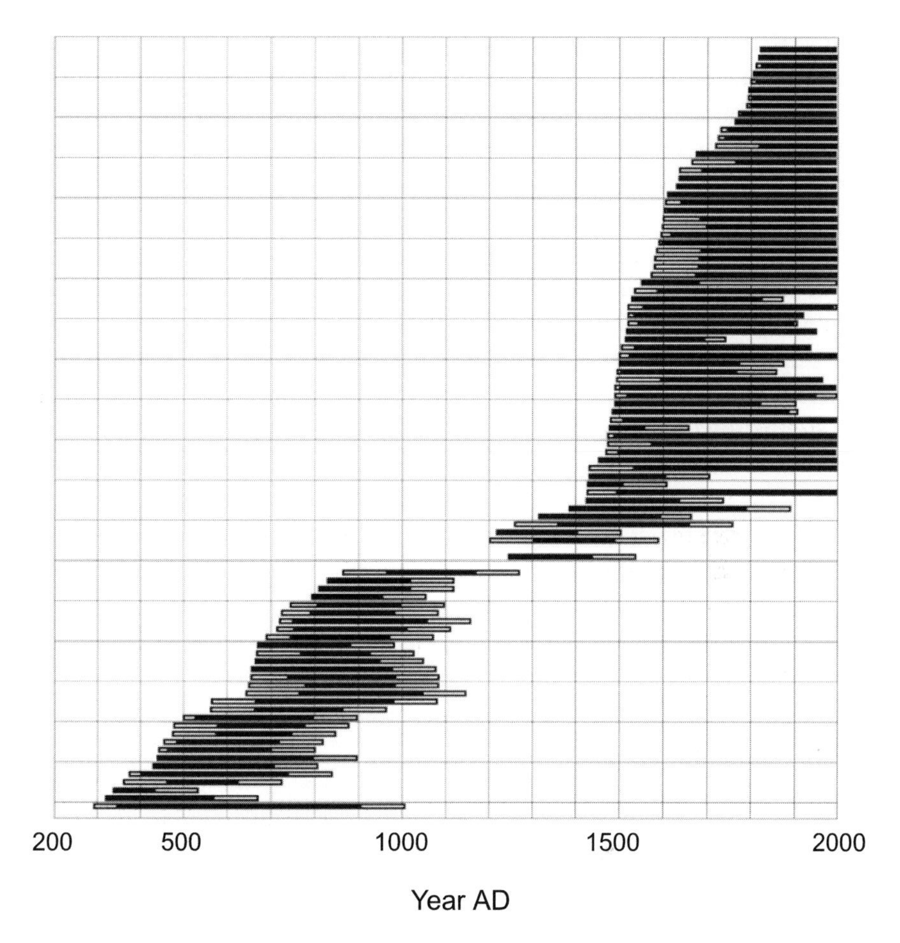

Figure 2 Life spans of pines (bars) as indicated by sampled (black) and estimated rings (white) at sample height. The log AD 1245-1536 is a subfossil pine from a lakeshore and is not included in the chronology

dating thus awaits verification. The longest documented life span was 558 years in a pine growing from at least AD 347-904. It probably germinated in the late 3rd century and died after the year 1000. The outermost rings of 11 pines dated to 950-1020, and the period 1056-1168 is represented by one pine only. The exact dates of these rings were determined by crossdating with the Torneträsk chronology.

The ring widths measured on the samples form two partial chronologies spanning 848 (AD 320-1167) and 775 years (AD 1220-1994), respectively, leaving a gap of 52 years (figure 3). The maximum internal replication is 27 radii from 17 trees in the early period and 68 radii from 38 trees in the late period, respectively. According to the EPS statistics (Expressed Population

Signal; Briffa and Jones 1990) for the period 1835-1991, a minimum of nine trees is required to represent 85 % of the tree-ring variability of the theoretical pine population. This requirement is met during 587-980 and 1507-1993 (figure 2). In these time windows, prolonged periods of poor growth occurred during 769-818, 1571-1620, 1786-1822, and 1865-1913. The narrowest rings were observed in the years 803/04, 1601, 1641/42/45, 1680, and 1837. Outside the EPS-85 % windows, there are noteworthy growth reductions around 542 and 1457. Good growth conditions of considerable duration prevailed during 587-627 (since 561 with EPS < 85 %), 872-946 and since 1914. Shorter undisrupted pulses of good growth occurred around 725, 1551-70, 1649-66, 1751-67, and 1943-62.

Radial growth was determined principally by July temperatures alone (figure 4). The correlation coefficient was r = 0.41 for the standard chronology and r = 0.58 for the residual chronology. Also late winter temperatures influenced growth; significantly in March, but just below the 95 % confidence

Photo 1 A solitary, probably more than 575 year old Scots pine in the subalpine birch forest. The innermost extracted ring dates to AD 1494, whereas another 10 cm of the central stem could not be sampled because of wood decay

Photo 2 Hollow stump of a fallen pine on shallow soil, revealing rings from AD 779-982. The pith date at 40 cm above ground was estimated to AD 650

limit in April. In contrast, precipitation was not related to growth. The multiple regression analysis revealed also that the first, third and fourth preceding rings significantly influence the current year's ring.

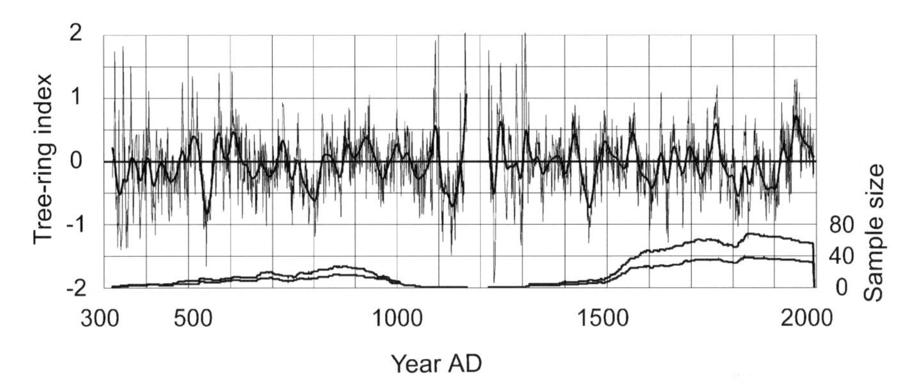

Figure 3 Tree-ring standard chronology: Annual and low-pass filtered variability of radial growth after detrending. The number of trees and radii is indicated in the lower part of the figure. Extreme ring-width indices in the centre of the figure are not shown

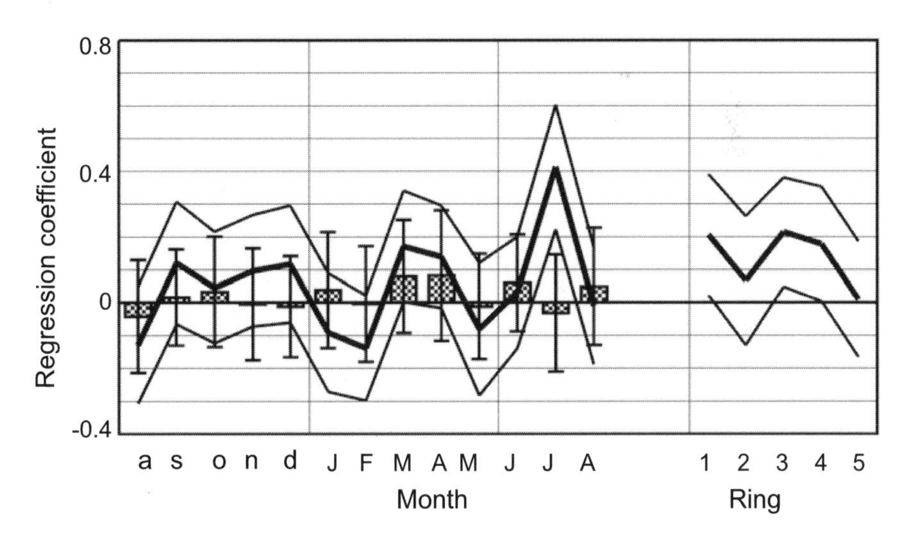

Figure 4 Response function for the standard ring-width chronology vs. Dividalen climate 1922-1992. Multiple bootstrap regression coefficients for temperature (lines) and precipitation (bars) with two standard deviations. Months from previous year's (a-d) August to current year's (J-A) August. 1-5: ring widths of the previous years

Based on the response functions, July temperature was chosen as the climate variable to reconstruct. The transfer function was calculated as

$$Temp_{(July)} = 5.564\ rw_{(t)} - 3.670\ rw_{(t+1)} + 11.235\ ,$$

where rw is the ring-width index of the current year (t) and the following year (t+1). The bootstrap multiple regression yielded correlation coefficients and standard deviations of 0.56 ± 0.07 (calibration) and 0.50 ± 0.14 (verification). Visual comparison of the 72 observed and reconstructed July temperatures 1921-1992 revealed good agreement in the year-to-year temperature fluctuations (figure 5). This was confirmed by the following statistics which were significant at the 95 % confidence level: correlation coefficient ($r = 0.54$), Reduction of error ($RE = 0.29$), t-value ($t = 3.9$), sign-test (47 positive products) and test of first differences (54 correct cases). However, the explained variance R^2_{adj} is only 28 %. High temperatures before 1949 (such as 1922, 1925, 1935 and 1937) were underestimated and cool summers after 1948 (such as 1949, 1951, 1962, 1964, 1965, 1968 and 1975) were overestimated. Thus the standard deviation of the estimated temperatures is far lower (1.13°C) than that of the observed temperature record (2.08°C).

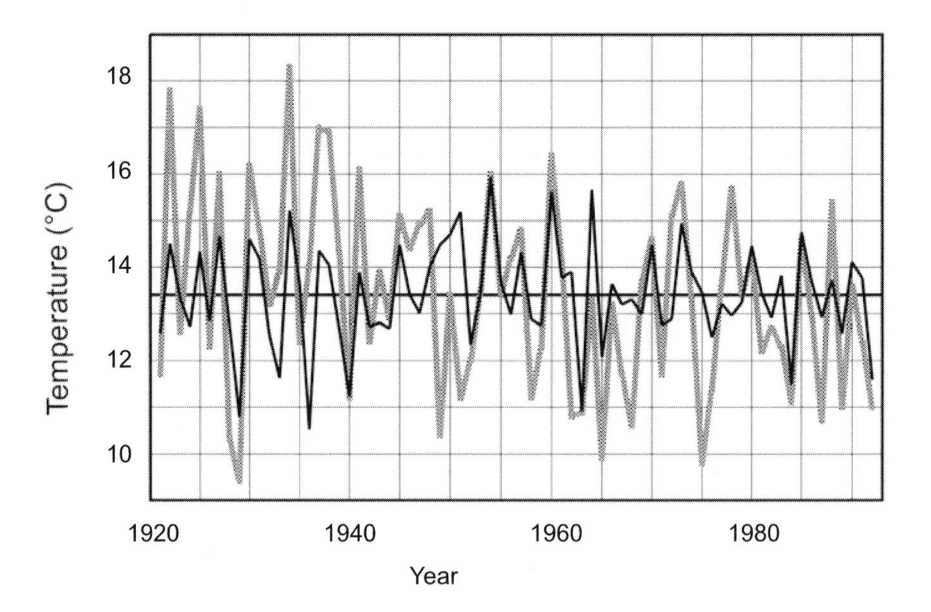

Figure 5 Observed (grey) versus reconstructed (black) July temperatures for Dividalen 1922-1992

Table 1 The warmest and coolest mean July temperatures for single years, 10- and 30-year periods as reconstructed for the Dividalen climate station. All single-year values differ by at least two standard deviations from the mean of 13.1°C (corresponding 10- and 30-year values underlined)

cool years	°C	warm years	°C
685	10.0	604	15.9
748	11.0	627	16.3
824	10.9	676	15.4
829	11.0	714	15.4
886	10.8	823	15.4
926	10.8	845	15.5
1574	10.7	885	16.1
1642	10.5	887	15.9
1645	11.0	925	16.3
1822	10.5	931	15.4
1825	11.1	934	16.4
1837	10.9	1629	16.0
1900	10.8	1694	15.4
1929	10.8	1739	15.5
1936	10.5	1799	15.4
1963	10.9	1954	16.0
		1960	15.6
		1964	15.7

cold 10-year periods	°C	warm 10-year periods	°C
788-797	12.5	604-613	13.9
803-812	12.3	725-734	13.7
1601-1610	12.4	876-885	13.7
1641-1650	12.3	925-934	13.7
1806-1815	12.4	1561-1570	13.7
1837-1846	12.4	1758-1767	14.0
1880-1889	12.4	1918-1927	13.7
1903-1912	12.4	1948-1957	14.1

cold 30-year periods	°C	warm 30-year periods	°C
640-679	12.9	600-629	13.5
788-817	12.6	714-743	13.4
1591-1620	12.8	917-946	13.5
1793-1822	12.7	1548-1577	13.4
1878-1907	12.7	1739-1768	13.6
		1945-1974	13.7

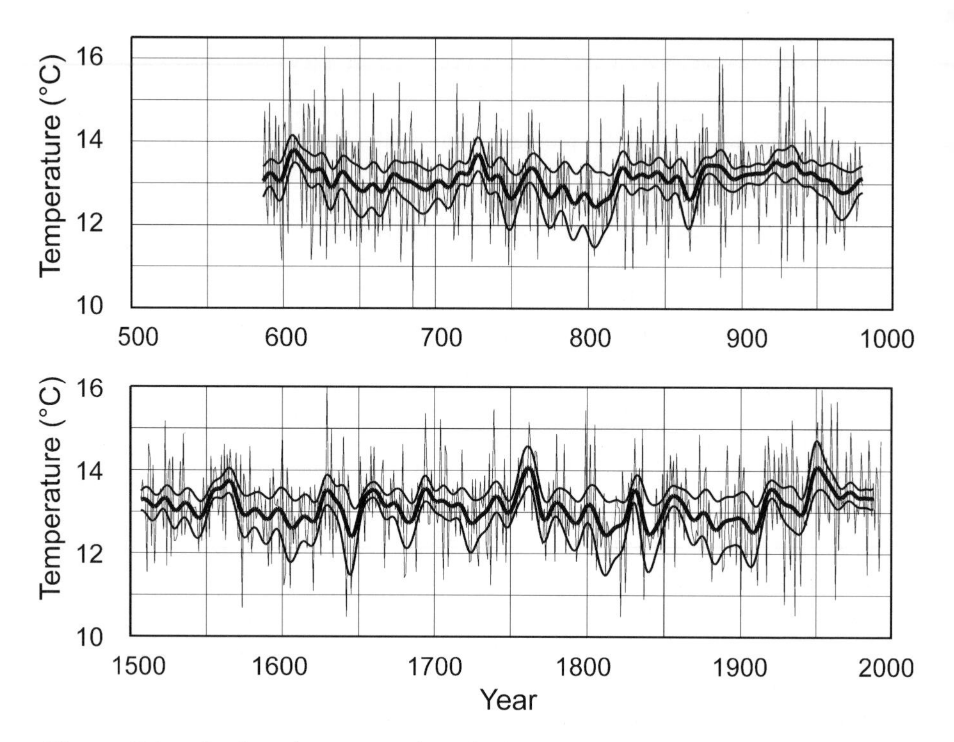

Figure 6 Tree-ring based reconstruction of July temperatures 587-980 and 1507-1993 for the Dividalen climate station, expressing climate fluctuations on time-scales of one year to about one century. Low-pass filtered series (10-year Gaussian filter) of the annual values and their 95% confidence limits are shown

The long-term mean of the July temperature reconstruction is 13.1°C (figure 6, table 1). The coldest and warmest summers were 685 (10.0°C) and 934 (16.4°C), respectively. For 10-year means, corresponding values were 1641-50 (12.3°C) and 1948-57 (14.1°C). The coldest and warmest 30-year periods were 788-817 (12.6°C) and 1945-74 (13.6°C).

5 Discussion

This investigation confirmed the impression that the pine tree-line forests in this part of Dividalen have a long continuity. The frequency and longevity of pine remains reflect both the cool-dry conditions and the low degree of disturbance by humans and fire. Both living and dead pine individuals show ages in the range of the oldest living pines elsewhere in Fennoscandia, which reach ages of 700-810 years (Sirén 1961; Engelmark and Hofgaard 1985).

The observed climate-growth response reflects the general limitation of tree growth by temperatures in the short summer season at high latitudes. In March and April, a positive response to temperature and the slightly positive, although non-significant relationship to precipitation might imply frost drought as a secondary growth-limiting factor. In contrast to many other response functions from northern Fennoscandia, July precipitation does not seem to affect pine growth. This probably is related to local climate. Whereas in the regional climate data, July temperatures and precipitation are negatively correlated ($r = -0.21$), they are weakly positively correlated in Dividalen ($r = 0.11$). Thus the response function might reflect that, on average, the growth-reducing effect of July precipitation (cooling) is outweighed or even exceeded by its growth enhancing effect, through preventing drought stress in warm summers. Signals of drought stress in Scots pine at dry sites in northernmost Norway have been shown previously (Kirchhefer 1998).

Given the dominant influence of July temperatures on radial growth, it is surprising that the explained variance of the July temperature reconstruction is only 28 %. Probable reasons are a) deviations between decadal trends of ring widths and July temperatures and b) the underestimation of temperature extremes. During the temperature optimum of the 1930ies, growth was retarded, and the growth optimum is delayed until around 1950. Mild winters in the 1930ies and increased spring precipitation have been proposed as candidate factors (Kirchhefer 1999). Furthermore, the relatively short calibration period of 72 years, as well as site-specific growth patterns, certainly contribute to the low explained variance. A higher explained variance (38 %) was achieved by combining several site chronologies from the valleys Dividalen and Målselvdalen with the longer regional climate series for northern Norway 1896-1992 (Kirchhefer 2000). Furthermore, when interpreting the presented dendroclimatic reconstruction, it must be stressed that its low-frequency variability is restricted to about one century. Because the chronology is composed of, on average, 235-year long individual ring-width series, long-term fluctuations longer than about one third of this length probably are removed in the process of tree-ring detrending (Cook et al. 1995). Therefore, the absolute values of the temperature series as reconstructed for the Dividalen meteorological station, particularly in terms of extreme temperatures and long-term trends, should not be taken literally. In spite of these limitations, the presented tree-ring data provide a wealth of information on past temperatures and forest-line dynamics.

The abundance of pine remains implies good conditions for forest growth from the Dark Ages to early Medieval times (figure 2). Although quantification is difficult due to sampling design and wood decay, one can assume that the altitude and density of the forest line stands were comparable to todays. Between AD 600 and 1000, solitary pines grew at about 460 m a.s.l., close to recent individuals, implying that the early Medieval pine-tree line was situated at the same elevation as today, or even higher. The early 7th century was the third warmest of the whole reconstruction, with summer temperatures of 0.4-0.9°C above average (figure 6, table 1). After another temperature peak around 728, a cooling trend led towards the second coolest decade of the reconstruction, 803-812 (-0.8°C). Thereafter, warm summers prevailed until ca. 950, only interrupted by a cool event in the 860ies. In terms of pine growth and distribution, the first part of the Medieval Warm Period thus resembles the 20th century after 1915. Further analysis of the tree-ring data and comparisons with other palaeoclimate data are necessary to reveal if the Medieval Warm period in Dividalen was warmer than the 20th century or not.

In spite of good growing conditions pine recruitment apparently came to a halt around 900, and consecutively remarkably few pine remains cover the period 1000-1450 (figure 2). This period conventionally would be assigned to the Medieval Warm Period (9th-14th century; Hughes and Diaz 1994), so the unfavourable conditions for pine growth in Dividalen are surprising. Apparently at some point after AD 1000, growth conditions changed dramatically and caused massive pine mortality at the pine forest line. The gap in the material is seen throughout all investigated sites, i.e. from the tree line at 460 m down to the closed forest at 320 m a.s.l. Considering that the outermost ring of 11 pine remains date to 950-1020, and assuming that between 100 and 200 rings have been eroded from the stems' surface, it is likely that this retreat took place in the early 12th century. The only tree preserved from this time grew slowly after 1100, particularly during the 1130ies (figure 3). This event is also prominent in the chronologies from Torneträsk and northern Finland (Lindholm and Eronen 2000; Grudd et al. 2002). Strong reductions in sample size around 1200 are also seen at Torneträsk and in central Sweden (Gunnarson and Linderholm 2002). However, at the Swedish localities, the pine forests had recovered by the 14th century, whereas the impact in Dividalen lasted much longer. In northern Finland and at the Norwegian coast (Kirchhefer 2001) on the other hand, forest regeneration was apparently continuous. Thus, the maximum impact occurred in the Scandes.

According to Sirén (1961), pine regenerated successfully in northern Finland, e.g., in 1140-50, 1200-1210, 1285-1300 and 1310-1325. Assuming that the same applied for Dividalen, the cold event around 1460 (figure 3) would be a strong candidate for eliminating a young, up to 200-year old pine forest. This cooling event is documented from Fennoscandia and large areas of the Northern Hemisphere (Eronen et al. 2002; Esper et al. 2002; Grudd et al. 2002; Gunnarson and Linderholm 2002). Trunks of trees that obviously died at a young age are present at the site, but only a few of those were sampled. Most of these were not dateable due to few rings and juvenile wood. Others might have decayed early because of the high portion of sapwood and juvenile wood. Radiocarbon dates and a continuous, local master chronology based on regularly grown, mature trees might help to place such samples in time, and show that there was pine recruitment in the 250 years before the proposed cold events of 1460 and, analogously, 1130.

Although the long gap in the pine material most probably is related to low summer temperatures, in other, continuous dendroclimatic reconstructions this period does not appear colder than the Little Ice Age (Lindholm and Eronen 2000; Grudd et al. 2002; Helama et al. 2002). This is supported by the fact that around 1200, two pines still grew at ca. 420 m a.s.l. in the present birch forest with only scattered pines, about 1 km east of the present pine forests. Interestingly, the habitat differs significantly from the dry rock outcrops where much of the older material was collected. One individual probably survived until at least 1268 not far from a willow mire, and the other germinated ca. 1220 on a peaty lakeshore (therefore it was not included in the chronology). The two pines that germinated in the 13th century also grew in slightly mesic habitats characterised by *V. myrtillus* and mosses. Therefore, drought potentially also played a role in delaying the re-advance of the pine forest. Fire scars in logs, which could indicate drought conditions and provide dates of forest fires, however, have not been detected yet. So far, scars have been detected on one stump only, caused by mechanical stress exerted to a sapling of ca. 3 cm diameter during 740-850. Charred logs are observed frequently, but cannot be dated by dendrochronology and might just be related to bonfires.

Immediately after the cooling event around 1460, a regeneration pulse of pine initiated a denser pine forest and a re-advance of the pine forest-line (figure 2). A similar increase in the number of trees is seen at the coast (Kirchhefer 2001). This also corresponds with a period of rapid tree-ring growth in Dividalen, at the coast of northern Norway, and elsewhere in northern Fennoscandia (figure 3). July temperatures during the 30-year period 1548-1577 were 13.4°C, which is 0.3°C above the reconstruction mean, and the 10-year mean 1561-79 was even 0.6°C above average (figure 6, table 1).

Fewer pines originate from ca. 1600-1750. This is referred to as the coldest period of the Little Ice Age (Bradley and Jones 1993). In Dividalen, the years 1570-1620 were coolest, whereas temperatures after 1650 were undulating around the average (figure 6). The coldest spells occurred around 1605, 1645, 1680 and 1720, interrupted by warm intervals around 1630, 1660 and 1695.

On average, this early part of the Little Ice Age appears to be warmer than the later part, ca. 1770-1910. This could, however, be an artefact of the standardisation technique and the dominance of old trees in the chronology.

A remarkable growth and presumably temperature maximum occurred around 1760. This probably triggered another regeneration pulse that is represented by those eight pines with innermost rings around 1800. Because sampling concentrated on old trees, this 200-250 year cohort presumably is underrepresented. Simultaneously in the 1770-1780ies, increased regeneration was observed in northern Sweden (Zackrisson et al. 1995).

Comparisons with the chronologies from coastal northern Norway (Kirchhefer 2001), northern Finland (Eronen et al. 2002) and Torneträsk (Grudd et al. 2002) reveal coast-inland differences as well as growth patterns constrained to the Scandes (i.e., Dividalen and Torneträsk) or just Dividalen. Of the conspicuous growth peaks in Dividalen about 1550, 1750, and 1950, for instance, none is found at the coast, and only the 1750ies' maximum shows in Finland and significantly strongly at Torneträsk. A pattern occurring only in the Scandes is the early onset (ca. 1570) of the early 17th century's cold interval. In the first millennium AD, contrasting increment rates between the Scandes and northern Finland were observed in the early 7th century (Scandes high, Finland low) and in the late 8th century (vice versa). Thus, growth patterns of Scots pine restricted to the intra-alpine valleys of the northern Scandes occurred also before the 20th century. These may be caused by strong regional gradients in summer temperature, related to anomalies in atmospheric circulation above northern Fennoscandia, or in sea surface temperatures. However, particularly during warm periods such as the 20th century, non-linear climate responses to secondary growth controlling factors certainly will represent a challenge for future climate reconstructions based on this material.

6 Conclusions

1. In Dividalen, abundant remains of Scots pine (*Pinus sylvestris* L.) from the Dark Ages are preserved. Many living trees are 500 years old. The elevation of the pine tree- and forest-line during the Dark Ages apparently was similar to the present situation.
2. Few logs are preserved from the period 1000-1450, indicating poor conditions for pine growth and regeneration during the later part of the Medieval Warm Period. Cooling events around 1130 and 1460 probably

triggered a tree-line depression of at least 50 m and its late recovery, respectively.
3. The tree-ring record for Dividalen now covers the periods AD 320-1167 and 1220-1994.
4. July temperatures were reconstructed for the time windows 587-980 and 1507-1993. The low explained variance (R^2_{adj} = 28 %) partly is due to deviations in decadal trends of July temperatures and ring widths in the calibration period 1922-1993.
5. Shorter spells of warm summers occurred around 607, 728, 1565 and 1762, and longer warm periods during the Viking Ages (ca. 819-957) and since 1915. Cold summers prevailed ca. 749, 769-818, ca. 866, 1573-1624, ca. 1645, 1785-1826, ca. 1842 and 1864-1914.

Acknowledgements

This study was financed by the Norwegian Research Council grants 107733/720 and 148791/720. Among field assistants I'd like to thank particularly Annette Furnes, University of Bergen, and Dr. Elisabeth Cooper, University Centre on Svalbard. Danny McCarroll, University of Wales Swansea, made valuable comments on the manuscript.

References

Bartholin TS (1983) Dendrokronologi i Lapland AD 436-1981. Meddelanden från Det Dendrokronologiska Sällskapet 5: 3-16

Bradley RS, Jones PD (1993) 'Little Ice Age' summer temperature variations: their nature and relevance to recent global warming trends. The Holocene 3(4): 367-376

Briffa KR, Jones PD (1990) Basic chronology statistics and assessment. In: Cook ER, Kairiukstis LA (eds) Methods of Dendrochronology: applications in the environmental science. Kluwer Academic Publishers, Dordrecht, pp 137-52

Cook ER, Briffa KR, Meko DM, Graybill DA, Funkhouser G (1995) The 'segment length course' in long tree-ring chronology development for palaeoclimatic studies. The Holocene 5(2): 229-237

Engelmark O, Hofgaard A (1985) Sveriges äldsta tall (The oldest Scots pine, *Pinus sylvestris*, in Sweden). Svensk Botanisk Tidskrift 79: 415-416

Eronen M, Zetterberg P, Briffa KR, Lindholm M, Meriläinen J, Timonen M (2002) The supra-long Scots pine tree-ring record for Finnish-Lapland – Part 1: chronology construction and initial inferences. The Holocene 12(6): 673-680

Esper J, Cook ER, Schweingruber FH (2002) Low-frequency signals in long tree-ring chronologies for reconstructing past temperature variability. Science 295: 2250-2253

Grudd H, Briffa K R, Karlén W, Bartholin TS, Jones PD, Kromer B (2002) A 7400-year tree-ring chronology in northern Swedish Lapland: natural climatic variability expressed on annual to millennial timescales. The Holocene 12(6): 643-65

Gunnarson BE, Linderholm HW (2002) Low-frequency summer temperature variation in central Sweden since the tenth century inferred from tree rings. The Holocene 12(6): 667-671

Helama S, Lindholm M, Timonen M, Merilaïnen J, Eronen M (2002) The supra-long Scots pine tee-ring record for Finnish Lapland: Part 2, interannual to centennial variability in summer temperatures for 7500 years. The Holocene 12(6): 681-687

Holmes RL, Adams RK, Fritts HC (1986) Tree-ring chronologies of western North America: California, eastern Oregon and northern Great Basin, with procedures used in the chronology development work, including users manuals for computer programs COFECHA and ARSTAN. VI, Laboratory of Tree-Ring Research, University of Arizona, Tucson

Hughes MK, Diaz HF (eds) (1994) The Medieval Warm Period. Climatic Change 26(2-3), pp 342

Jensen C, Kuiper JGJ, Vorren K-D (2002) First post-glacial establishment of forest trees: early Holocene vegetation, mollusc settlement and climatic dynamics in central Troms, North Norway. Boreas 31: 285-301

Kirchhefer AJ (1998) Climate and site effects on tree-rings of *Pinus sylvestris* L. in northernmost Norway – an exploratory pointer-year study. In: Griffin K, Selsing L (eds) Dendrokronologi i Norge. AmS-Varia 32: 15-28

Kirchhefer AJ (1999) Dendroclimatology on Scots pine (*Pinus sylvestris* L.) in northern Norway. Dr. scient.-thesis, University of Tromsø, Norway

Kirchhefer AJ (2000) The influence of slope aspect on radial increment of *Pinus sylvestris* L. in northern Norway and its implications for climate reconstruction. Dendrochronologia 18: 27-40.

Kirchhefer AJ (2001) Reconstruction of summer temperature from tree rings of Scots pine, *Pinus sylvestris* L., in coastal northern Norway. The Holocene 11(1): 41-52

Lindholm M, Eronen M (2000) A reconstruction of mid-summer temperatures from ring-widths of Scots pine since AD 50 in northern Fennoscandia. Geografiska Annaler 82A (4): 527-535

Moen A, Lillethun A, Odland A (1999) Vegetation. In: Lillethun A (ed) National Atlas of Norway. Hønefoss, Norwegian Mapping Authority

Ording A (1941) Årringaanlyser på gran og furu. Meddelelser fra Det norske skogforsøksvesen VII(25): 101-354

Ruden T (1987) Hva furuårringer fra Forfjorddalen kan fortelle om klimaet i Vesterålen 1700-1850. Norsk institutt for skogforskning, NISK rapport 4/87

Schweingruber FH (1985) Dendro-ecological zones in the coniferous forests of Europe. Dendrochronologia 3: 67-75

Sirén G (1961) Skogsgränstallen som indikator för klimatfluktuationerna i norra Fennoskandien under historisk tid. Communicationes Instituti Forestalis Fenniae 54(2): 1-66

Thun T, Vorren K-D (1996) Short dendroseries from northern Norway reflect oceanic and subcontinental climates. In: Frenzel B, Birks HH, Alm T, Vorren K-D (eds) Holocene treeline oscillations, dendrochronology and palaeoclimate. Paläoklimaforschung 20: 120-126

Till C, Guiot J (1990) Reconstruction of precipitation in Morocco since 1100 A.D. based on *Cedrus atlantica* tree-ring widths. Quaternary Research 33: 337-351

Vorren K-D, Jensen C, Kirchhefer AJ, Alm T, Mørkved B, Karlsen SR, Kunzendorf H (2000) Regional Report 1. Northern Sub-oceanic. In: Hicks S, Jalkanen R, Aalto T, McCarroll D, Gagen M, Pettigrew E (eds) FOREST (Forest Response to Environmental Stress at Timberlines). Sensitivity of northern, alpine and mediterranean forest limits to climate – Final Report. EU-contract ENV4-CT95-0063. Oulu, University of Oulu, pp 11-34

Zackrisson O, Nilsson M-C, Steiljen I, Hörnberg G (1995) Regeneration pulses and climate-vegetation interactions in nonpyrogenic boreal Scots pine stands. Journal of Ecology 83: 469-483

Woodland Recolonisation and Postagricultural Development in Italy

Pietro Piussi

Abstract

The principal change in Italy's rural landscape over the past fifty years has been the spontaneous return of woodlands on land previously used for agriculture and cattle-raising. This secondary succession, together with re-afforestation projects carried out by man, is an ongoing process. The spontaneous recolonisation by woodland takes place in many areas, from agricultural land by the sea, in the Mediterranean environment, to the high altitude pastures of the Alps. This colonisation, which involves numerous native and exotic tree species, is sometimes preceded by a more or less long phase of shrub dominance; in other cases, the shrub phase is absent, and trees colonise the land immediately after agricultural use has ceased or when grazing pressure is reduced. The pattern of recolonisation is often influenced by the presence of remnant fodder or fruit trees or pre-existent agricultural artefacts, such as terraces or walls. Already existing woodland types, such as stone pine woods, expand, and new stand types are being formed, like ash-sycamore stands that in the past did not exist in the Italian forest landscape. Many broad-leaved tree species invade old-established sweet chestnut (*Castanea sativa*), cork oak (*Quercus suber*) and umbrella pine (*Pinus pinea*) plantations. The ecological and social consequences of this woodland expansion are, so far, virtually unknown. The social perception of this phenomenon is rather complex: some believe that a degradation of the rural territory is taking place and that the spread of woodlands close to villages and along the roads should be halted. Others, for various reasons, consider the 'renaturation' of the rural landscape as a positive factor.

1 Introduction

The most relevant land-use change in Italy over the past fifty years has been the increase in wooded area which has replaced agricultural land, partly as a result of afforestation by man, but principally due to processes of secondary succession such as the recolonisation of abandoned agricultural land by woodland. This phenomenon is present in many European countries (Piussi 2000), as well as in the United States, where it marks a very significant moment in the history of land use. In these countries, the proportion of wooded area was at its lowest during the 18th and 19th centuries, before the process was reversed by afforestation and woodland recolonisation. By contrast, on a global scale, the deforestation process is still ongoing and of great importance (Mather 1992).

In Italy, the decline of agriculture and subsequently the spontaneous recolonisation of these areas by woodland started in the first half of the 19th century, increased in the second half of that century and during the 1920ies and 30ies, was interrupted by a resumption of agricultural land use during war-times and other critical periods, and saw its most dramatic development during the 1950ies and 60ies. However, woodlands had become established on temporarily abandoned fields also in earlier times, in a cyclical alternation of: woodland – clearance – agricultural use – abandonment – woodland. Archaeological evidence, ancient maps and woodland soil profiles frequently testify a different distribution of woodland cover and cultivated land in the past. Statistical data on the wooded area which, according to Italy's National Statistical Bureau (ISTAT), over the last fifty years increased by 21.4 % (from 5.4 to 6.8 million ha) provide information only at a small scale, whereas the analysis of woodland stands and soils may explain the origin of some woods (Guidi et al. 1994). Historical land registers may also furnish ambiguous information since certain terms of land use had different meanings in different times (Giovannini et al. 2001). Not only did secondary succession take place on fields, meadows, and pastures, but also in open stands of cork oak (*Quercus suber*), sweet chestnut (*Castanea sativa*) groves, and alpine meadows interspersed with larches (*Larix decidua*) (Magini and Piussi 1966; Mondino 1991). During this process, a species change often took place in pre-existent woodlands, leading to the formation of denser stands, frequently with a different structure than the original ones. The cessation of cultivation does not necessarily exclude other, usually extensive and occasional, uses, such as grazing, hunting or the gathering of berries, nor does it reduce the danger of fire.

This study outlines the state of knowledge regarding the recolonisation of abandoned agricultural land in Italy, and considers this phenomenon principally in relation to events which took place over the last fifty years and within the confines of Italy. Documentation on the spread of woodlands in the last two or three centuries is very scarce: there are sporadic indications in Musoni's (1915) work in eastern Friuli, in Del Noce's (1849) study in Tuscany, and occasional references in the research on the depopulation of Italian mountain regions, undertaken by the National Institute of Agricultural Economics (Istituto Nazionale di Economia Agraria) in the 1930ies. Pelleri and Sulli (1997) have given a detailed description of a case in the Lombard Pre-Alps where recolonisation began in the 18th century and is still ongoing. Dendrochronological research undertaken on the upper timberline both of the western and eastern Alps (Piussi 1994; Motta and Nola 2001) revealed the gradual expansion of woodland into pastures from the latter 19th century onwards. Evergreen mediterranean vegetation, which has taken over olive groves in the Maremma region near Grosseto in Tuscany, can also be traced back at least to the second half of the 19th century.

2 Results and Discussion

From the 1960ies onwards, there is evidence of secondary succession and the formation of new woodland or scrub in many parts of the peninsula, particularly in the Alps, Pre-Alps, Karst, and in the central north Apennines. Until now, research has focused on a single or few communities in well defined areas so that a regional overview is not available (figure 1).

There is no reliable estimate of the total area affected by woodland recolonisation. The data available on a regional or national level do not

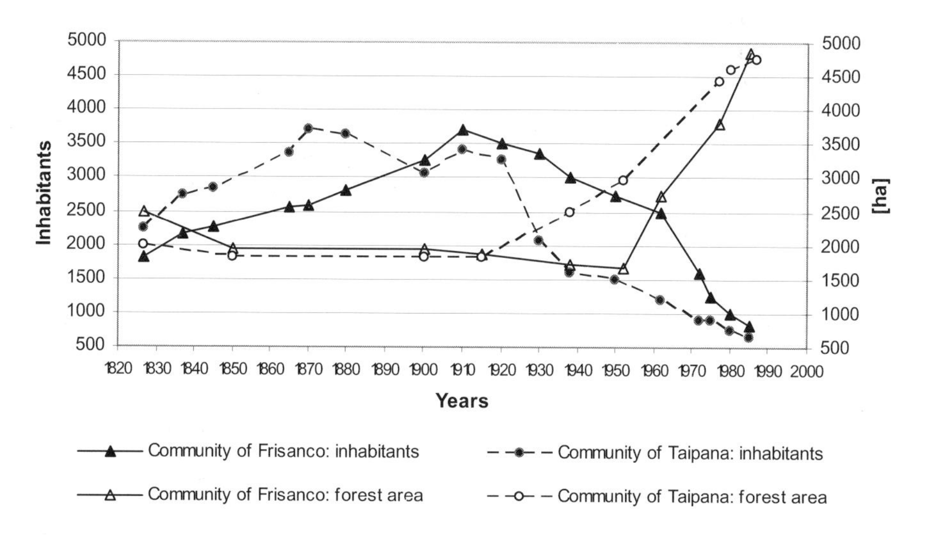

Figure 1 Variations of population and forest area [ha] in two municipalities of the Friuli-Venezia Giulia Region (Guidi et al. 1990)

distinguish between recolonisation and afforestation, and even data on the total wooded area in Italy vary considerably depending on the source and the survey methods employed. The Coordination of Information on the Environment system (CORINE) puts 'evolving woody vegetation, both young forests on former non-forest land, and degraded stands' into the same category (map unit 3.2.4 Transitional woodland/shrub) which covers 1.6 million ha, and, therefore, offers no possibility for distinction either.

It would also be useful to be able to date the beginning of this recolonisation process, even if only approximately. On the other hand, there is agricultural land which had been abandoned and invaded by woody vegetation in the past, but was subsequently cleared and put back to agricultural use; in those cases, woodland recolonisation constituted a brief episode in the site's history. In Italy, every year thousands of hectares of agricultural land are abandoned, and spontaneous woody vegetation can establish there, at least for a short time. The species composition in the newly formed woods is extremely varied, depending on the climate and on soil characteristics. Stands of larch, Swiss stone pine (*Pinus cembra*), dwarf mountain pine (*Pinus mugo*), and alder (*Alnus* sp.) become established in high-altitude alpine pastures. Norway spruce (*Picea abies*), common beech (*Fagus sylvatica*), Austrian pine (*Pinus nigra*) and Scots pine (*Pinus sylvestris*) invade medium-altitude meadows in the Alps. Common ash (*Fraxinus excelsior*) and sycamore (*Acer pseudoplatanus*) on nutrient rich and moister sites, as well as hop-hornbeam (*Ostrya carpinifolia)* and manna ash (*Fraxinus ornus)* on dry sites prevail in the Pre-alpine and Apennine areas. Medium-altitude meadows and pastures in the Apennines are occupied by Turkey oak (*Quercus cerris*), downy oak (*Quercus pubescens*), sweet chestnut, common beech and other broad-leaved trees, generally interspersed with shrubs. Holm oak (*Quercus ilex*) and various evergreen shrubs got established in abandoned olive groves and vineyards in the areas closest to the sea. The black locust (*Robinia pseudoacacia*) is one of the most widespread non-native species in fields and meadows, as well as in open and degraded woodland. Frequently, particularly in the Apennine region, there are shrub formations which include Spanish broom (*Spartium junceum),* common broom *(Citysus scoparius),* blackthorn (*Prunus spinosa*), juniper (*Juniperus communis*) and hazelnut (*Corylus avellana*), which are sometimes interspersed with occasional trees (figure 2).

Relevant changes are also taking place with regard to woodland stand types: Swiss stone pine invades areas once dominated by larch; ash-sycamore stands which are nowadays common throughout the pre-alpine region, were rare in the past; black locust has become a significant part of the forest landscape, old chestnut groves are more and more being substituted by mixed oak communities.

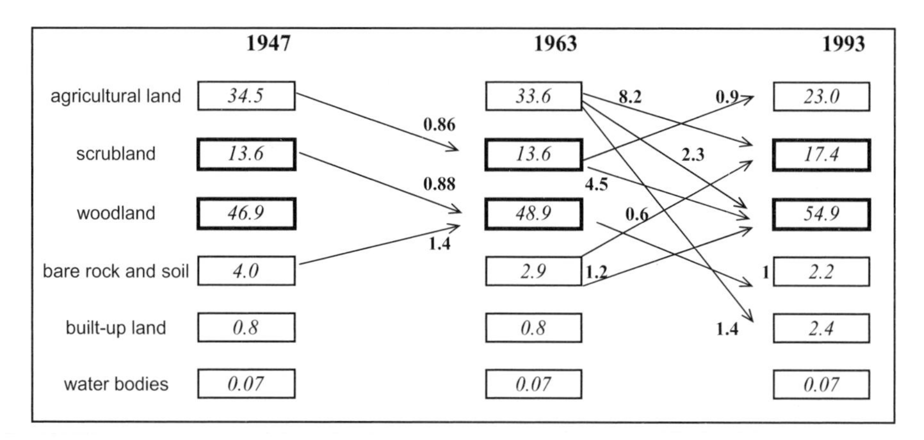

Figure 2 Direction and intensity of changes over 46 years. Boxes indicate % of area for each cover class; arrows indicate % of change per year. Changes less than 0.5 % (i.e. 42.5 ha) are not reported. Firenzuola (Tuscany) study area (Torta 1997)

Since the recolonisation process is so gradual, it is often difficult to determine the environmental conditions prevailing at the time of a woodland's establishment, or the precise moment when it became established, unless this process is testified by written documents or maps. Some of these stands are already productive, and several oak coppices produce firewood. At the same time, this kind of vegetation, particularly if dominated by shrubs, easily becomes subject to fire, which may spread to pre-existing woodlands.

3 The Recolonisation Process

In order to determine the structure and dynamics of these new woods, the process of establishment of woody species in its different phases and the environmental factors that may affect it, were studied by researchers whose findings are summarised below.

3.1 Fallow Land

Recolonisation by woodland takes place especially in mountainous or hilly terrain, where previously marginal arable land, meadows and pastures have been abandoned. The recolonisation process is faster on small plots of land, especially if these are located in the vicinity of wooded areas or are separated from other plots by walls or hedges which often acted as property boundaries. The signs of landscape modification by man are often evident: in the terracing of slopes (Ghidotti and Piussi 1999; Monser et al. 2003) or the deliberate or involuntary creation of specific substrates, such as piles of stones removed from the fields, boundary walls or ruins (Guidi and Piussi 1993; Gandolfo 1994).

3.2 Stand Formation: Beginning and Duration of the Recolonisation Process

Land use may cease suddenly and completely, or gradually. Arable fields may become meadows, then pastures; whilst on pastures the number of grazing animals may reduce progressively, or different grazing species may be introduced. Enquiries regarding the moment of establishment of woody species, how long this process took (the 'window of time'), and the factors which determined its beginning and end have shown that the results vary in the few examples analysed.

For Abruzzo, Peroni et al. (2000) described a temporal sequence consisting of annual species, followed by herbaceous perennials, shrubs and finally trees. In the vicinity of Fabriano (Central Apennines), Ballerini et al. (1997) observed that the land was colonised by Spanish broom, which was followed by manna ash. Similar situations, in which the invasion of woody species was preceded by a rather long period of shrub cover, were described in other areas of the Pre-Apennines and Apennines, and in Sardinia (Boccotti 1997).

Thus, in many cases woody species come in gradually, several years after land use has ceased. The distribution of some shrubs is aided by animals, others propagate vegetatively forming dense, extensive polycorms which enable them to better survive grazing and trampling. It is, however, not clear whether, and in which cases, shrubs hinder the establishment of trees, or whether they represent a pioneer stage within a more complex successional process.

By contrast, in the eastern Pre-Alps and in parts of the Apennines, the colonisation by trees starts a few years after the fields were abandoned and lasts some decades. Chiesi (1998) described the emergence of several shrub and tree species in the year following the last cut of hay from a meadow.

Torta (1997), who studied an area in Firenzuola, northern Apennines, covering the period 1947-1993, found that the 'window' for Turkey oak seems to be limited to a few years, perhaps because a developing layer of herbaceous perennials and shrubs gradually occupied the site, thereby 'inhibiting' (Connell and Slatyer 1977) any further oak colonisation.

Also in the eastern Pre-Alps, Monser et al. (2003) surveyed recolonisation in three different sites where land use had been given up fifteen to thirty years previously, and found that the introduction of new plants had, in fact, continued under the canopy of a well-established tree community, with the greatest density of seedlings on the edge of the tree canopy.

In a terraced, mountainous area in the western Alps, Siniscalco et al. (1995) observed that a climax grove of Scots pine may become established twenty to thirty years after abandonment, unless the soil disturbance caused by deer and wild boar periodically returns the succession to earlier stages.

3.3 Terraced Fields

In Italy, as in most mediterranean countries, steep terrain, anywhere from the sea level to the mountains, was commonly levelled out by constructing terraces whose land could be cultivated. With the onset of mechanisation in agriculture, the terraces were the first agricultural land to be abandoned and were recolonised by shrubs and trees, or they collapsed and became eroded. Local differences in the rocky material used for building the terraces, in the degree of slope, climate, grazing activities or wildlife expansion influence the successional process and the stability of the terraces. Trees often establish within their retaining walls or amongst piles of stones dumped there by farmers to ‘clean up’ the surrounding land, and often they did so well before the land was abandoned. Presumably, on these sites, above and below ground competition is reduced or absent.

It is very rarely possible to determine the germination substrate and soil cover before recolonisation, the pattern of dissemination and the early recolonisation process. If recolonisation started when meadows and pastures were still in use, the seedling stems were removed by mowing or grazing; in arable fields, the seedlings were uprooted. Usually, any seedling growing near a terrace wall was uprooted or cut, as it competed with the crops grown on the terraces, and later its roots would have endangered the stability of the wall. Their ability to produce suckers has, however, allowed some of these plants to survive (Gandolfo 1994), which now constitute a sort of ‘stump bank’ whose shoots will grow undisturbed only once hay-making or grazing has been given up completely.

By contrast, Monser et al. (2003) observed that the soils on the calcareous terraces built on the Bernadia plateau (Friuli-Venezia Giulia), after having remained uncultivated for decades, are now decalcified: The top soil pH varies from strongly acid to neutral and becomes neutral or weakly alkaline at greater depths. The surface horizon consists of well-humified organic matter. In Val Seriana (Orobic Pre-Alps, Lombardy), on terraces built in areas where the geological substrate consists of calcareous rocks, sycamore-ash woodlands are widespread, whereas in contiguous non-terraced sites the most common species are manna ash and hop-hornbeam (Ghidotti and Piussi 1999).

Di Pietro and Filibeck (2000) also emphasised the role of terraces in secondary succession: the terraces in the coastal Aurunci mountain range (Lazio) are often recolonised rapidly, mainly by holm oak, pubescent oak and hop-hornbeam. In the Ausoni mountains (Lazio), a fine-grained structure of vegetation was described by Blasi et al. (2000) on terraces formerly occupied by olive groves. Vegetation communities in the centre of the terrace, where the soil is deep and there is virtually no protection from grazing, differ from those near the back wall of the terrace, where the soil is shallow and disturbance from grazing is reduced.

3.4 Mechanisms of Propagation

Seed dispersal by wind or ingestion by animals are the best mechanisms for recolonising a vast territory. The seeds of some species are carried externally by animals, which is an important method of seed dispersal: this is the case for Swiss stone pine and, less often, for several oak species and for umbrella pine (*Pinus pinea*).

In species with large seeds, such as oak, beech and umbrella pine, seed dispersal by gravity is the predominant mechanism, but the seed is not carried very far. By contrast, dispersal by animals, which, however, works only for a limited amount of seed of these species, covers a distance of several dozen and sometimes (Swiss stone pine) several kilometres. This gives rise to random seed distribution, linked to the existence of 'safe sites' (sensu Harper 1977) which offer conditions favourable to seed survival, and to seedling germination and growth. Once a tree is established, new plants may grow in its vicinity: from the seed it produces, by vegetative propagation, due to the more favourable microclimate created by the tree or by the seed distributed by birds which use isolated trees as a perch.

Vegetative propagation by root suckers is important only in a few tree species, such as black locust and, to a lesser extent, aspen (*Populus tremula*) and elm (*Ulmus* sp.); it is more relevant for several shrub species including blackthorn, which sometimes extends along the entire woodland edge.

According to Torta (1997) Turkey oak, pubescent oak and sweet chestnut, whose seeds are dispersed by animals or gravity, regenerate within fifteen metres from the woodland edge, and their seedling density decreases with distance in a negative exponential function. For field maple (*Acer campestre*), on the other hand, a species with wind-borne seeds, regeneration distribution is randomised.

Centrifugal expansion takes place around isolated oak trees in meadows and fields, a common situation in the hills of central Italy; the same applies to shrubs which propagate vegetatively and produce polycorms. Nocentini and Mori (1991) studied the clonal dispersion of a blackthorn shrub made up of 168 individuals: its suckers, aged from one to eleven years, arose from thirty-two main roots and covered an area of 9.3 m^2. In a study of *Juniperus communis,* Urbinati et al. (1995) also described a population with an aggregated distribution: the age structure indicated a few older specimen, several individuals of intermediate age, and no young plants.

Trees growing on abandoned farmland were often introduced by previous agricultural practices. Potential seed include oaks used for seed production, beeches and larches in alpine pastures, fodder trees, common walnut (*Juglans regia*), field maples planted as hedges or as living support for vines, common alders planted as soil fertilisers (Piussi 1998). Woody species, and usually trees, are common in traditional Italian rural landscapes, from the dry Apulia and Sicily to the cold alpine regions.

Some non-native species, usually introduced in re-afforestation projects, have become naturalised and now play a relevant role in secondary succession. Black locust in most Italian regions and black cherry (*Prunus serotina*) in Lombardy have become weeds over wide areas and have also invaded many abandoned sweet chestnut groves, which had been devastated by the chestnut cancer (*Cryphonectria parasitica*). Several cases of colonisation have been documented, involving Austrian pine, Douglas fir (*Pseudotsuga menziesii*), tree of heaven (*Ailanthus altissima*) and, more recently, Italian alder (*Alnus cordata*) whose narrow original range has expanded by re-afforestation.

4 Influence of Secondary Woodland on Soil

The variation of carbon content in the soil has been studied in the forest of San Martino di Castrozza, Trentino (Thuille et al. 2000), in several Norway spruce stands on fields that were abandoned in different periods since the 1940ies. There is a gradual increase in carbon (0.36 Mg ha^{-1} $year^{-1}$) in the organic horizon, caused by an accumulation of scarcely decomposed or undecomposed litter.

In the central Apennines (Ballerini et al. 1997), soil analyses in an area covered by Spanish broom growing on calcareous parent material, have shown that the amount of organic matter accumulating in the soil increases proportionately to the broom's density and development, irrespective of stand age.

Vos and Stortelder (1992) studied landscape development in the Solano basin (Arezzo province). In secondary woodlands, a moder humus built up.

Half a century without agricultural activity diversified the soil horizon, enriched the mineral topsoil (greater content of C, N, and P), lowered the pH, reduced the amount of N, P and cations in the deeper Ahp horizon below, and generally favoured a progressive transformation of the profile by leaching.

There are also long-term and far-reaching consequences when soil is no longer cultivated. On the Tyrrhenian coast, near the Ombrone River estuary in Tuscany, between the Etruscan era (about 5th century BC) and the 19th century, the river delta moved out to sea by about 7 km, but from the mid-19th century onwards there has been a retreat which, in the 20th century, reached about 800 m (Pranzini 1994). This process can be explained by changes in solid transport by the river which are due to several factors: land reclamation by the diversion of muddy flows, gravel dredging to provide construction material, a reduction in the water flow caused by an upstream diversion of water for irrigation purposes, and also reduced soil erosion. In the past, the clearing of woodlands for the purpose of land cultivation increased soil erosion and the transport of solid material, and thus contributed considerably to the formation of beaches. By contrast, the natural recolonisation by woodland or re-afforestation projects which have taken place during the last decades, have had the opposite effect.

Torta (1997) noted that following the agricultural abandonment of land in Firenzuola, there was an increase in landscape heterogeneity, and a fragmentation of former fields and meadows by the invasion of shrubs. The number of patches (fields or woodlands) and their area have increased and their forms become more irregular. At the same time, in Abruzzo, Peroni et al. (1997) noticed an increase in the surface area of shrub patches, decreased heterogeneity (as a result of reduced human activity), and less fragmentation.

Vos and Stortelder (1992) stated that the landscape pattern changed between 1935 and 1985: the mosaic of cultivated fields, pastures and small woods was reduced, and landscape diversity decreased as a result.

5 The Social Perception of Recolonisation

Hardly any information exists on the local population's opinion regarding spontaneous woodland regeneration. Nutini (1998) conducted a thorough survey among the inhabitants of Firenzuola where, although large tracts of land have been abandoned by farmers, agriculture and animal husbandry are still being practised. Firenzuola's youth often has a positive, almost romantic image of this abandonment - farmhouses partially ruined, 'wild-looking' spontaneous vegetation - in part as a memory of the living conditions of the older generations, in part as a site of games and imagination. For adults not employed in agriculture, the image is negative, it being perceived as a lack of care and interest in the conservation of the land. For the people still employed in agriculture, on the other hand, the abandoned fields are considered useful grazing land.

The inhabitants, nowadays mostly elderly people, of Natisone Valley villages (Friuli-Venezia Giulia), where the invasion by trees proceeds quite rapidly, have asked the local council to help reduce the expansion of woodlands near villages and along roads. Trees shade out buildings, impede road visibility, increase the danger from ice formation during the winter, and their leaves make the roads slippery.

In the past, the woods were cleared to provide sufficient arable land and grazing to satisfy the population's basic food requirements. Nowadays, the presence of woods is considered in more abstract terms: as protection of the environment, as a symbolic value of 'nature', and as a recreational resource. However, these values are generally expressed by the urban population that is far more numerous than the rural one, where the majority of people no longer benefits economically from woodlands and timber production, rarely uses woods for recreational purposes (Scrinzi et al. 1996; Paci and Cozzi 2000) and has a different relationship to the land than farmers and shepherds once had.

6 Conclusions

Most studies on secondary succession were carried out in mountainous and hilly areas in central and northern Italy, whereas in southern Italy and on the big islands woodland recolonisation has rarely been dealt with. Recolonisation by woodland constitutes a great change of the Italian rural landscape. The wooded area expands, and trees colonise previously unwooded or partly wooded areas where woodlands traditionally did not characterise the landscape, or land on which in the past scattered trees produced fodder, cork, chestnuts, acorns and so on, and often the species composition changes. The transformation of the landscape on a grand scale began about fifty years ago, it is still continuing and the area occupied by secondary woodland is increasing. Detailed research provides information only on very limited areas, although some results may be applicable to other areas.

The research work carried out up to now covers only a few years so that it has been impossible to analyse the factors affecting different stages of the succession, in particular starting from the moment when cultivation ends and, at least potentially, a woody vegetation begins to establish itself. Because of their large scale, old aerial photographs are of limited use where information on the early phases of succession is required, or detailed studies of small plots are carried out. There is virtually no data available concerning the relationship between seed rain and the development of herbaceous vegetation on the site which will be colonised by shrubs and trees; there is also little knowledge of the role of wildlife whose numbers have increased dramatically over the last decades, severely affecting the woodland vegetation. Information on the reduction of landscape diversity is also scarce.

Human influence on vegetation and soils, such as grazing of domestic animals, fire, hunting or recreational use, before and during the initial phases of recolonisation by tree species in a secondary succession, has considerable weight and a long-term impact, visibly affecting the spontaneous regeneration process. These direct or indirect human activities also affect the early years of a tree's life. On the other hand, spontaneous woodland recolonisation, considered to be a process of 'renaturation' of the landscape, is indeed the result of a series of natural processes (seed dispersal, survival and germination, seedling establishment, interspecific competition, predation etc.), although it takes place in an environment (vegetation, fauna and soil) that is largely determined by past human land use or by disturbances such as fire or acid deposits. One should, therefore, distinguish between the 'naturality' of the successional process and the 'non-naturality' of the vegetation make-up.

In low and medium altitude mountainous areas, two different kinds of successional processes can be distinguished: in the Alps and in some parts of the northern Apennines, tree species become established soon after a field has been abandoned, whilst in the Apennine and pre-Apennine mountains a shrub stage of fairly long duration comes first which is frequently dominated by common broom (*Cytisus scoparius*) or Spanish broom (*Spartium junceum*), blackthorn (*Prunus spinosa*) and bracken (*Pteridium aquilinum*). There are, however, exceptions: direct colonisation by common beech (*Fagus sylvatica*) in Abruzzo and of Austrian pine (*Pinus nigra*) in the Sila mountains of Calabria, as well as extensive hazelnut (*Corylus avellana*) groves in the Pre-Alps. The dynamics of natural processes in hot and dry environments are still unclear: on the one hand, holm oak (*Quercus ilex*) is the pioneer species in abandoned vineyards on the isle of Elba and olive groves in Sardinia, on the other hand, there is evidence of desertification in some southern Italian regions and on the larger islands (Sardinia and Sicily). It can be assumed that factors of disturbance such as the grazing of domestic animals or fire (which occur to a certain extent also in central and northern Italy) slow down the woodland colonisation process, favouring instead a process of regression under harsh environmental conditions.

Abandoned agricultural land and spontaneous recolonisation by woodland form part of the history of rural communities, confronting today's rural and urban society. There are virtually no studies on the social impact of these changes; it is, therefore, impossible to determine the repercussions of recolonisation by woodland on a psychological, administrative, economic and agricultural level. The return of woodlands testifies to a change in Italian society which has passed to the industrial and post-industrial phase. It may now be asked in what way these new woodlands affect society, and what they represent for both urban and rural inhabitants. The few data available suggest that both sides have different, contradictory, perceptions. Agricultural land may also be perceived as having shaped social life in its various forms: family ties, work, friendships. Vice versa, this agricultural land was created by the relationships between people and their environment (Tilley 1994; Magnaghi 2000); consequently, the present day's changing landscape indicates a historical kind of community on its way to extinction. The new generations that remain, or settle, in the rural area, lack a vision of the spatial context in which past social life took place.

References

Ballerini V, Biondi E, Calandra R (1997) Structure and dynamics of a *Spartium junceum* (L.) population in the Central Apeninnes (Italy). Colloques Phyto-sociologiques 27: 1071-1096

Blasi C, Di Pietro R, Fortini P (2000) A phytosociological analysis of abandoned terraced olive grove shrublands in the Tyrrhenian district of Central Italy. Plant Biosystems 134 (3): 247-259

Boccotti F (1997) Man and forest in Monte Arci. News of Forest History (IUFRO) Subject group S 6.07.00) 25/26: 111-118

Chiesi M (1998) Insediamento ed affermazione del novellame di specie legnose in rimboschimenti spontanei sottoposti a ceduazione nelle Prealpi Orientali (prov. di Udine). Unpublished thesis, pp 106

Connell JR, Slatyer RO (1977) Mechanisms of succession in natural communities and their role in community stability and organisation. Amer Natur 111: 1119-1144

Del Noce G (1849) Trattato istorico scientifico ed economico delle macchie e foreste del gran-ducato toscano. Firenze, Ducci, pp 340

Di Pietro R, Filibeck G (2000) Terrazzamenti abbandonati e recupero della vegetazione spontanea: il caso dei Monti Aurunci. Informatore Botanico Italiano 32 (1-3): 17-30

Gandolfo G (1994) Indagine storico-biologica sull'insediamento della vegetazione forestale in conseguenza della cessata coltivazione delle pendici terrazzate nel Ponente ligure (Valle Arroscia, Imperia). Unpublished thesis, Firenze, pp 95

Ghidotti N, Piussi P (1999) Rimboschimento spontaneo di coltivi abbandonati nelle Alpi Orobiche. Atti del Congresso SISEF 'Applicazioni e Prospettive per la Ricerca Forestale Italiana', pp 23-26

Giovannini G, Sulli M, Zanzi A (2001) Bosco e campo a Carmignano in Età moderna e contemporanea. In: Contini A, Toccafondi D (eds) Carmignano e Poggio a Caiano. EDIFIR, Firenze 125-142, pp 207

Guidi M, Piussi P (1993) The Influence of Old Rural Land-management Practices on the Natural Regeneration of Woodland on Abandoned Farmland in the Prealps of Friuli, Italy. In: Watkins C (ed) Ecological Effects of Afforestation. CAB, Wallingford, pp 57-67

Guidi M, Salbitano F, Stiavelli S (1990) I rimboschimenti spontanei nelle Prealpi Carniche e Giulie. In: Atti del Convegno sulle Possibilità di sviluppo della arboricoltura nelle zone collinari e montane delle Alpi orientali. Cividale del Friuli 40-50

Guidi M, Piussi P, Lasen C (1994) Linee di tipologia forestale per il territorio prealpino friulano. Annali Accademia Italiana di Scienze Forestali XLIII: 221-285

Harper J (1977) Population biology of plants. Academic Press, London, New York, San Francisco, pp 892

Magini E, Piussi P (1966) Insediamento spontaneo di specie arboree nei castagneti abbandonati: considerazioni sulle conseguenze pratiche del fenomeno. Atti del Congresso Internazionale sul castagno. Cuneo 293-294

Magnaghi A (2000) Il progetto locale. Bollati Boringhieri, Torino, pp 256

Mather AS (1992) The forest transition. AREA 24 4: 367-379

Mondino GP (1991) Caratteristiche dei boschi di sostituzione e loro tendenze evolutive. In: Ferrari C, Bagnaresi U. (eds) I boschi italiani. Valori naturalistici e problemi di gestione. Società Italiana Pro Montibus et Silvis 53-61

Monser U, Piussi P, Albani M (2003) Woodland recolonization of abandoned farmland in the Julian Pre-Alps. Gortania Atti Mueso Friul. di Storia Naturale 25 207-231

Motta R, Nola P (2001) Growth trends and dynamics in sub-alpine forest stands in the Varaita Valley (Piedmont, Italy) and their relationship with human activities and global change. Journal of Vegetation Science 12: 219-230

Musoni F (1915) La popolazione in Friuli. Nuove ricerche antropogeografiche nelle Prealpi del Natisone. Annali Istituto Tecnico Zanon, serie II, anno 32: 1-112, Udine

Nocentini G, Mori P (1991) Primo contributo alla conoscenza di dispersione clonale del Prunus spinosa. Monti e Boschi 3: 49-54

Nutini R (1998) Abbandono colturale e rimboschimento spontaneo nell'Appennino Settentrionale. Università degli Studi di Firenze, Facoltà di Magistero, aa 1997-98

Paci M, Cozzi F (2000) La percezione del bosco nella società contemporanea: risultati di un'indagine svolta nelle province di Prato e Firenze. L'Italia forestale e montana, XV, pp 97-100

Pelleri F, Sulli M (1997) Campi abbandonati e avanzamento del bosco. Un caso di studio nelle Prealpi lombarde (Comune di Brinzio, Provincia di Varese). Annali, Istituto Sperimentale per la Selvicoltura, Arezzo 28: 89-126

Peroni P, Ferri F, Avena G (2000). Temporal and spatial changes in a mountainous area of central Italy. Journal of Vegetation Science 11: 505-514

Piussi P (1994) Mixed *Pinus cembra* Stands on the Southern slope of the Eastern Alps. Workshop Proceedings on Subalpine Stone Pine and Their Environment: The Status of Our Knowledge. Forest Science INT-GTR 309: 261-268

Piussi P (1998) Piantagioni di ontano nero in prati falciabili nel Friuli Orientale. Annali di San Michele 11: 215-230

Piussi P (2000) Expansion of European Mountain Forests. In: Price M, Butt N (eds) Forests in Sustainable Mountain Development: a State of Knowledge Report for 2000. CABI-IUFRO, pp 19-25

Pranzini E (1994) Bilancio sedimentario ed evoluzione storica delle spiagge. Il Quaternario 7 (1): 197-204

Scrinzi G, Floris A, Flamminj T, Agatea P (1996) Un modello di stima della qualità estetico-funzionale del bosco. Trento, ISAFA, comunicazione di ricerca, 95/2

Siniscalco C, Barni E, Montacchini F (1995) Dinamismo di vegetazione in coltivi abbandonati della Valle di Susa (Alpi occidentali). Allionia 33: 259-270

Thuille A, Buchmann N, Schulze E-D (2000) Carbon stocks and soil respiration rates during deforestation, grassland use and subsequent Norway spruce afforestation in the Southern Alps, Italy. Tree Physiology 20: 849-857

Tilley C (1994) A phenomenology of landscape. Berg, Oxford, Providence, pp 221

Torta G (1997) Cambiamenti del paesaggio e dinamica della vegetazionein coltivi abbandonati dell'Appennino Settentrionale (Comune di Firenzuola, Firenze). Unpublished PhD thesis, Università degli Studi di Padova

Urbinati C, Carrer M, Rosa F (1995) Dinamismo spaziale e cronologico di Juniperus communis (L) in campi abbandonati nelle prealpi orientali. Linea ecologica 2: 13-19

Vos W, Stortelder A (1992) Vanishing Tuscal Landscapes. Pudoc, Wageningen, pp 404

Mountain Ecosystems

Studies in Treeline Ecology

Regional Treeline Studies in Asia

the highest variability in summer precipitation of all places in Mongolia (between 8 and 162 mm in August), with very unreliable timing. About 70 % of rainfall events yield less than 1 mm of water (Weischet and Endlicher 2000).In the mountain ranges of the area conditions are considerably wetter in summer, due to convective precipitation centred around higher altitudes (cf. Retzer 2003).

In the Ih Bogd, rainfall levels are roughly similar to those in the Dalandzadgad region. Like in other regions, summer conditions start and end very abruptly (Hilbig et al. 1984). From its geographical location within the climatic gradients described above, one might assume slightly wetter and cooler conditions than those observed at Dalandzadgad and in the Dzüün Sayhan.

2.3 Vegetation

Desert steppe and dry steppe vegetation surround the Govi Altay Mountains, but vegetation changes noticeably with altitude. On the northern declivity of the Ih Bogd range the following units are observed (bottom up): *Stipa glareosa – Anabasis brevifolia*-dominated semi desert and desert steppe associations (< 2000 m a.s.l.), mountain steppe with *Stipa krylovii*, *Koeleria gracilis* and *Agropyron cristatum* (2000-2700 m a.s.l.), alpine mats of *Kobresia* (2700-3100 m a.s.l.), and high alpine rock cushion vegetation with *Rhodiola quadrifida*, *Saxifraga* spp. and *Arenaria meyeri* (> 3100 m a.s.l.) (Hilbig et al. 1984).

In a similar form the same is encountered in the Dzüün Sayhan. Desert steppe vegetation with *Stipa* and *Allium* spp. reaches up to 1800 to 2000 m a.s.l., mountain steppe covers the upper part of the range (Dasch and Tschimedregsen 1996). In few locations surrounding the forest sites meadow steppe can be encountered, regarded as a replacement community for forest in northern and central Mongolia (Sommer 1998; Sommer and Treter 1999; Wesche et al. 2003). As the Dzüün Sayhan barely reaches 2800 m, the alpine belt is missing almost entirely, only a few mats of *Kobresia myosuroides* are present in the highest locations (Miehe 1996; Wesche et al. 2003).

The striking feature of the Ih Bogd and Dzüün Sayhan are woodlands of *Betula* (birch) and *Salix* (willow) trees and shrubs, found on north-facing slopes. While on south-facing slopes, the grass-dominated vegetation of the mountain steppe is frequently interrupted by large patches of *Juniperus (J. pseudosabina* in the Ih Bogd range, *J. sabina* in the Dzüün Sayhan), woodland of *Betula* and *Salix* trees and shrubs are found in parts of this zone in northern exposure. The latter occur on north-exposed slopes, and in north-facing rills and gullies (Hilbig et al. 1984). They are the subject of the present paper.

3 Methods

In order to achieve the aim of the present study, i.e. to assess the position of the forests in the framework of human impact and climatic changes, a range of methods was applied. Combining vegetation surveys, dendroecological studies, and radiocarbon dating of charcoal findings the authors tried to portray the natural history and the present-day state of the forests, trace possible reasons for their current isolation and identify ecological trends regarding their potential for persistence.

3.1 Vegetation

Vegetation surveys followed the Zürich-Montpellier School of Braun-Blanquet (1932), as described by Müller-Dombois and Ellenberg (1974). Almost all recent studies of the vegetation of Mongolia employ the Braun-Blanquet phytosociological approach, so that comparability of results was ensured by this choice of methodology. In accordance with this approach, floristically complete vegetation surveys were conducted in subjectively selected locations. Sites considered in this investigation were located along forest – steppe gradients, i.e. forest centre, forest fringe inside, forest fringe outside, and open steppe vegetation. Plots of the last type were also surveyed on slopes resembling those with forest cover with respect to slope, altitude and exposure.

Species ground cover was estimated in percent rather than on the Braun-Blanquet scale to ensure compatibility with previous surveys in the area (e.g. Miehe 1996). Plant specimens were collected for reference and exact identification.

Identification of all plant specimens collected in the field was carried out at the herbarium at the University of Halle, Germany and using the Flora recently published in English (Grubov 2001). Plant names and authorities were assigned based on Gubanov (1996).

Statistical analysis of the data was carried out using the packages PC-Ord 3.15 (McCune and Mefford 1999) and SPSS 11.0.0 (SPSS Inc. 2001). Identification of communities was based on Two Way Species Analysis (TWINSPAN) (Hill 1979). In accordance with Braun-Blanquet's plant-sociological approach it was run on a presence / absence-based matrix rather than abundance. Since it is objective in terms of repeatability this approach was given preference over the traditional tabular rearrangement by hand.

3.2 Dendrochronology

For the dendrochronological approach 100 m^2 plots were chosen on the fringes and in the centres of the forests. On these plots all trees larger than 1 centimetre diameter at breast height were cored (as smaller trees might be fatally damaged by the coring) using standard techniques (Schweingruber 1983; Fritts 2001). One increment core was taken at each tree's bole base for determining its age and two at breast height for further dendroecological analysis. Where the pith was not reached, the missing years were estimated according to the method of Bräker (1981). Though the encountered tree species can grow single-stemmed and do so on the sites, most of the trees grow multi-stemmed – ranging from two stems up to 50 stems per individual in some cases. In multi-stemmed trees only the major stem was cored. On one plot all stems were sampled for reference.

The sampling at breast height excluded reaction wood. Sample processing followed Pilcher (1990). Tree-ring width measurement and crossdating were conducted using the TSAP 3.6 package (Rinn 2000). Statistical tests for crossdating were Gleichläufigkeit values and Student's t-test. The single curves of 10 dominant and codominant willows were combined into tree mean-curves and a site chronology for the Dzüün Sayhan Forest. The low-frequency variability in individual tree ring series attributable to tree ring ageing, forest stand development or differences in the vitality of individual trees was removed by standardising the data using the moving-average trend index that is included in the TSAP package. The annual mean increments of the site-chronology were correlated with monthly mean temperatures and monthly mean precipitation sums of each month, starting with May of the preceding year until September of the year the tree ring was formed. The meteorological data used is from Dalandzadgad, located about 30 km southwest of the Dzüün Sayhan Forests at ca. 1400 m a.s.l.

Additional information such as damage by fraying, ramming and browsing, tree height and diametre at breast height was collected.

3.3 Others

In soil examinations accompanying the vegetation surveys layers of charcoal were found in different soil depths. Genus and ^{14}C-age of the specimens were determined.

In support of the ecological studies a small number of semiformal interviews were conducted with nomads, to reveal further aspects of the distribution and utilization of the forests.

4 Results and Discussion

4.1 The Forests and Their Present-Day State

4.1.1 Vegetation

In the Dzüün Sayhan, a large forest described by Miehe (1996; 1998) ('Dzüün Sayhan A' photo 1) and a small one in a neighbouring valley ('Dzüün Sayhan B') were studied. In the Ih Bogd range a forest in Bitüüt valley was concentrated on ('Ih Bogd' photo 2). Interviews with locals at the mountain ranges between the Dzüün Sayhan and Ih Bogd as well as at the latter two revealed that today there are no other *Betula-Salix* forests than the ones studied.

The three forests investigated differ greatly in extent, covering 0.5, 30 and 50 ha, respectively. They are composed of *Betula* and *Salix* spp., growing in more or less dense stands on north- to north-east facing slopes at altitudes roughly between 2400 and 2700 m a.s.l. Position and area figures of all three forests are summarized in table 1. The smallest of the forests, Dzüün Sayhan B, may still be referred to as a forest in accordance with the definition put forward by the FAO Forest Resources Assessment, which assumes an area of at least 0.5 ha with a tree crown cover of at least 10 % (FAO 2001).

Photo 1 *Betula-Salix* Forest in the Dsüün Sayhan Mountains

Photo 2 *Betula-Salix* Forest in the Ih Bogd Mountains

The dominant tree species are *Betula microphylla*, *B. platyphylla* and *Salix taraikensis*. On average they are around 5 m high; the highest trees reach 7.5 m.

Two Way Indicator Species Analysis (TWINSPAN) of the entire vegetation dataset from all forests allowed the vegetation to be classified into communities. The first two levels of the two-way ordered TWINSPAN table are followed up in the following remarks.

Table 1 Study sites

	Dzüün Sayhan A	Dzüün Sayhan B	Ih Bogd
Coordinates of centre	43° 29.15' N 104° 05.59' E	43° 29.75' N 104° 04.64' E	44° 58.51' N 100° 19.60' E
Total area (estimate)	30 ha	0.5 ha	50 ha
Altitudinal range (m a.s.l.)	2380-2580	2470-2550	2510-2680
Protected area	Yes	Yes	No
Slope exposure	NNW to NNE	NE	NNE

The Ih Bogd vegetation quadrats are identified as the same association as the four species-richest plots of the Dzüün Sayhan. These lie notably above 2500 m a.s.l. The herbaceous vegetation of these plots and the Ih Bogd plots can be described as alpine *Kobresia* meadow (Hilbig 1995), a community generally found at higher altitudes. The main difference between these Ih Bogd and Dzüün Sayhan locations is the presence of trees and shrubs in the Ih Bogd, making them a different sub-association. Since, according to Hilbig (1987), potential forest sites reveal themselves by a herbaceous vegetation which largely resembles that of forest understoreys, the Dzüün Sayhan meadow steppes may be seen as possible forest sites. Indeed, as Hilbig (1995, p. 127) remarks, 'most meadow steppes on mountain slopes used to be forest'.

Within the forest and forest fringe quadrats of the Dzüün Sayhan, the second association identifiable from the TWINSPAN table, a clear separation is possible between plots encroached by *Juniperus sabina* and those with a rich shrubby understorey including *Rosa acicularis*, *Grossularia acicularis*, *Spiraea media*, *Juniperus sabina*, *Ribes rubrum*, *Cotoneaster* spp. and *Lonicera* spp.

4.1.2 Climatic Constraints

Water availability is the crucial factor limiting the growth of forests in the mountain forest steppe zone of central and northern Mongolia. Although this zone is significantly more humid than the Govi Altay, forests are already very near their dry limits and can only survive due to a very complex system of water conservation where evaporation is kept to a minimum. Forests in the mountain forest steppe zone are strictly confined to north-facing slopes. It would appear that this is due to the reduced radiation in these locations, resulting in lower temperatures and evapotranspiration levels (Treter 1996). High levels of mean sensitivity (table 2) and the good crossdatability of the willow curves of the Dzüün Sayhan forests are distinctive indicators for a strong dependency of annual increment to water availability in contrast to for example nutrient supply or temperature regime.

However, in accordance with observations made further north (Haase 1983), the seasonal timing of moisture availability seems to be of much greater ecological importance than total precipitation sums. Firstly, most rain falls in summer when the evaporation levels are highest. Secondly, forests are not supplied with rain water before the summer rains of July and August. When the snow melts early in the year, temperatures are still too low to support plant growth (Richter et al. 1963; Treter 1996).

Table 2 Descriptive statistics of the willow ring width chronology

	Dzüün Sayhan
Beginning [yr]	1801
End [yr]	2001
No. of years	201
Missing rings [%]	0
Tree ring width	
Mean [1/100 mm]	41
Median [1/100 mm]	43
Standard Deviation	120.33
Mean sensitivity	36

The solution to both problems is 'home-made'. When rainwater becomes available, it quickly infiltrates into the forest soils, and with the forest canopy in place direct evaporation is kept to a minimum. At the same time, snow cover remains in place for a longer time, its moisture becoming available to forest plants in spring when it is needed most. This was confirmed by an account of a Dzüün Sayhan National Park ranger who reported snow cover as late as May in the Dzüün Sayhan forest area.

These findings are supported by results from the climate-growth correlation between yearly increment and meteorological data from the Dalandzadgad meteorological station shown in figure 3. One can see that even though January precipitation only combines a small portion of the annual precipitation sums it does predict annual increments significantly on a 95 % level.

During summer convective processes lead to pronounced local differences in precipitation levels, especially in mountainous areas. This may explain the low predictive power the climatic data from Dalandzadgad has for the annual increment on the Dzüün Sayhan forests.

The strong summerly convective influence in the mountains was affirmed by meteorological measurements by Retzer (2003) in the nearby Dund Sayhan Mountains.

Tree physiognomy also indicates a significant role of snow. Photo 3 shows a typical sabre-formed stem bases found in areas with high snow pressure due to heavy snowfall or creeping snow (e.g. Timell, after Schweingruber 1996; Holtmeier 2000; 2003). One could mistake bends due to creeping hills but they typically show accumulation of material above the tree and erosion of material on the lower side. Neither is predominant on the researched sites. Although snow constitutes only a small part of total anual precipitation it appears to play an important ecological role. One must assume that snow is

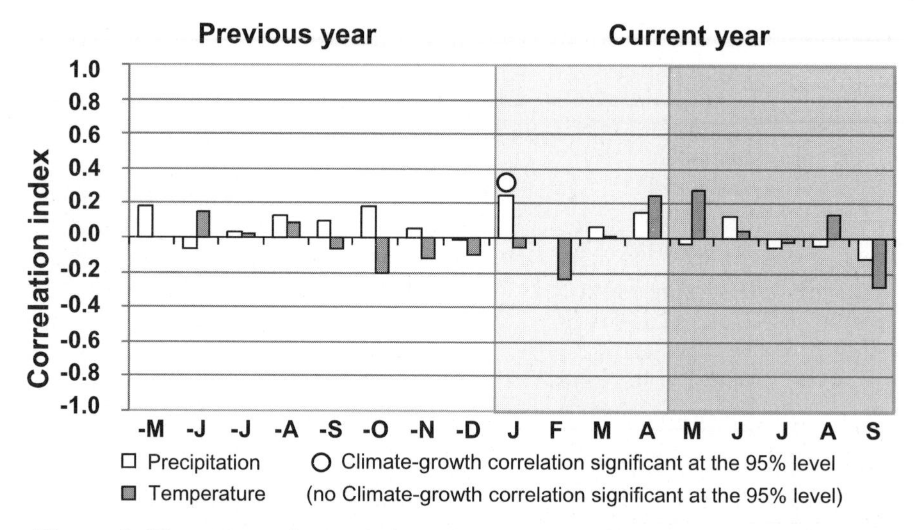

Figure 3 Climate-growth correlation of the willow chronology from the Dsüün Sayhan Forest with monthly mean temperatures of the years 1968-1999 and monthly precipitation sums of the years 1938-1999 from Dalandzadgad meteorological station. The annual increments of each year are being correlated with monthly climate data starting with May of the previous year until September of current tree ring year. The correlation index ranges between -1 and 1

blown to the forests from the adjacent steppe. The protective forest canopy supports snow accumulation.

Another interesting aspect is the presence of a frozen soil layer from as little as 80 centimetres downwards. Given the unusually hot and dry conditions prevalent in southern Mongolia during the summer of 2001 (cf. Retzer 2003), one may assume a more permanent and possibly perennial presence of this frost layer.

In northern Mongolia the presence of a permanent frost layer has been described as a function of forest cover: Canopy and organic soil layers protect the soil from solar radiation in summer. While air temperatures show no marked differences between steppe and forest areas, soil temperatures do (Treter 1996). Where the forest is cleared, the permafrost layer soon disappears; snow thaws early and the conditions for forest growth are no longer given. Considering discussion about the mentioned permafrost layers in the forests of northern Mongolia this presumed permafrost layer has important ecological functions in terms of water preservation and cooling (Haase 1983; Brzezniak and Pacyna 1989). In the north the permafrost is usually found at a depth between a few decimetres and 150 cm in summer. In winter the frost stratum extends to include the upper soil layers. The forest topsoil freezes later than the topsoil of 'open' vegetation, leaving more time

Photo 3 Typical Sabre-Shaped Stem Bases, Dzüün Sayhan

for the absorption of water from early snow (Bernatzky 1978). In the process of freezing the subsoil dries out and the moisture is concentrated in the topsoil. When temperatures begin rising in spring, the 'active layer' of the ground frost thaws and releases its moisture to the vegetation (Succow and Kloss 1978; Haase 1983). Correspondingly, the lowest water deficit in the forest steppe zone occurs in May, at the beginning of the growing period (Glazik 1999).

The discovery of the frozen layer in the particular situation outlined above may be interpreted as an indication of similar processes being in place in the Govi Altay. However, longer term observations would be needed to ascertain this.

Another adaptation to dealing with water stress can be concluded from the above-mentioned trend to build colonies: All of the birch and willow species encountered can grow as single-stemmed trees and sometimes do so on the researched sites. Nevertheless most of the specimens encountered were multi-stemmed. This phenomenon has been observed in many forests at the alpine timberline and has been interpreted as an adaptation to different ecological constraints (Holtmeier 2000; 2003).

Figure 4 shows the annual increment of three stems of the same tree. As can be seen, stem 1 is the oldest of the stems and has had the predominant annual increment until 1990. From that time on the stems' growth rate declines

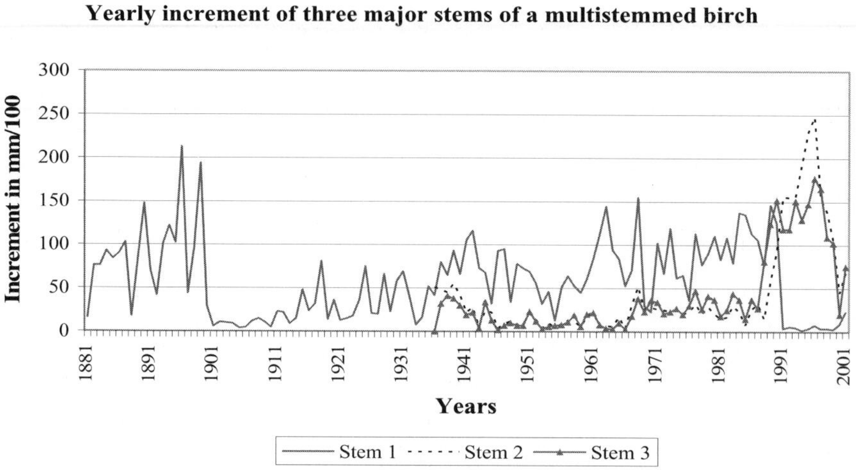

Figure 4 Yearly increment of three major stems of a multi-stemmed birch. Declining growth of the major stem in 1990 is accompagnied by increasing growth of the two minor stems

rapidly and it is to be expected that the specimen is going to die in the near future. At the same time the other two stems 'take over' – the tree survives. The important factor is that the already established root system is still in place – crucial water access is thus granted and a decisive advantage in competing with other individuals or even other species is given. This strategy would also explain why no clearings and in turn no seedlings were found within the forest whereas seedlings are present at the forest fringes.

4.1.3 Human and Livestock Impact

Livestock droppings, trampling and traces of grazing and browsing were encountered in most vegetation plots. Some timber was found near the Ih Bogd forest. In 1957, a major earthquake occurred in the Govi Altay. With its epicentre in the Ih Bogd and Baga Bogd ranges, it resulted in ruptures over a length of 260 km ('Bogd rupture') and extensive rockfalls, landslides and tension cracks in the mountain ranges (Kurushin et al. 1997).

A vast landslide occurred in the upper reaches of the Ih Bogd, in the vicinity of today's forest. A block of rock of about 140 x 10^6 m^3 broke off and slid down to the valley ground. The material now covers the valley along a stretch of more than 3 km with a maximum width of about 1.2 km. According to local herders the valley used to be inhabited seasonally before the landslide. Since then however the main entrance route has been blocked for horse-back and cattle access, thus today only a few hunters come to the upper reches of the valley.

This 'remoteness' leads to one of the major differences between the forests in the Dzüün Sayhan and Ih Bogd mountains. While in the Ih Bogd almost no signs of browsing, ramming or fraying could be found – only 6 % of the trees showed such signs – evidence of strong disturbance at the forest fringes is omnipresent in the Dzüün Sayhan. In these plots up to 93 % of the sampled trees were partly heavily damaged by large herbivores. Without doubt the yak population is capable of contributing to forest decline in this area. But as with the microclimate the forest seems to protect itself by being so dense that large herbivores cannot access its centre easily so that only few signs of animal damage - on 4 % of the trees - can be observed there.

4.2 Environmental History

4.2.1 Short Term Environmental History

The age structure of the three Ih Bogd plots (figure 5) clearly indicates a young and expanding forest with an age gradient from the forest centre plot to the outer plots. Outposts of the emerging forest too small for coring could be observed on adjacent north-facing slopes as well. Overgrown tree stumps as well as solitary 'older' trees (115 years in the outer plot, 85 and 82 years in the forest centre) witness that there were trees even before this new expansion process.

Dendrochronology promises to supply exact dates for changes in forest dynamics. Looking at the forest centre plot where the process of expansion most likely started out one can observe that most trees are between 40 and 60 years of age – in other words: their germination took place between 1940 and 1960. At a first glimpse this would contradict the hypothesis that forest expansion is a direct result of the 1957 earthquake as mentioned above. But one has to keep in mind that tree growth is extremely slow under present conditions – taking 20 years to reach 1 cm diameter in breast height. Thus assuming that the use of timber was one of the major threats to forest growth before the earthquake, it seems plausible that an array of trees with an age of about 20 years would have been there when the 1957 earthquake happened and in the consequence the forest was left to itself. At the forest fringe all the trees germinated after the earthquake and no old stumps were encountered. This is a clear hint that the area had been cleared of forests before and that the abrupt decline of grazing after the earthquake made the succession possible.

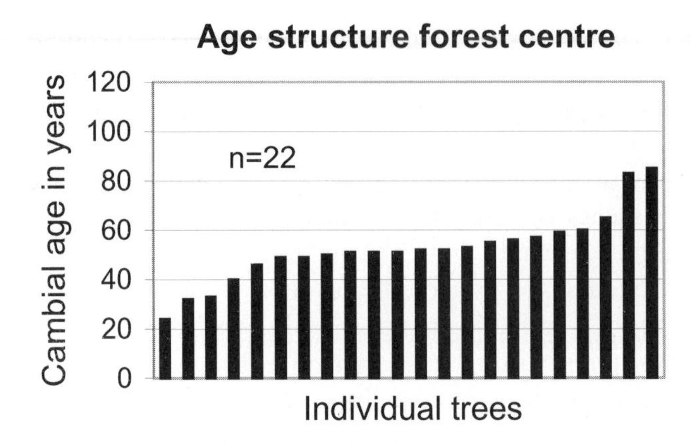

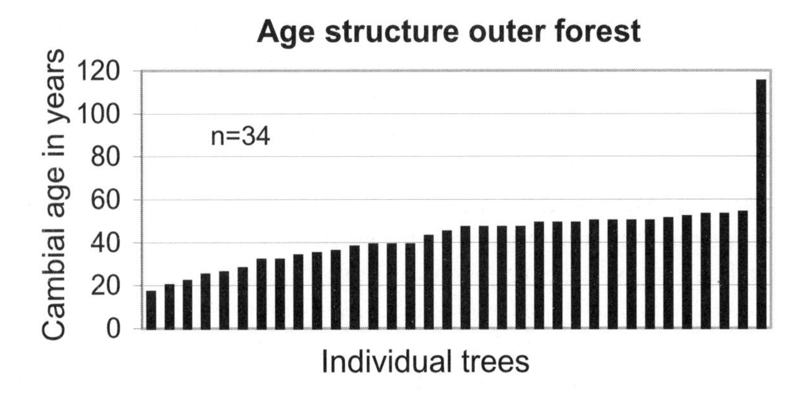

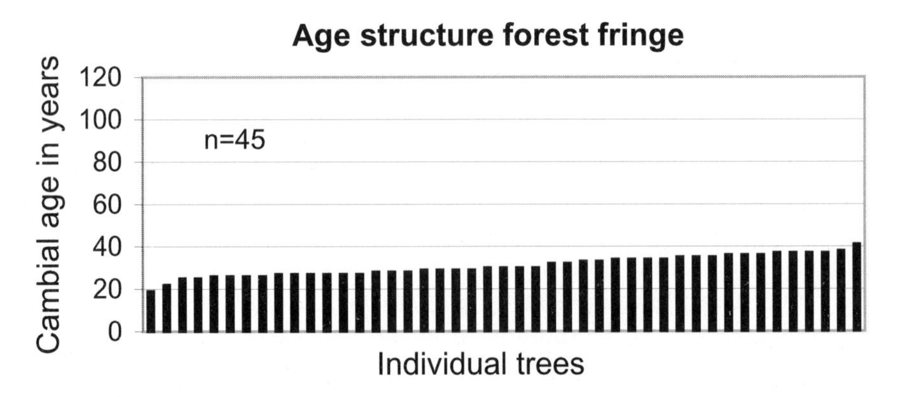

Figure 5 Tree age structure on three plots from centre to forest fringe. Tree ages are represented by cambial ages

Tree ages in the Dzüün Sayhan forests witness a totally different situation. All the plots sampled show a much more diversified age structure – which means that no directed expansion as in the Ih Bogd can be observed here. On the contrary, it seems that this little forest patch at least within today's fringes must have been stable for at least 150 to 220 years, as the oldest cored trees are that age and large decaying trunks of deadwood are found throughout the forest.

4.2.2 Long Term Environmental History

In several locations under forest and under adjacent steppe charcoal was found at different soil depths. These have been identified as *Betula, Salix,* and *Juniperus* and in one case *Picea* or *Larix* (no further discrimination possible from charcoal, Schoch 2002, pers. comm.). The *Picea/Larix* has been ^{14}C dated to 3743 $\pm$ 51 years (2296 BC – 2014 cal. BC, Erl-5517). The presumably oldest *Betula*-specimen has been ^{14}C dated to 4335 $\pm$ 53 years (3098 BC – 2878 cal. BC, Erl. 5518). These findings correspond well with dated macrofossil findings of *Larix*, *Picea* and *Abies* at Bayan Sayr, the Tsakhir Khalgyn Mountains and Uert valley (Dinesman et al. 1989, cited in Dorofeyuk 2000). This underlines the likeliness of a closed forest belt at some time in the Holocene with forest cover in locations that are forest-free at the moment. This belt may have appeared during a more humid phase between 5000 and 2000 years BP, which has been suggested for the north before (Klimek 1980). This conjecture is further corroborated by floristic features encountered in and around the forests, such as *Viola* spp. and *Paeonia*. Not being anemochores, these species are known to be slow migrators (Jäger 2002; pers. comm.).

5 Conclusions

All evidence collected about the three *Betula* - *Salix* forest islands of the Govi Altay supports the hypothesis that today's isolated forests are the last remnants of a former forest belt spanning the entire Govi Altay. Charcoal findings and floristic evidence link the present-day forest islands to periods past.

The reasons underlying the retreat of a former forest belt to today's few forest islands can only be guessed upon. Today's islands suggest that even though the climate developed unfavourably after the establishment of a forest belt in the Altay the climatic conditions would support an already established forest ecosystem. Charcoal findings in different soil layers – or in other words:

forest fires at different times, the strong pressure of livestock, and well documented current timber utilization can surely account for a large portion of the forests' decline. As the situation in the Dzüün Sayhan forests suggests - once the forest cover was broken up it was at least very difficult for forests to once again penetrate the steppe.

The picture presented today differs for the two investigated areas. While all evidence in the Ih Bogd mountains points to an expanding forest on the north-facing slopes as a consequence of people and their livestock leaving the area after the 1957 earthquake, the Dzüün Sayhan forests seem to be under strong pressure from livestock. This seems especially challenging since these forests must be regarded as 'true' remnants: Since the forests ensure their own survival today, their eradication under today's climatic circumstances would most certainly be final.

Acknowledgements

We would like to thank Werner Schoch of the Laboratory for Quaternary woods, who determined the charcoal specimens. Andreas Rigling from the Swiss Federal Institute for Forest, Snow and Landscape Research who assisted with the dendrochronological analysis. Thomas Hennig of the University of Marburg, who assisted us in the field-campaign and Christiane Enderle for preparing some of the graphs.

The Bundesministerium für Wirtschaftliche Zusammenarbeit (BMZ) and the German Research Council (DFG) supported the research with financial grants.

References

Barthel H (1983) Die regionale und jahreszeitliche Differenzierung des Klimas in der Mongolischen Volksrepublik. Studia Geographica 34: 1-91

Bernatzky A (1978) Tree Ecology and Preservation. Elsevier, Amsterdam

Bräker OU (1981) Der Alterstrend bei Jahrringdichten und Jahrringbreiten von Nadelhölzern und sein Ausgleich. Mitt. Forstl. Bundes-Vers.anst. Wien 142: 75-102

Braun-Blanquet J (1932) Plant Sociology: The Study of Plant Communities. Facsimile 1965, 1. edn, Hafner, New York

Brzezniak E, Pacyna A (1989) Types of the Vertical Plant Zonality in the Mountains of Mongolia Against a Background of the Climate. Prace Botaniczne 18: 7-19

Dasch D, Tschimedregsen L (1996) Forschungsbericht der Pflanzengeographischen Untersuchungen im komplexen Schutzgebiet Gobi-Gurvan-Saichan. Akademie der Wissenschaften der Mongolei, Ulaanbaatar

Dorofeyuk NI, Tarasov PE (2000) Rastitel 'Nost' Sapadnoy i Yuzhnoy Mongolii v Poesdnem Pleystozene i Golozene. Botanicheskiy Zhurnal 85, 2: 1 – 17

Ellenberg H (1979) Man's influence on Tropical Mountain Ecosystems in South America. Journal of Ecology 67: 401-416

Food and Agricultural Organization of the United Nations (FAO) (2001) Global Forest Resources Assessment 2000 - Main Report. FAO Forestry Paper 140. FAO, Rome

Freitag H (1971) Die natürliche Vegetation Afghanistans. Vegetatio 22: 285-344

Fritts HC (2001) Tree Rings and Climate. The Blackburn Press, Caldwell, New Jersey

Glazik R (1999) Struktura Bilansu Wodnego na Obszarze Mongolii. Acta Universitatis Nicolai Copernici 103: 257-268

Grubov VI (2001) Key to the Vascular Plants of Mongolia (With an Atlas). Science Publishers, Enfield

Gubanov IA (1996) Konspekt flory vneshnei Mongolii: sosudistye rasteniia. Valang, Moscow

Haase G (1983) Beiträge zur Bodengeographie der Mongolischen Volksrepublik. Studia Geographica 34: 231-367

Hilbig W (1987) Zur Problematik der ursprünglichen Waldverbreitung in der Mongolischen Volksrepublik. Flora 179: 1-15

Hilbig W (1995) The Vegetation of Mongolia. Amsterdam

Hilbig W, Knapp HD (1983) Vegetationsmosaike und Florenelemente an der Wald-Steppen-Grenze im Chentej-Gebirge (Mongolei). Flora 174: 1-89

Hilbig W, Stubbe N, Dawaa N, Schamsran Z, Dorn M, Helmecke K, Bumzaa D, Ulykpan K (1984) Vergleichend biologisch-ökologische Untersuchungen in Hochgebirgen der Nordwest- und Südmongolei. Erforschung biologischer Ressourcen der Mongolischen Volksrepublik 4: 5-49

Hill MO (1979) TWINSPAN – a FORTRAN Program for Arranging Multivariate Data in an Ordered Two Way Table by Classification of the Individuals and the Attributes. Cornell, Department of Ecology and Systematics, Ithaca, NY

Holtmeier F-K (2000) Die Höhengrenze der Gebirgswälder. Arbeiten Institut für Landschaftsökologie, Westfälische Wilhelms-Universität 8

Holtmeier F-K (2003) Mountain Timberlines – Ecology, patchiness, and dynamics. Advances in Global Change Research, 14. Kluwer Academic, Dordrecht

Kessler M (2002) The „Polylepis Problem": Where do we stand? Ecotropica 8: 97-110

Klimek K (1980) Relief and Paleogeography of the Southern Khangai Mountains. Prace Geograficzne 136: 19-27

Kurushin RA, Bayasgalan A, Ölziybat, M, Enhtuvshin B, Molnar P, Bayarsayhan C, Hudnut KW, Lin J (1997) The Surface Rupture of the 1957 Gobi-Altay, Mongolia, Earthquake. The Geological Society of America, Boulder

McCune B, Mefford MJ (1999) PCOrd: Multivariate Analysis of Ecological Data. MjM Software Design, Gleneden Beach

Meteorological Survey of Mongolia (1996) Meteorological Data Dalandzadgad. Personal Communication
Miehe S (1996) Vegetationskundlich-ökologische Untersuchungen, Auswahl und Einrichtung von Dauerprobeflächen für Vegetationsmonitoring im Nationalpark Gobi-Gurvan-Saikhan. GTZ, Ulaan Baatar, Technical Report
Miehe S (1998) Ansätze zu einer Gliederung der Vegetation im Nationalpark Gobi-Gurvan Saikhan. GTZ, Technical Report, Ulaan Baatar
Miehe G, Miehe S (1994) Zur oberen Waldgrenze in tropischen Gebirgen. Phytocoenologia 24: 53-110
Miehe G, Miehe S, Huang Jian, Otsu Tsewang (1998) Forschungsdefizite und -perspektiven zur Frage der potentiellen natürlichen Bewaldung in Tibet. Petermanns Geographische Mitteilungen 142: 155-164
Miehe S, Miehe G, Huang Jian, Otsu Tsewang, Tuntsu Tseren, Tu Yanli (2000) Sacred Forests of South-Central Xizang and their Importance for the Restauration of Forest Ressources. In: Marburger Geographische Schriften 135: 228-249
Miehe G, Miehe S, Schlütz F, Schmidt J, Schawaller W (2002) Sind Igelheiden die zonale natürliche Vegetation altweltlicher Hochgebirgshalbwüsten? Vegetationskundliche Zoogeographische und palynologische Befunde aus dem tibetischen Nepal - Himalaya. Erdkunde 56: 268-285
Müller-Dombois D, Ellenberg H (1974) Aims and Methods of Vegetation Ecology. Wiley & Sons, New York
Murzaev EM (1954) Die Mongolische Volksrepublik: Physisch-geographische Beschreibung. VEB geographisch-kartographische Anstalt, Gotha
Pilcher JR (1990) Sample Preparation, Cross-Dating and Measurement. In: Cook ER and Kairiukstis LA (eds) Methods of Dendrochronology. Applications in the Environmental Sciences. Kluwer Academic, Dordrecht, pp 40-51
Reading RP, Amgalanbaatar S, Lhagvasuren L (1999) Biological Assessment of Three Beauties of the Gobi National Conservation Park, Mongolia. Biodiversity and Conservation 8: 1115-1137
Retzer V (2003) Grazing Ecology of the Gobi Gurvan Sayhan, South Gobi, Mongolia: Carrying Capacity and Forage Competition Between Livestock and a Small Mammal (*Ochotona pallasii*) in a Non-Equilibrium Ecosystem. University of Marburg, unpublished PhD thesis
Richter H, Haase G, Barthel H (1963) Besonderheiten des Periglazials unter kontinentalen Klimaverhältnissen Zentralasiens. Wissenschaftliche Zeitschrift der Technischen Universität Dresden 12: 1153-1158
Rinn F (2000) TSAP 3.6. Heidelberg, Rinntech
Schlütz F (1999) Palynologische Untersuchungen über die holozäne Vegetations-, Klima- und Siedlungsgeschichte in Hochasien (Nanga Parbat, Karakorum Nyanbaoyeze, Lhasa) und das Pleistozän in China (Qinling Gebirge, Gaxun Nur). Dissertationes Botanicae 315
Schweingruber FH (1983) Der Jahrring: Standort, Methodik, Zeit und Klima in der Dendrochronologie. Paul Haupt, Bern

Schweingruber FH (1996) Tree Rings and Environment. Dendroecology. Paul Haupt, Bern
Sommer M (1998) Die Bärentrauben-Lärchenwälder im Charchiraa-Gebirge (Mongolei). Feddes Repertorium 109: 129-138
Sommer M, Treter U (1999) Die Lärchenwälder der Gebirgswaldsteppe in den Randgebirgen des Uvs Nuur Beckens. Die Erde 130: 173-188
SPSS Inc. (2001) SPSS for Windows, Release 11.0.0
Succow M, Kloss K (1978) Standortverhältnisse der nordmongolischen Waldsteppenzone im Vorland des westlichen Chentej. Archiv für Acker- und Pflanzenbau und Bodenkunde 22: 529-542
Treter U (1996) Gebirgs-Waldsteppe in der Mongolei: Exposition als Standortfaktor. Geographische Rundschau 48: 655-661
Troll C (1959) Die tropischen Gebirge. Bonner Geographische Abhandlungen 25
Walter, H (1974) Die Vegetation Osteuropas, Nord- und Zentralasiens. Stuttgart
Walter H, Medina E (1969) Die Bodentemperatur als ausschlaggebender Faktor für die Gliederung der subalpinen und alpinen Stufe in den Anden Venezuelas (vorläufige Mitteilung). Berichte Deutsche Botanische Gesellschaft 82: 275-281
Weischet W, Endlicher W (2000) Regionale Klimatologie – Teil 2: Die Alte Welt. Teubner, Stuttgart, Leipzig
Wesche K (2002) The High-Altitude Environment of Mt. Elgon (Uganda/Kenya): Climate, Vegetation, and the Impact of Fire. Ecotropical Monographs 2
Wesche K, Miehe S, Miehe G (submitted): Desert and Semi-Desert Vegetation in Southern Mongolia: the Plant Communities of the Gobi Gurvan Saikhan National Park (South Gobi Aimag, Mongolia). Phytocoenologia
Zhang Jinwei (ed) (1988) The Vegetation of Xizang. Beijing. In Chinese.

The Upper Timberline in the Himalayas, Hindu Kush and Karakorum: a Review of Geographical and Ecological Aspects

Udo Schickhoff

Abstract

Based on comprehensive evaluations and analyses of existing literature and data sources, a review of geographical and ecological aspects of the upper timberline in the Himalayan mountain system is presented. Upper timberline elevations increase along two gradients: a NW-SE gradient corresponds to higher temperature sums at same elevations along the mountain arc. However, mean temperatures of the warmest month are higher at timberlines in the NW, which develop at lower elevations than expected since extreme winter cold, later snow melt, and shorter growing seasons overcompensate the advantage of higher summer temperatures. A second gradient is developed in peripheral-central direction from the Himalayan south slope to the Great Himalayan range and the Tibetan highlands. Increasing timberline elevations along this gradient are related to the combined effects of continentality and mass-elevation, both leading to higher temperature sums. *Juniperus tibetica* stands in S-Central Tibet even reach 4600-4800 m, the most elevated timberline in the northern hemisphere.

Potentially natural timberline elevations are higher at south-facing slopes compared to north-facing slopes. The difference in altitudinal position is up to several hundred meters. The pronounced effects of exposure to solar radiation result in a much higher utilization pressure at sunny slopes, in particular with regard to pastoral use. South-facing slopes have since long been subjected to massive human impacts throughout the mountain arc so that natural conditions are hard to reconstruct. The depression of upper timberline may amount to more than 500 m, largely depending on the local/regional utilization potential of alpine pasture areas. With regard to physiognomy, high coniferous forests give way to medium-sized broadleaved tree stands and finally to a krummholz belt. This is the dominant timberline pattern on shady slopes, whereas remnant open coniferous forest stands on sunny slopes dissolve into isolated patches or single crippled trees higher up. Along the NW-SE gradient, north-facing slopes show a floristic change from deciduous *Betula*- to evergreen *Rhododendron*-dominated upper timberlines, which must be attributed to decreasing winter cold and strongly increasing humidity levels. *Juniperus* spp. are the principal timberline tree species on south-facing slopes throughout the mountain system.

Relationships of Himalayan timberlines to other ecological conditions and processes such as carbon balance, freezing and frost drought, soil temperatures, wind, snow cover, soils, regeneration, etc., are still largely unexplored. The state of the art is summarized in this paper. More systematic, interdisciplinary timberline research in the Himalaya is strongly needed to better understand how complex ecological and socio-economic processes are expressed in present spatial and physiognomic timberline structures.

1 Introduction

The upper timberline is the most conspicuous physiognomic boundary within the altitudinal zonation of vegetation and at the same time one of the most fundamental ecological boundaries. It represents a significant limit in the continuous change of ecological conditions with increasing elevation, expressed by more or less abrupt alterations of dominating life forms and plant communities. At a global scale, natural alpine timberlines are mainly caused by heat deficiency, i.e. insufficient air and soil temperatures during growing season and insufficient length of growing season, which adversely affects growth, regeneration, cold hardiness and survival of trees at their upper distribution limit. Inadequate summer warmth as the main causal factor for worldwide positions of upper timberlines has to be seen in connection with other factors and processes important on a local scale such as extreme climatic events, freezing and frost drought, wind, snow and ice, soil physical and chemical properties, topography, carbon balance, regeneration, growth forms and forms of tree stands, etc. (Holtmeier 1989; 2000; 2003; see also Tranquillini 1979; Arno 1984; Wardle 1993; Körner 1998a).

It must be stressed that at a local scale the upper limit of mountain forests is not only a thermally induced boundary or governed by temperature-related ecophysiological conditions for tree growth. In his comprehensive treatment of the upper timberline, Holtmeier (2000; 2003) explained in detail that causes of the current position of uppermost tree stands are in general extremely complex. The integration of timberlines into high mountain terrain as well as the physiognomic and structural appearance of timberlines are the result of a great variety of interrelated factors and processes, spatially manifested in a small-scale mosaic of differing site conditions. Factor complexes include topoclimate (radiation, temperature, length of growing season and snow cover), topography (slope inclination, relief forms), ecology of tree species (regeneration, seed dispersal, successional stage), site history (climate oscillations, fire, human impact, insect calamities), current biotic (browsing, trampling, diseases and insect pests) and anthropogenic influences (burning, logging, grazing, recreation and tourism).

Since the upper timberline occurs everywhere high mountains rise above forested altitudinal belts, it has since long attracted geographers and ecologists to do research on the phenomena it encompasses. Imhof (1900), Brockmann-Jerosch (1919), and Däniker (1923) were prominent among the pioneers of systematic timberline research, which has started in the European Alps. Hermes (1955) presented an extensive survey of the altitudinal position of upper timberline in the world's high mountains, following and later followed by smaller contributions to global comparisons (Troll 1948; 1973; Ellenberg 1966; Wardle 1974; Hinckley et al. 1985; Holtmeier 1989; 1993a; Plesnik 1991; Körner 1998b; Jobbágy and Jackson 2000; Miehe and Miehe 2000) and by Holtmeier's (2000; 2003) recent comprehensive reviews. Holtmeier followed a complex landscape-ecological approach in his numerous timberline studies in Europe and North America (e.g., 1965; 1966; 1973; 1974; 1985; 1987; 1989; 1993a, b; 1994; 1995; 2000; 2003; Holtmeier and Broll 1992), the results of which largely contributed to our current understanding of the various factors and processes at work in the timberline ecotone. Ecophysiological aspects of tree species at timberline were summarized by Tranquillini (1979). Arno (1984) provided a comprehensive overview of North American timberlines. Wardle (1971; 1993), Stevens and Fox (1991), and Körner (1998; 1999; 2003) attempted at general explanations and hypotheses for timberline formation based on recent ecological and ecophysiological research. Meanwhile, profound knowledge of upper timberlines in Europe, North America and New Zealand has been accumulated. Considerable research deficits, however, still remain as far as timberlines on other continents, especially in subtropical and tropical high mountains, are concerned. Recent overviews of upper timberlines in the tropics were provided by Miehe and Miehe (1994; 1996; 2000), for tropical/subtropical oceanic islands see Leuschner (1996), for South and East Asia Ohsawa (1990), for Middle Asia Zimina (1973), for North Asia Malyshev (1993).

The growing prominence of global environmental change issues has stimulated new scientific interest in the upper timberline, concomitant to increasing awareness of the essential role mountain ecosystems play in the biosphere and to the efforts to establish mountains as a research priority (e.g., Chapter 13 of Agenda 21, Rio Earth Summit 1992; International Year of the

Mountains 2002; cf. Messerli and Ives 1997; Price and Messerli 2002). For instance, the question of potential vegetation shifts in response to global warming provoked numerous scenarios and modelling studies of the future position of altitudinal vegetation zones and upper timberline (e.g. Peters 1990; Ozenda and Borel 1991; Franklin et al. 1992; Halpin 1994; Lischke et al. 1998; Bolliger et al. 2000; Grace et al. 2002). The majority of these studies are based on the assumption that present-day elevation limits are in balance with current climate, and on linear extrapolations of the relationships between tree growth, metabolism, regeneration and climate conditions at present-day ecotones according to predicted increases in temperatures. However, Holtmeier (1995; 2000) showed that the timberline ecotone has to be seen as a space-time-related phenomenon that does not respond linearily to an altitudinal shift of any isotherme, and that the actual position of upper timberline reflects site history (extreme events, fire, insect pests, human impact, inertia of tree populations, biotic interactions, etc.) rather than present climate. He concluded that a synchronous adjustment of environmental conditions and timberline elevation to changing climate is not to be expected. This view is corroborated by recent empirical studies showing enhanced tree growth and an increase in tree populations within the timberline ecotone during the past century, but either no timberline advance at all or merely minor shifts lagging behind actual warming (e.g. Hättenschwiler and Körner 1995; Szeicz and MacDonald 1995; Paulsen et al. 2000; Klasner and Fagre 2002). Recently, single seedlings were reported far above current timberline (e.g. Dubey et al. 2003; Kullman 2004); their fate remains uncertain for the time being.

Considerable uncertainties still exist as far as our knowledge of upper timberline conditions in the Hindu Kush-Himalayas (HKH mountain system) is concerned. This holds true for both basic and applied aspects of timberline research. Neither basic relationships of upper timberline to ecological conditions and processes nor recent timberline fluctuations in response to global warming have been sufficiently investigated so far. The aim of this study is to review and compile the existing knowledge regarding geographical and ecological aspects of the upper timberline in this vast mountain system. The focus lies on the following questions: What are the regional patterns of timberline elevation, physiognomy and floristics? What is known about ecological conditions and processes at timberline elevation that might control timberline position? How far is the timberline ecotone influenced by anthropogenic disturbances? Is there any differentiation of human impacts at various spatial scales? Based on this review of the state of the art, current research deficits are highlighted and a future intensification of timberline research is proposed considering changing ecological and socio-economic environments.

In this review, the term 'upper timberline' is used sensu Holtmeier (2000; 2003), i.e. it refers to the upper elevational limit of forests in high mountains, that may occur as an abrupt boundary, but usually forms a transition zone, a more or less wide ecotone, between closed contiguous forests and single, often crippled individuals of tree species higher up. In this review I follow this definition, which unifies the terms 'timberline', 'treeline', and 'tree species line' since it stresses the ecotone aspect of the upper limit of mountain forests.

2 Regional Differentiation of Upper Timberline Position, Physiognomy and Floristics

From Afghanistan in the NW (ca. 36°N and 70°E) to Yunnan in the SE (ca. 26°N and 100°E), the Himalayan mountain system as delimited by Schweinfurth (1957a) crosses several horizontal climatic zones as indicated by the vegetation of its immediate forelands: Along a pronounced humidity gradient, the vegetation changes from subtropical semi-desert and thorn steppe formations in the NW to tropical evergreen rain forests in the SE. Likewise, a steep S-N humidity gradient is developed across the mountain system being a climatic barrier for moisture-bearing monsoonal air masses as well as for extremely cold and dry air currents from Central Asia. Accordingly, at a regional scale the densely forested Himalayan south slope sharply contrasts with steppe lands on the Tibetan Plateau. The staggered high mountain ranges itself create a great variation of annual precipitation and thermal conditions along and across the mountain system, i.e. a complex pattern of local and regional climatic and related edaphic and biotic conditions. Corresponding to this marked three-dimensional landscape-ecological differentiation of the HKH-region, the altitudinal position of the upper timberline considerably varies (figure 1), and, moreover, the timberline encompasses considerable differences in species composition.

Generally, the upper timberline increases from NW to SE and from S to N within the mountain system. At the northwestern distribution limit of monsoon-dependent coniferous forests (central Hindu Kush), upper timberline merely lies between 3000 and 3300 m. In the SE, the upper limit of forest and tree growth attains an elevation of 3900 m in the southern and may reach up to 4600 m at north-facing slopes in the mountain ranges of Yunnan and Sichuan (cf. figure 1). However, this general pattern is not consistent throughout the mountain arc. E.g., the more or less linear ascent from S to N is often modified

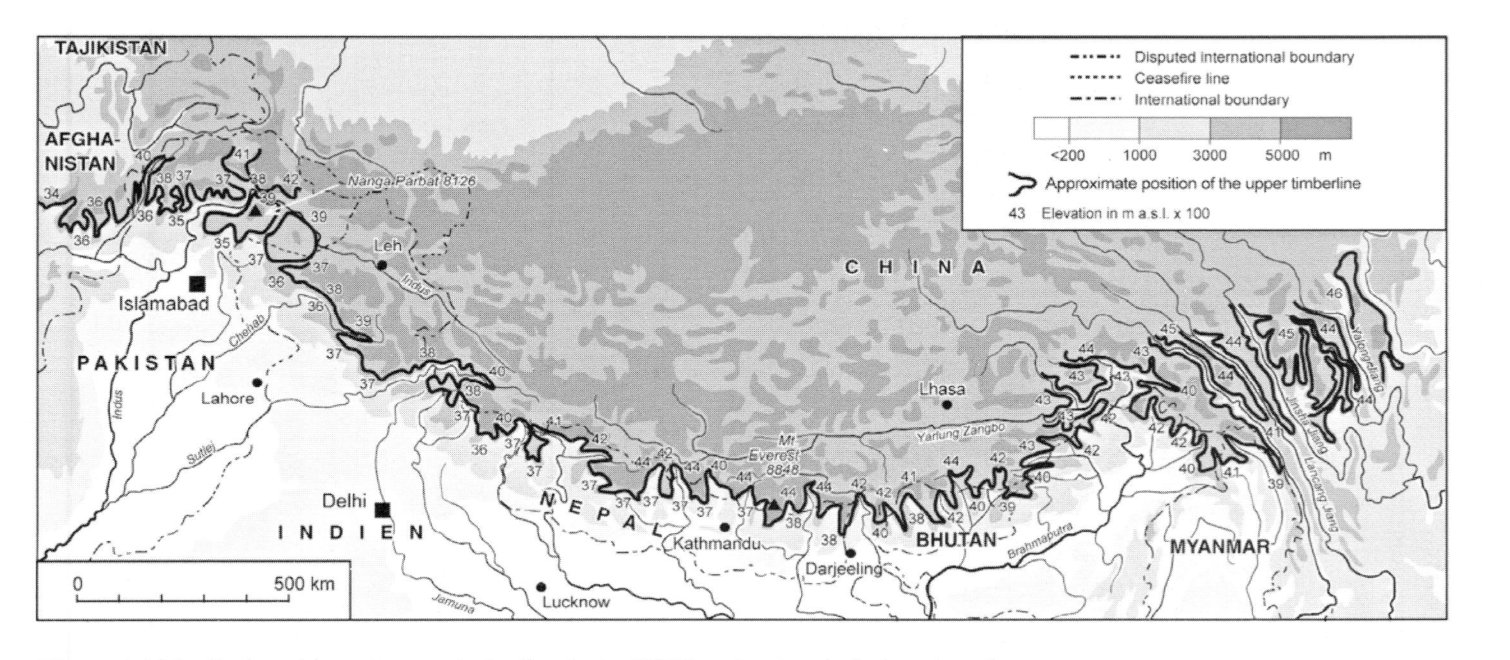

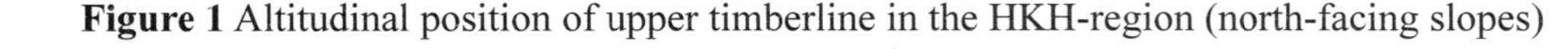
Figure 1 Altitudinal position of upper timberline in the HKH-region (north-facing slopes)

at single high mountain massifs, where the upper timberline might be higher than in mountain ranges further north. The upper timberline position also varies according to aspect and related topoclimatic variations. As a rule, S-facing slopes show an elevated upper timberline compared to N-facing slopes. However, this pattern is to a locally different extent disturbed by anthropogenic interferences.

As for general floristic and physiognomic patterns at upper timberline, natural vegetation of the subalpine belt in the NW and W Himalayas primarily consists of coniferous (*Abies pindrow*, *Pinus wallichiana*, *Picea smithiana*) and birch forests (*Betula utilis*). The birch is dominant constituent of the upper subalpine forest belt on N-facing slopes, where coniferous forests already have a considerable proportion of birch and finally turn into a narrow belt of pure birch stands. *Rhododendron campanulatum* and *Salix* spp. are principal understory species. Further upslope, *Betula utilis* grows scattered in a crippled growth form and merges into a *Rhododendron* and/or *Salix* krummholz belt, which delimits the timberline ecotone and thus the subalpine belt (photo 1). Evergreen trees and large shrubs of the genus *Rhododen-*

Photo 1 *Salix denticulata* krummholz delimiting the upper subalpine belt at 3500 m in lower Kaghan V./W Himalaya (Schickhoff, 1990-06-02)

dron (*Rh. campanulatum*, *Rh. barbatum*, *Rh. wightii*, *Rh. fulgens*, *Rh. lanatum* a.o.) replace the birch belt under less severe winter cold and increasing influence from tropical monsoon moisture in the E Himalayas. Uppermost *Rhododendron* forests are interspersed with single conifer species (*Abies* spp., *Picea* spp., *Larix griffithiana*), ascending from dense forest stands in the lower subalpine and upper montane belt. The krummholz belt higher up is often formed by the same *Rhododendron* species, frequently associated with *Salix* spp. Several juniper species (*Juniperus excelsa*, *J. turkestanica*, *J. recurva*, *J. wallichiana*, *J. indica*, *J. tibetica* a.o.) are characteristic for the uppermost tree growth on S-facing slopes along the entire mountain arc (photo 2), occurring from fairly dense stands to very open woodlands and finally to single isolated individuals. S-facing slopes are often lacking a distinct and easily recognizable sequence of altitudinal belts, especially in the drier NW of the HKH-region, due to the homogenizing effect of irradiation exposure.

Photo 2 Actual upper timberline at a south-facing slope in Bagrot V./Karakorum, formed by *Juniperus turkestanica* at 3700 m (Schickhoff, 1992-08-04)

Scattered junipers may change their growth form from tree to shrub (e.g. *J. indica, J. recurva*) at higher altitudes and often ascend up to an elevation, which corresponds to the alpine belt of shady slopes. As long as it is not clear, whether they indicate a potentially forested altitudinal belt, they are less suitable for the delimitation of the upper timberline ecotone and the sub-alpine belt. Potential natural stand structure of juniper groves is hard to assess since S-facing slopes are heavily impacted by man and his animals for many centuries.

In this paper, no distinction is made between *Juniperus excelsa* M Bieb., *Juniperus excelsa* ssp. *polycarpos* (K. Koch) Takht. (syn. *J. macropoda* Boiss., *J. polycarpos* K. Koch, *J. seravschanica* Kom.), and *Juniperus semiglobosa* Regel (for recent nomenclature see Farjon et al. 2000) since the information on species names is very confusing in the evaluated literature and correct determination is often doubtful. The respective taxa are subsumed under *J. excelsa*. *Betula utilis* ssp. *jacquemontii* and *Betula utilis* ssp. *utilis* are subsumed under *Betula utilis*.

Before ecological conditions and processes at upper timberline are discussed in subsequent chapters, general patterns will be examined more closely by means of several S-to-N cross-sections along the mountain arc (figure 2). The timberline ecotone is depicted in all cross-sections (figures 3 - 13) with a homogeneous width of 300 m for illustrative reasons. Naturally, ecotone width varies according to timberline physiognomy, topography, topoclimate, exposure, climatic continentality, etc. The cross-sections follow the NW-SE gradient:

Paktia – Safed Koh Mts. – Nuristan – Chitral

In the far NW of the HKH-region, upper timberline is developed at comparatively lowest elevations (figures 2 and 3). Still, open woodlands of *Juniperus excelsa* attain elevations of 3500-3600 m on S-facing slopes in the mountains between Khowst and Gardez (33°34'N/69°31'E) (Rathjens 1969, Freitag 1971). Here, coniferous forests are delimited downslope by a drought-related lower timberline. Further northeast, in the Safed Koh Mts., upper timberline occurs at slightly higher elevations. At Mt. Sikaram (4761 m), *Picea smithiana* and *Juniperus excelsa* were observed to be the primary timberline trees on shady slopes at 3500-3600 m (34°03'N/69°54'E) (Breckle 1973, 1975). The occurrence of *Betula utilis* at timberline and *Rhododendron* dwarf shrubs higher up (cf. Aitchison 1881; Breckle 1972) indicates higher humidity and already resembles W Himalayan timberline patterns. Aitchison (1881) and Fischer (1970) mentioned *Abies pindrow* as associated with *Picea*

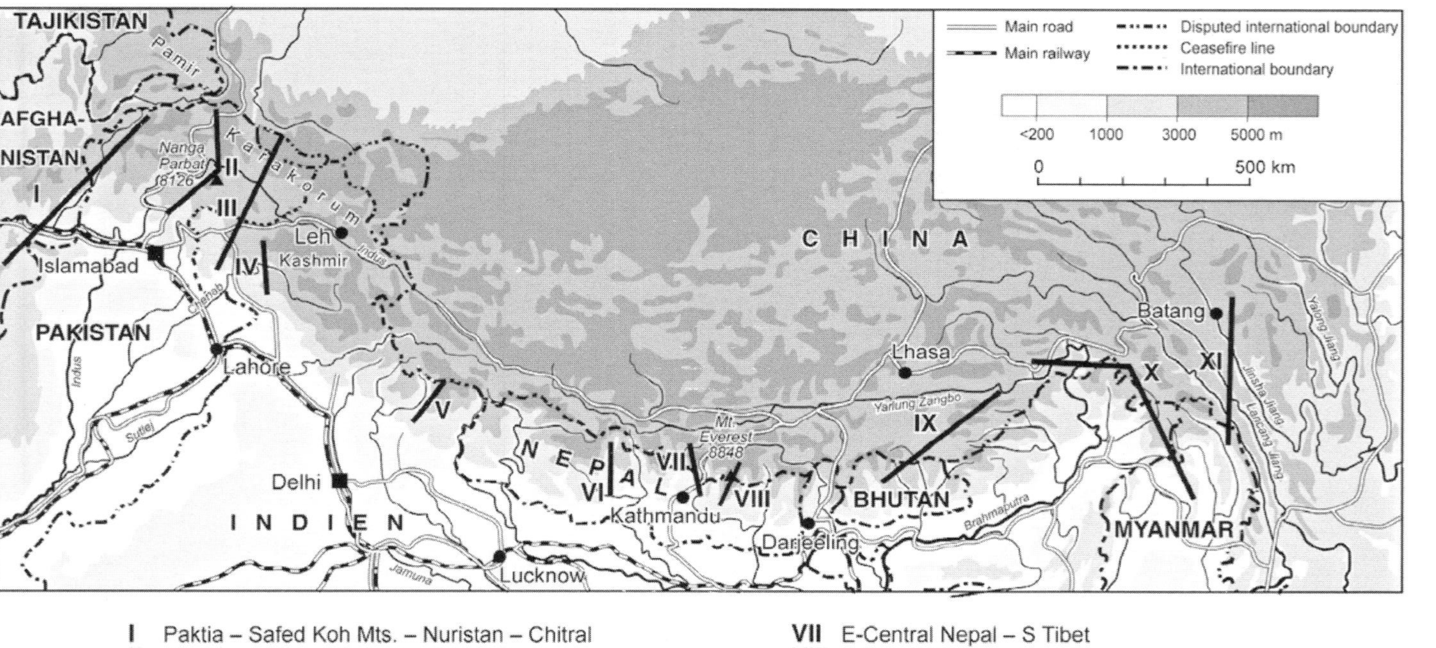

Figure 2 Location of timberline ecotone cross-sections described in the text

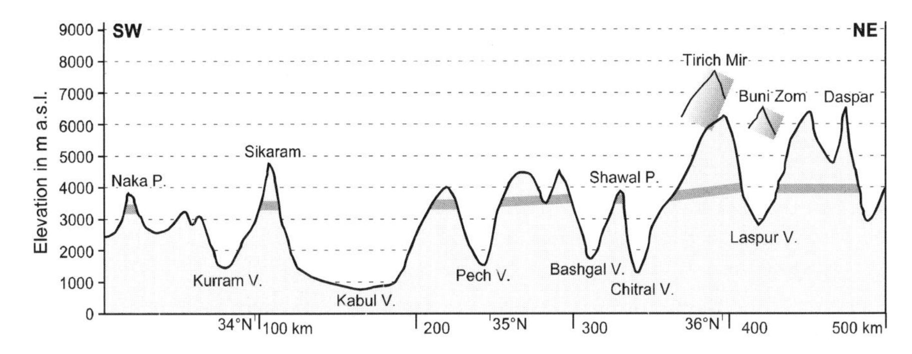

Figure 3 Altitudinal position of timberline ecotone on north-facing slopes along the cross-section Paktia – Safed Koh Mts. – Nuristan – Chitral

and *Juniperus*, the latter found *J. excelsa* ascending to 3700 m on south-facing slopes.

Across the wide valley of Kabul River to the NE, upper timberline elevations do not increase for a while. Elevations between 3300 and 3600 m were recorded at northern aspects in E Nuristan, the principal tree species being *Betula utilis*, *Abies pindrow*, *Pinus wallichiana*, *Picea smithiana*, and *Juniperus excelsa* (cf. Voigt 1933; Linchevsky and Prozorovsky 1949; Volk 1954; Kitamura 1960; Fischer 1970; Freitag 1971; Breckle and Frey 1974; Rathjens 1974). In dry inner valleys, *Pinus gerardiana*, increasingly common further west, may be the principal upper timberline species at 3300-3400 m (Linchevsky and Prozorovsky 1949; Breckle and Frey 1974). Similar elevations (3400-3500 m) are reported from northern aspects in lower Chitral, where *Betula utilis* and *Juniperus excelsa* dominate at upper timberline (Haserodt 1980; 1989; own observations). However, if a krummholz belt (*Salix denticulata*, *Salix karelinii*, *Betula utilis*) is well developed, its upper limit may occur as high as 3800 m in S Chitral (cf. Nüsser and Dickoré 2002). Scattered junipers may reach exceptional elevations even in this section: Breckle (1973) found *Juniperus excelsa* growing at 4400 m in E-exposed rock crevices in Suyengal V. (upper Nuristan; 35°52'N/71°06'E), attaining a height of 2 m. In upper Chitral, uppermost birch stands are reported to ascend to 3850 m, followed by *Salix* krummholz up to 4150 m (Pargham Gol; 35°57'N/72°15'E; Haserodt 1980). However, because of the dry climate these stands are progressively confined to azonal sites such as water surplus habitats, and they are patchy and isolated from forests below since monsoonal influence and humidity is no longer sufficient for the growth of closed sub-humid coniferous forests in the montane and subalpine belts (cf. Schickhoff 2000a). South facing slopes are covered by montane/alpine dwarf shrublands and steppes, which are, in some distance away from settlements, sprinkled with isolated juniper groves or single individuals. Uppermost *Juniperus excelsa* trees/shrubs were recorded at 4000 m above Barum Glacier (36°13'N/71°58'E; Wendelbo 1952) and at 4050 m at Khot Pass (36°31'N/72°38'E; Haserodt 1980). Summarizing the altitudinal position of upper timberline along this cross-section, the ascent amounts to 300-400 m in northern and southern aspects over a distance of about 400 km. Considering the exceptional finding of *Juniperus excelsa* at 4400 m (see above), the S-N increase in elevation is still higher.

Proceeding from this cross-section to the west, upper timberline elevations further decrease, in particular on N-facing slopes. According to Kerstan (1937), Neubauer (1954/55) and Breckle (1973) the SE-slope of the central Hindu Kush has upper timberlines at 3000-3200 m in the side valleys of Pech and Alingar rivers (Nuristan/Laghman). In the lower Panjshir V., *Pinus gerardiana* does not ascend to more than 3000 m (Fischer 1970). However, single crippled junipers (*J. excelsa*) may still be found at 4000 m (Kotal-i-Spé Pass, 35°05'N/ 69°50'E; Frey and Probst 1982). This holds true for the NW-slope of the Hindu Kush as well: Juniper trees (*J. excelsa*) attain elevations of 3800-4000 m in the Parshui, Farkhar, and Jokham V. (Grötzbach 1965; Breckle and Frey 1974; Frey et al. 1976). Towards central Afghanistan, where severe summer droughts exclude juniper trees, upper timberline is made up by *Amygdalus* spp. at 2700-2900 m (Freitag 1971).

Kaghan – Nanga Parbat – Hunza Karakorum

In the outer ranges of NW Himalayas (lower Kaghan, Kohistan, Swat), present-day upper timberline occurs around 3500 m at N- and S-facing slopes (figures 2 and 4) (Ogino et al. 1964; Schickhoff 1993; 1994; 2000a; Rafiq

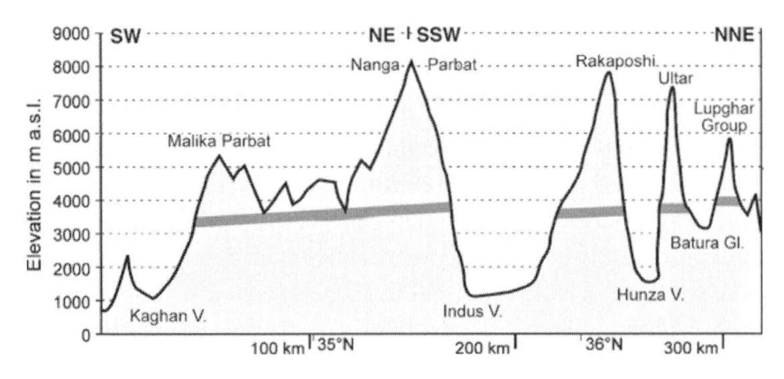

Figure 4 Altitudinal position of timberline ecotone on north-facing slopes along the cross-section Kaghan – Nanga Parbat – Hunza Karakorum

1996). *Betula utilis* is the principal tree species on shady slopes, forming a narrow forest belt between coniferous forests (*Abies pindrow*, *Pinus wallichiana*) below and *Salix denticulata* krummholz above (photo 3). Southern aspects at timberline elevations are covered by uppermost stands and crippled individuals of *Pinus wallichiana* with some *Juniperus excelsa*. Towards the inner ranges, upper timberline considerably increases in elevation to 3750 m (N-exp.) and 3800 m (S-exp.) (e.g. in upper Kaghan; 34°54'-56'N/73°39'-43'E). Here, *Betula* and *Salix* are more and more confined to strictly northern aspects, and *Artemisia* dwarf-shrublands and alpine steppes wooded with very open *Juniperus* stands become the distinctive feature on S-facing slopes (Schickhoff 1993; 1996a). Immediately north of the Himalayan crestline (Babusar Pass; 35°08'N/74°02'E), upper timberline

Photo 3 Altitudinal zonation of timberline vegetation at a north-facing slope in lower Kaghan V./W Himalaya (3200-3500 m): *Abies pindrow* forests give way to *Betula utilis* stands, followed upslope by *Salix denticulata* krummholz (Schickhoff, 1990-05-30)

potentially lies at about 3700 m (*Betula utilis*), but is anthropogenically depressed to 3500 m (*Pinus wallichiana*) (own observations).

The considerable increase in timberline elevations continues north of the main Himalayan range, especially around Nanga Parbat massif (8126 m). Elevations of 3800-3900 m (*Betula utilis, Salix karelinii*) were recorded on shady slopes in Diamir, Rupal, Raikot and Chilim V. (Schlagintweit 1854-57; Finsterwalder 1938; Wagner 1962; Repp 1963; Schickhoff 1996b; Dickoré and Nüsser 2000). *Salix karelinii* krummholz even extends up to 4420 m in Kosto Gah (35°20'N/74°27'E), decreasing in height from 250 cm in moist ravines lower down to below 30 cm at its upper distribution limit (Nüsser 1998). However, since this willow species has an altitudinal range from 2100 m up to 4500 m (Polunin and Stainton 1984), it is not very suitable for the delimitation of altitudinal vegetation belts (see also Hartmann 1968). Scattered trees of *Juniperus excelsa* and *J. turkestanica* make up the timberline in southern aspects at 4100-4200 m in Diamir (35°16'N/74°27'E) and Rupal V. (35°12'N/74°40'E) (Wagner 1962; Dickoré and Nüsser 2000; own observations). It is quite remarkable and points to a considerable human impact in recent decades that Troll (1939) observed much higher timberline elevations in some of the same localities sixty years ago (*Betula utilis*: 4150 m; *Juniperus excelsa*: 4250 m).

Towards the Hunza Karakorum further NE, upper timberline shows a slight decrease in elevation immediately north of the Nanga Parbat region before it again increases a little in the northern high mountain ranges of the NW Karakorum. In the valleys of Bagrot, Naltar, Chaprot, and on the Rakaposhi North Slope (Nilt, Minapin) upper timberline develops between 3700 and 3800 m in northern aspects, made up by *Betula utilis* and *Salix karelinii* krummholz (photo 4), whereas south-facing slopes have uppermost *Juniperus turkestanica* trees between 3800 and 4000 m (Paffen et al. 1956; Miehe et al. 1996; Richter et al. 1999; Schickhoff 2000a; 2002; Reineke 2001). Proceeding further northeast, single dwarf juniper trees (*J. turkestanica, J. excelsa*) attain elevations around and above 4000 m, e.g. above Batura and Pasu Glacier (4000 m; Visser 1928; Paffen et al. 1956; Dickoré and Miehe 2002; Eberhardt 2004; own observations; photo 5), in Morkhun V. (3900-4000 m; 36°35'N/74°58'E; Klötzli et al. 1990; Klötzli 2004; Esper 2000; own observations), in Lupghar V. and upper Chupursan V. (3980 m; 36°44'N/74°35'E, 36°50'N/74°17'E; Eberhardt 2004), above Sarat (4100 m; 36°20'N/74°50'E; Braun 1996), and in Hispar V. (4350 m; 36°10'N/75°03'E; Workman 1910). North of Batura Glacier, even north-facing slopes may show upper birch stands (*Betula utilis*) around 4000 m with *Salix* krummholz reaching 4200 m (Paffen et al. 1956). Recently, Eberhardt (2004) reported uppermost

Photo 4 *Betula utilis* at upper timberline in Nilt V./Rakaposhi North Slope, Karakorum at 3650 m (Schickhoff, 1992-07-19)

Betula trees from 4320 m above Batura Glacier (Wartom V.; 36°38'N/ 74°37'E). However, like in upper Chitral these stands are patchy and isolated and are surrounded by subalpine/alpine steppes and montane dwarf shrublands because of the dry climate (see above). Further to the north, on the outer slopes of W Kun Lun (Kongur-Oytagh; 38°53'N/74°20'E), upper timberline (*Juniperus pseudosabina*, *Picea schrenkiana*) decreases again in elevation to 3300-3700 m (Dickoré and Miehe 2002).

Along this entire cross-section, upper timberline elevations increase by about 500-600 m in northern aspects and 600-700 m in southern aspects over a distance of ca. 300 km. However, nearly all of this increase already occurs along the first 150 km towards the main Himalayan range and the Nanga Parbat massif, whereas upper timberline elevations do not change much in the Transhimalayan part of this cross-section. This pattern seems to be valid also for natural conditions, although human impact has, of course, modified timberline elevations in many localities.

Photo 5 One of the last remaining near-natural *Juniperus turkestanica* stands at upper timberline (3610 m) above Batura Glacier/Karakorum (Schickhoff, 1994-07-10)

Jammu – Kashmir – Baltistan/Central Karakorum

The outer Himalayan ranges of Jammu and Kashmir have upper timberlines of 3500-3600 m in all aspects as recorded for the Wardwan/Chenab V. (33°48'N/75°50'E; Drew 1875; Meebold 1909; Meusel and Schubert 1971), Bhadarwah Hills (32°55'N/75°40'E; Kaul and Sarin 1974), or the Pir Panjal Range (figures 2 and 5) (Drew 1875; Singh 1929; Wright 1931; Polunin 1956-57; Rao 1961;

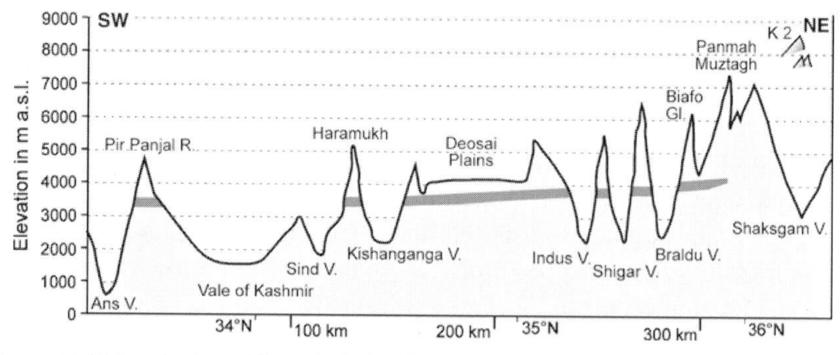

Figure 5 Altitudinal position of timberline ecotone on north-facing slopes along the cross-section Jammu – Kashmir – Baltistan/Central Karakorum

Gupta and Kachroo 1983; Singh and Kachroo 1983). In contrast to the regions adjoining to the northwest, *Rhododendron campanulatum* now appears to be a major component of subalpine forests on north-facing slopes pointing to a stronger monsoonal influence in Jammu and Kashmir. It occurs in the understory of *Betula utilis* and *Abies pindrow* forests, but may also form pure stands or is associated with *Salix* and *Juniperus* krummholz further upslope. South-facing slopes are often covered with *Pinus wallichiana* forests giving way to juniper stands (*J. turkestanica, J. excelsa*) at the treeline ecotone. Elevations are still around 3500-3600 m towards the main Himalayan Range, e.g. at Zoji La (35°17'N/75°24'E; Drew 1875; Polunin 1956-57; Billiet and Leonard 1986; Hartmann 1990), in Liddar V. (34°09'N/75°16'E; Duthie 1894-98; Meebold 1909; Dhar and Kachroo 1983), Dachhigam woodland (34°10'N/75°01'E; Singh and Kachroo 1976), Dagwan V. (34°15'N/75°00'E; Herzhoff and Schnitzler 1981) and at Burzil (34°56'N/75°05'E) and Tragbal Pass (34°29'N/74°38'E; Troll 1937). Drew (1875) found *Betula utilis* growing up to 3600-3700 m at Kamri Pass (34°44'N/74°57'E). Mani (1978) gives a general regional figure of 3600 m.

A considerable increase in elevation is to be witnessed beyond the main Himalayan range in Baltistan, where the upper timberline of *Betula utilis* and *Salix karelinii* krummholz lies between 3700 and 3900 m at north-facing slopes in the valleys west of Skardu (own observations). Webster and Nasir (1965) found *Betula utilis* (attaining a height of up to 5 m) and *Sorbus tianshanica* at 4000-4100 m in upper Hushe V. (35°31'N/76°24'E). Uppermost junipers grow between 3800 and 4000 at south-facing slopes, e.g., near Marol (Suru/Indus confluence; 34°46'N/76°13'E) (Meebold 1909), and in the Skardu-Khaplu area (Miehe et al. 1996; Esper 2000; Schickhoff 2000a; own observations). Entering the Central Karakorum Range, the increase in elevation continues up to 4000-4200 m in northern aspects at Biafo Glacier (35°50'N/75°48'E), where Hartmann (1966) observed *Juniperus excelsa* and *Salix karelinii* krummholz. At the Central Karakorum north slope, Dickoré (1991) found *Juniperus turkestanica* growing at almost 4000 m. Further north, upper timberline elevation considerably decreases again. In the Kun Lun, forest line (*Picea schrenkiana*) develops at about 3400 m on north-facing slopes, whereas junipers (*J. turkestanica, J. centrasiatica, J. excelsa*) reach 3600 m in southern aspects (Zheng 1990; Zhang 1995).

Summing up, current upper timberline elevations increase 500-600 m along a cross-section of about 300 km. Exceptional high records, which are reported in older sources, are not considered in this calculation. Meebold (1909) found *Betula utilis* at 4200 m near Burji La (Satpara/Deosai; 35°07'N/75°33'E). Calciati (1914) even observed a dead *Juniperus excelsa* specimen at 5000 m at Kaberi Glacier (35°25'N/76°38'E), which is probably a relic from warmer periods in the Holocene.

Himachal Pradesh – Zanskar

This cross-section from the mountains around Beas River and Dhaola Dhar Range across Chamba and Lahul into Zanskar shows an upper timberline increase of ca. 400 m along a distance of ca. 100 km (figures 2 and 6). North-exposed timberline of *Betula utilis*, *Rhododendron campanulatum*, *Sorbus aucuparia* and *Salix* spp. develops at 3500-3600 m in the outer ranges, e.g. at Beas River (32°01'N/77°25'E; Bruce 1914; Gorrie 1933), and in Uhl V. (Dhaola Dhar Range; 32°19'N/76°51'E) (Chandra 1949). Wynter-Blyth (1951) encountered the upper timberline at ca. 3700 m in Kulu V. (32°12'N/77°09'E). According to Rau (1974), upper timberline (*Betula utilis*, *Rhododendron campanulatum*, *Juniperus* spp.) even attains 4200 m in Brahmaur V. (Chamba; 32°40'N/77° E), a figure which has to be verified since it seems to be very exceptional. Nevertheless, towards the inner ranges timberline elevations considerably increase to 3800-4000 m as observed in inner Sutlej V. (Thomson 1852; Gorrie 1933; Schweinfurth 1957a), where locally *Pinus wallichiana* or *Quercus semecarpifolia* may form an upper forest belt. However, *Betula utilis* and *Rhododendron campanulatum* usually dominate the uppermost forest and krummholz belts. Drew (1875) encountered uppermost birch stands at ca. 3800 m in upper Bhutna V. (33°30'N/76°15'E). Seybold and Kull (1985) found uppermost *Juniperus excelsa* trees on both sides of Shingo La (pass between Lahul and Zanskar; 32°58'N/77°02'E) at about 3900 m. The same altitude is given by Gupta (1994) for junipers in Zanskar and Ladakh. Further north at Khardung (Ladakh; 34°23'N/77°40'E), Thomson (1852) still saw uppermost junipers at 4250 m in northern aspect.

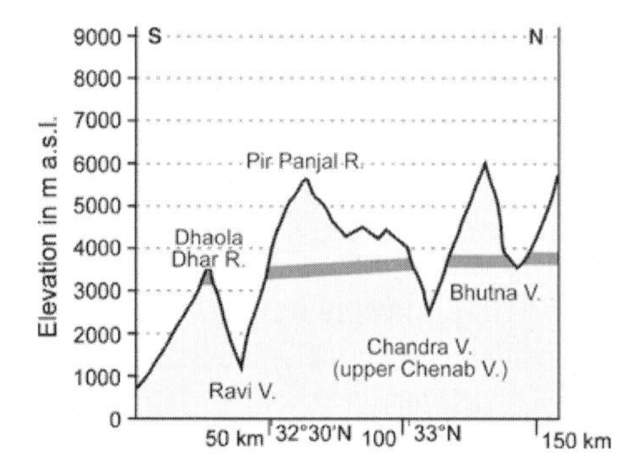

Figure 6 Altitudinal position of timberline ecotone on north-facing slopes along the cross-section Himachal Pradesh – Zanskar

Uttaranchal

Timberline elevations in the outer Himalayan ranges of Uttaranchal differ broadly between 3500 and 3700 m (Singh and Singh 1992) and tend to be slightly higher compared to the regions further northwest due to moister, warmer summer conditions (figures 2 and 7). Timberline observations are almost exclusively restricted to north-facing slopes and include the mountains in outer Tehri Garhwal (Heske 1932), around Nainital (Meusel and Schubert 1971), Pindari/Sarju V. at the southern slopes of Nanda Devi massif (7816 m) (30°07'-10'N/79°48'-80°04'E; Kalakoti et al. 1986; Rawal and Pangtey 1993, 1994; Rawal and Dhar 1997; Maikhuri et al. 1998), and Tungnath (30°30'N/79°15'E; Sundriyal 1994). Principal timberline species are again *Betula utilis* and *Rhododendron campanulatum*, sometimes associated with *Abies pindrow*, *Juniperus recurva*, *Pyrus foliolosa*, *Betula alnoides*, *Rhododendron arboreum*, *Rh. barbatum*, *Sorbus* spec. and/or *Quercus semecarpifolia*.

A more or less continuous and quite steep increase in elevation towards the inner ranges and the Great Himalaya can be observed. Timberline lies around 3700-3800 m in upper Yamuna V. (31°15'N/78°19'E; Heske 1932, 1937; Tyson 1954; Puri et al. 1989), Yumnotri (31°03'N/78°28'E; Gupta 1983), in the Valley of Flowers National Park (30°41'-30°48'N/79°33'-79°46'E; Kala et al. 2002), and in the Nanda Devi Biosphere Reserve (30°17'-41'N/79°40'-80°50'E; Longstaff 1907; 1908; 1928; Champion 1936; Garkoti and Singh 1994). In the latter, it may occasionally ascend to 3900-4000 m, e.g. in Kuhti Yankti V. (Schmid 1938; Heim and Gansser 1939), or in Bishanganga and Dhauliganga V. (Osmaston 1922). Even higher timberlines were found at Gangotri Mountain (*Betula utilis* at 4100 m; N-exp.; 30°55'N/78°52'E; Dudgeon and Kenoyer 1925; Heske 1932), in Tons V. (*Betula utilis*

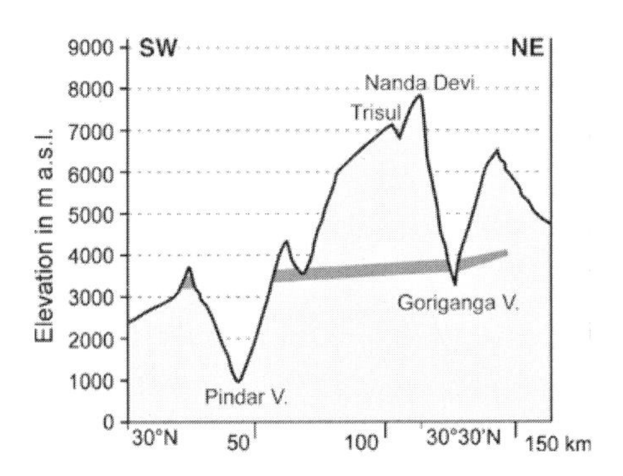

Figure 7 Altitudinal position of timberline ecotone on north-facing slopes along the cross-section Uttaranchal

and *Rhododendron campanulatum* at 4000-4100 m; N-exp.; 31°11'N/78°13'E; Gupta and Singh 1962), and at southern aspects in upper Bhagirathi V. (*Juniperus* spec. at 4250 m; 31°06'N/78°58'E; Gupta 1983). Beyond the Indian-Tibetan border in this region (upper catchment of Sutlej River/Langchen Zangbo), humidity considerably decreases in the rain shadow of the Great Himalaya and is no longer sufficient to support forest growth. Along this cross-section, the increase in timberline elevations amounts to ca. 500 m in northern aspects and 500-600 m in southern aspects at a horizontal distance of ca. 100 km.

W-Central Nepal (Dhaulagiri/Annapurna Himal)

Proceeding from the Himalayan foothills and midlands to the north, mountain ranges surpass timberline elevation for the first time northwest of Pokhara at the southern escarpment of Dhaulagiri and Annapurna Himal, where timberline lies between 3600 and 3700 m (figures 2 and 8) (Goodfellow 1954; Schweinfurth 1957a; Miehe 1982; Metz 1998; own observations). Upper subalpine forests in northern aspects are composed of *Betula utilis* and *Abies spectabilis*, followed upslope by *Rhododendron campanulatum* krummholz. As soon as one enters these huge high mountain massifs, timberline elevations show a remarkable increase. Along the western slope of Annapurna (8091 m), upper timberline at north-facing slopes (*Betula utilis*, *Rhododendron campanulatum*) ascends to 4000-4100 m above Ghasa (28°36'N/83°37'E) and Marpha (28°45'N/83°40'E) in the Thak Khola gorge (Kali Gandaki), and even to 4400 m on Nilgiri north slope (28°43'N/83°45'E) (Dobremez and Jest 1971; Grey-Wilson 1974; Miehe 1982). Similar altitudes are reached at Annapurna north slope, e.g., in Kone Khola (*Betula utilis*; 4300 m; 28°43'N/

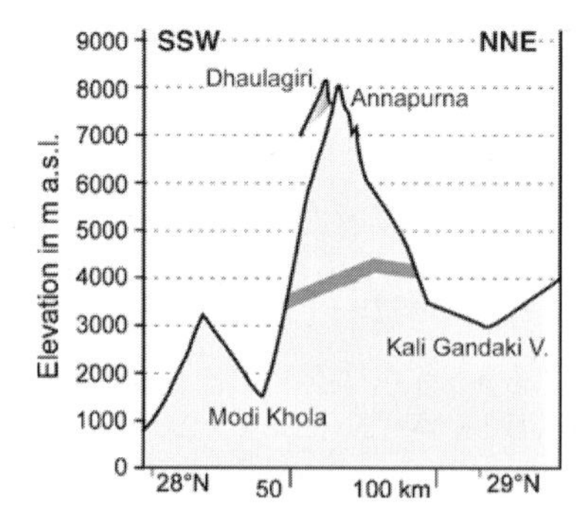

Figure 8 Altitudinal position of timberline ecotone on north-facing slopes along the cross-section W-Central Nepal (Dhaulagiri/Annapurna Himal)

83°59'E; Miehe 1984), and in Jargeng Khola (upper Marsyandi V.; 28°40'N/ 83°56'E), where *Betula utilis* was found growing in southern aspects up to 4500 m, in northern aspects up to 4200 m, mixed with *Rhododendron*, *Salix* and *Juniperus* spp. (cf. Imanishi 1954; Nakao 1955; Kawakita 1956; Schweinfurth 1957a). With increasing distance from Annapurna massif to the north, timberline elevations slightly decrease again – a similar pattern as at Nanga Parbat massif in NW Himalaya. Around Kagbeni and Muktinath (28°45'-51'N/83°45'-55'E) and in Cha Lungpa V. (28°54'N/83°38'E), uppermost *Betula utilis* stands reach 4150 m, whereas southern aspects have *Juniperus indica* groves between 4000 and 4300 m (cf. Tilman 1952; Kawakita 1956; Dobremez and Jest 1971; Miehe 1982; Miehe et al. 2002). In the semiarid upper Kali Gandaki, the landscape has already Tibetan character and is almost devoid of trees, that are replaced by thorn shrublands. According to Miehe et al. (2002), the drought limit of *Juniperus indica* stands corresponds to an annual precipitation of ca. 250 mm in northern Mustang (Ghami area).

The Dhaulagiri massif (8172 m) exerts a similar influence on upper timberline elevations. Miehe (1982) observed *Betula utilis* and *Juniperus recurva* krummholz ascending to 3850 m in upper Mayangdi Khola at 28°40'N/83° 27'E, southwest of Dhaulagiri I. Immediately northwest of Dhaulagiri massif, in Barbung Khola (upper Bheri V.; 28°50'N/83°15'E), upper timberline formed by *Betula utilis* and *Juniperus indica* krummholz rises to 4300-4500 m (cf. Polunin 1950a; Tichy 1954; Stainton 1972; Kleinert 1974). This cross-section shows the steepest increase in upper timberline elevations. Over a very short distance of only 20-30 km, from the deeply incised valleys at the southern escarpment of Dhaulagiri and Annapurna Himal to the high valleys in the rain shadow areas on the northern declivity, upper timberline rises by 700-800 m.

Proceeding further NW and WNW respectively, timberline elevations do not exceed 4200 m, a figure which is not surpassed in the entire NW Nepal according to Schweinfurth (1957a), Dobremez (1976), and Shrestha (1985). However, Kleinert (1974) observed *Pinus wallichiana* and *Betula utilis* at above 4200 m near Shey Gompa (Kanjiroba north slope; 29°17'N/82°55'E). Bishop (1990) mentioned *Betula utilis* at northern slopes and *Juniperus wallichiana* in southern exposures growing up to 4000-4100 m in the upper Mugu Karnali V (29°39'N/82°40'E). For the Humla-Jumla area and Dozam Khola (30°04'N/82°05'E), elevations of 4150 m (N-exp.; *B. utilis*, *Rh. campanulatum*) and 4200 m (S-exp.; *Juniperus indica*) were given by Williams (1953) and Dobremez and Shrestha (1980).

E-Central Nepal – S Tibet

A comparably steep gradient in timberline elevations exists in eastern Central Nepal when crossing the main Himalayan range via Helambu and Langtang (figures 2 and 9). In the outer ranges, upper timberline lies at about 3700 m in northern (*Rhododendron campanulatum, Betula utilis*) and southern aspects (*Juniperus recurva*) as observed at Chyochyo Danda (27°55'N/85°44'E) by Schmidt-Vogt (1990a) and at Yangri Danda (27°56'N/85°41'E; photo 6) by Schickhoff and Klaucke (1988). Towards the main Himalayan range upper timberline already rises to 4020 m (N-exp.) and 4100 m (S-exp.) at Dupku (Helambu; 28°05'N/85°35'E) (Miehe 1990). Further to the west, in Shiar Khola/Buri Gandaki V. (28°33'N/85°05'E), *B. utilis* and *Rh. campanulatum* are mixed with *Rhododendron barbatum, Larix griffithiana*, and *Juniperus* spp. at slightly lower elevations (3800-3930 m) (Nakao 1955; Kawakita 1956). Entering Langtang Himal, upper timberline attains similar altitudes as in Dhaulagiri and Annapurna Himal. At northern exposures in Langtang V., upper timberline is often around 4000 m (Malla et al. 1997). However, stands of *B. utilis, Rh. campanulatum*, and *Sorbus microphylla* may occur between 4200 and 4400 m (cf. Polunin 1952; Miehe 1990). *Rhododendron* and *Sorbus* form a second tree layer in *Betula* forests, which are followed by *Rhododendron* krummholz stands further upslope. These stands are again superseded by dispersed, multi-stemmed *Sorbus microphylla* trees that attain a height of 2 m within a matrix of *Rhododendron anthopogon* dwarf scrub (Miehe 1990). *Juniperus recurva* inhabits southern slopes up to 4540 m (at Tikeapsa; 28°12'N/85°38'E) (Miehe 1990). In Chilime V. (right tributary of Trisuli River; 28°23'N/85°15'E), upper subalpine forests (*B. utilis, S. microphylla, Rh. campanu-latum*) may even reach 4500 m in sheltered locations (cf. Lloyd 1950; Polunin 1950b). Upper timberline elevation decreases again to 4300 m

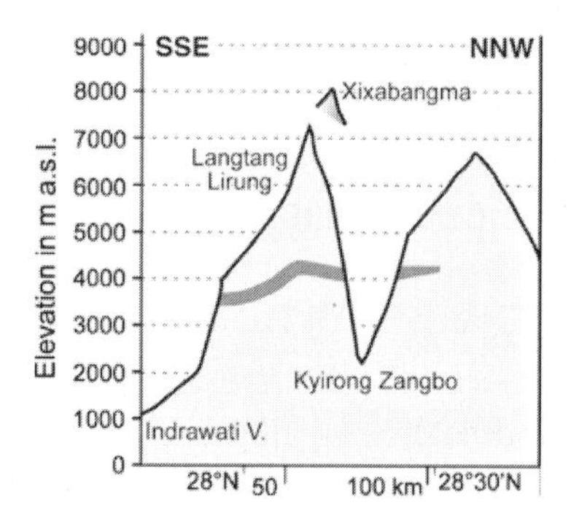

Figure 9 Altitudinal position of timberline ecotone on north-facing slopes along the cross-section E-Central Nepal – S Tibet

Photo 6 Upper timberline (3700 m) at Yangri Danda (Helambu, Central Himalaya) made up by *Rhododendron campanulatum* (Schickhoff, 1986-09-10)

in the upper Trisuli catchment (28°32'N/85°16'E) beyond the Tibetan border (Kyirong Zangbo), where Miehe and Miehe (2000) observed dwarf trees of *Sorbus microphylla* with *Lonicera obovata* and *Rhododendron campanulatum* within the treeline ecotone.

Along this cross-section, increase in upper timberline elevations amounts to 600-700 m in northern exposures and 700-800 m at southern slopes over a horizontal distance of only 30-40 km.

E Nepal (Khumbu Himal)

About 100-150 km further to the east, upper timberline elevation in the outer Himalayan chains continues to increase due to more favourable climatic conditions at same elevational levels (figures 2 and 10). Approaching the Mt. Everest region from the south, Haffner (1979) reported lowest upper timberline

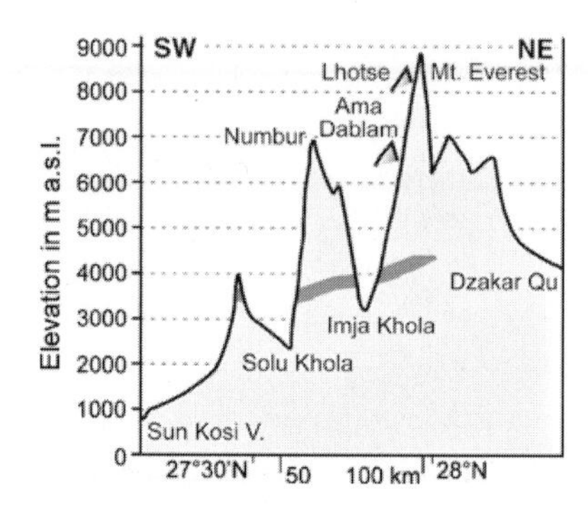

Figure 10 Altitudinal position of timberline ecotone on north-facing slopes along the cross-section E Nepal (Khumbu Himal)

elevations of 3700 m (N-exp.) at the southern escarpment of Numbur-Kang range (6959 m) (*Abies spectabilis*, *Rhododendron campanulatum*). However, Numata (1967) and Yoda (1967) observed timberline elevations of 3900-4000 m (*Juniperus indica*, *Rhododendron lanatum*, *Rhododendron campanulatum*) at Mt. Numbur south slope (Solu Khola; 27°40'N/86°30'E). A general figure of 3800 m for the outer ranges of E Nepal was given by Dobremez (1976), who mentioned *Betula utilis* and *Rhododendron* spp. as principal timberline species. Proceeding towards the inner ranges, a considerable increase in elevation is confirmed by several sources. Heuberger (1956) described *Juniperus recurva* krummholz in the Nangpo V. (27°54'N/86°35'E), NW of Namche Bazar, ranging from 3800 to 4200 m. Haffner (1967) observed here *Betula-Abies-Rhododendron* forests up to 4000 m at NE-aspect. Sakai and Malla (1981) noted the same forest type near Tengpoche (27°49'N/86°47'E) at about 3900 m. Above Pangpoche (Imja Khola, NE of Namche Bazar; 27°52'N/86°48'E), timberline further rises to 4200-4300 m on northern slopes, above Chulungche (27°52'N/86°49'E) to 4400 m, primarily formed by *Betula utilis*, *Rhododendron campanulatum*, and *Sorbus microphylla* (cf. Shipton 1952; Troll and Schweinfurth 1968; Donner 1972; Haffner 1967; 1972; 1979; Miehe 1991a; b). Southern exposures at this location support *Juniperus recurva* groves up to 4440 m (Miehe 1989; 1991a), the same figure (4400 m) is given by Stainton (1972) for the upper juniper tree level in inner valleys of E Nepal.

In upper Barun Khola (upper Arun catchment; 27°49'N/87°05'E) adjoining to the east, upper timberline elevations may even reach 4500 m, a krummholz figure provided by Ohsawa et al. (1973) for northern slopes (*Rhododendron* spp., *Juniperus* spp., *Salix* spp.), and by Byers (1996) for southern aspects (*Juniperus recurva*). Other sources for upper Arun catchment give figures of

4000-4200 m (Howard-Bury 1922a; Carpenter and Zomer 1996; Chaudhary and Kunwar 2002), which are also valid for Rolwaling Himal adjoining Khumbu Himal to the west (cf. Howard-Bury 1922b; Kikuchi 1991).

Like the E-Central Nepal cross-section, this cross-section is marked by an enormous increase in upper timberline elevations over a very short distance. Northern exposures show an increase of 600-700 m, southern slopes one of 700-800 m at a horizontal distance of only 50 km.

Sikkim/Bhutan – S-Central Tibet

The outer ranges of Sikkim and Bhutan Himalaya have lowest upper timberlines at about 3600-3700 m with increasing elevations towards the east (figures 2 and 11). Pradhan et al. (2001) described the upper timberline being composed of *Abies densa*, *Betula utilis*, *Arundinaria aristata*, and *Rhododendron campanulatum* at this altitude in their study of Singalila National Park (26°31'-27°31'N/87°59'-88°53'E). Older records give considerably higher elevations. Champion (1936) and Kanai (1963) reported several *Rhododendron* species (*Rh. campanulatum*, *Rh. wightii*, *Rh. lanatum*, *Rh. molle*) from the upper timberline at 3800-4000 m in the Singalila Range (27°20'N/88°10'E). *Juniperus pseudosabina* krummholz had even reached 4500 m on the northern ridge of Singalila Range according to Gammie (1894). East of Tista River, upper timberline in the outer ranges occur at elevations up to 4000 m as exemplified by *Abies densa* stands at the west face of Jelep La (27°20'N/88°55'E; Gammie 1894; Shebbeare 1934). *Rhododendron* krummholz attains 4200 m on northern slopes in the same area (above Karponang) (Rao 1963). Troll (1937; 1972) observed a rather abrupt timberline at 3900 m on a southern slope SE of Tsomgo Lake, made up of *Abies densa*

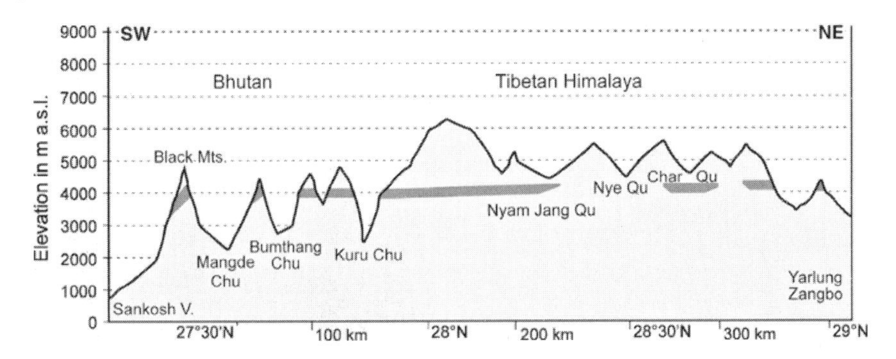

Figure 11 Altitudinal position of timberline ecotone on north-facing slopes along the cross-section Sikkim/Bhutan – S-Central Tibet

with an undergrowth of *Betula utilis* and *Rhododendron* spp. Trees in the upper subalpine forests here are clad with *Usnea* beardmosses indicating very high humidity. *Betula utilis*, already of minor importance in eastern Nepal (cf. Chaudhary 1998), has finally lost its predominance in the upper subalpine forest belt of Sikkim. Instead, *Rhododendron* spp. (photo 7) are prevalent throughout the eastern Himalayas. Ascending Natu La (4310 m; 27°25'N/ 88°46'E), *Abies* stands reach 3700-3800 m (Taylor 1947). *Rhododendron* spp. can grow up to 4 m high in the upper subalpine belt, and *Rhododendron* krummholz reaches 0.5 m even near the top of the pass (cf. Smith 1913; Buchanan 1919; Troll 1937).

Hooker (1855) already provided a general figure for the upper timberline in upper Sikkim (4100 m on north-facing slopes) and for most of the Kangchendzonga massif (8585 m). Subsequent researchers more or less

Photo 7 *Rhododendron* spp. (here SW of Annapurna) are prevalent at upper timberline throughout the eastern Himalaya (Schickhoff, 1984-03-16)

corroborated this figure. 4000-4200 m were observed in Zemu V. (27°45'N/88°10'E; *Rhododendron campanulatum*, *Rh. wightii*, *Juniperus* spec.) (Smith and Cave 1911; Smith 1913; Wien 1937), and above Tanggu (27°54'N/88°34'E) in the already drier upper Lachen V. (*Juniperus* spec.) (Petermann 1861). At southern exposures, *Juniperus* krummholz sometimes stretches to 4500 m, e.g., in Lhonak V. (27°47'N/88°21'E) (Smith and Cave 1911). According to recent observations, upper timberline on north-facing slopes in Kangchendzonga Biosphere Reserve lies at 3700-3800 m, made up of *Abies densa*, *Rhododendron barbatum*, *Rh. hodgsonii*, and *Betula* spec. (Singh et al. 2003).

In the outer Himalayan ranges adjoining to the east similar upper timberline elevations are to be observed. In lower Chumbi V. (in the V-shaped protrusion of Tibetan territory between Sikkim and Bhutan; 27°44'N/89°05'E), Champion (1936) and Ludlow (1937) found the upper timberline at 3700-3900 m, composed of *Rhododendron* spp., *B. utilis*, *Juniperus* spp., and *Salix* spp. However, timberline elevations can be significantly higher in outer ranges further east in Bhutan. Cooper (1933) reported an upper timberline of *Abies* spec., *Juniperus* spec., and *Rhododendron* spp. at 4000 m on a northern slope in the Black Mountains (27°17'N/90°29'E). Recently, Miehe and Miehe (2000) gave an account of a near-natural upper timberline of *Rhododendron* spp. and *Abies densa* at 4250 m in the same area (27°23'N/90°42'E). Gratzer et al. (1999) observed *Abies densa* stands with an understory of *Rhododendron hodgsonii*, *Rh. campylocarpum*, *Rh. campanulatum*, and *Rh. lanatum* up to 3700 m and 4100 m in high precipitation areas of the front ranges and in the somewhat drier interior, respectively. Sargent et al. (1985) gave a general figure of 4200 m for inner valleys of Bhutan and mentioned *Juniperus pseudosabina*, *Picea spinulosa*, *Abies densa* as principal species. Grierson and Long (1983) and Ohsawa (1987) provided information on the upper timberline in Sankosh Valley (27°40'N/90°E), where northern slopes have *Rhododendron campanulatum* and *Rh. cinnabarinum* at 4200 m, and southern exposures are inhabited by *Juniperus recurva* krummholz up to 4400 m. Further observations in inner valleys indicate upper timberlines of similar elevations, including Tremo La (27°43'N/89°15'E; > 4000 m; *Betula utilis*, *Rhododendron* spp.; Cooper 1933), and Pologong Chu (Bumthang V.; 27°30'N/90°35'E; 4000 m; *Rhododendron* spp.; Cooper 1933). In upper valleys of northern Bhutan, shrubby *Rhododendron* thickets (*Rh. bhutanense*, *Rh. aeroginosum*) form the undergrowth of uppermost *Abies densa* and *Betula utilis* stands (4400 m), and may reach up to 4800 m, reducing their height from 2-3 m to 40-50 cm (Miehe 2004).

In eastern Bhutan, timberlines at 3800-4000 m were observed at Narimthang (Kuru V.; 27°55'N/91°13'E) and at Kalong Chu (Manas V.; 27°54'N/91°32'E), dominated by *Rhododendron* spp. and *Abies densa* (Griffith 1847; Ludlow 1937). Information on adjoining westernmost Arunachal Pradesh were provided by Ward (1936; 1940a, b), who ascertained several *Rhododendron* spp. (*Rh. barbatum*, *Rh. hodgsoni*, *Rh. campanulatum*), *Abies*, *Picea* and *Arundinaria* spp. at 3900-4000 m around Se La (27°35'N/92°08'E) and Orka La (27°25'N/92°02'E; upper Bhareli watershed). Towards the Tibetan border (Mon Yul), krummholz of *Juniperus recurva*, *Rhododendron* spp., and *Salix* spp. may even climb to 4500 m (Chaudhuri 1992).

The increase in upper timberline elevations already amounts to several hundred meters from the outer ranges to the main Himalayan range. Nevertheless, upper timberline continues to rise in the Transhimalayan ranges of S-Central Tibet adjacent to the north. Compared to the Transhimalayan ranges further west, a higher monsoonal humidity influx via the SE declivity of the Tibetan plateau and more favourable thermal conditions (mild winters) enable forest growth, at least in isolated patches, quite far beyond the main Himalayan crestlines. These forest patches, dominated by *Juniperus tibetica*, reach their western drought limit at about 90°E. Juniper stands are more or less confined to southern exposures, whereas shady slopes are dotted with *Salix* and *Rhododendron* scrub or with birch stands under more humid conditions (Miehe et al. 2000). Corresponding to the gradient of increasing humidity towards the east, SE Tibet is more extensively forested E of ca. 93°E (cf. Zhang 1988). E.g., the upper Tsari V. (28°43'N/93°23'E) is thickly forested up to 4000 m (Ludlow 1944). Likewise, in the southern tributaries of Yarlung Tsangpo (Ne Chu, Lilung Chu) and in upper Siyom V. around 94°E, dense coniferous forests were observed between 3500 and 4000 m below *Salix*/*Rhododendron* scrub (Ludlow 1944; Taylor 1947; Schweinfurth 1957b).

Juniperus tibetica stands in S-Central Tibet are obviously remnant stands in an old cultural landscape. They occur as isolated, often sacred and thus protected groves, and form the upper timberline between 4600 and 4800 m, often superseding *Juniperus convallium* stands of lower altitudes (Miehe et al. 1998; Miehe et al. 2000). This is the most elevated timberline on earth – with the exception of *Polylepis* stands in the Andes. As an extension of this cross-section from Sikkim/Bhutan northward to the western Nyainqentanglha Shan, the following upper timberline locations, documented in Miehe et al. (1998) and Miehe et al. (2000) can be added: above Tsamtschü (E of Nagarzê; 28°58'N/90°28'E; 4650 m, see also Wardle 1981; Bräuning 1999), above Pamtschü (29°16'N/91°57'E; 4600 m), E of Nyemo River (29°20'N/90°00'E;

4800 m), above Porong Ka Monastery (29°47'N/91°06'E; 4600 m), in upper Kyi Chu catchment (30°07'N/92°08'E; 4600 m), at Reting Monastery (30°18'N/91°31'E; 4750 m), and SW of Damxung (30°22'N/90°51'E; 4280 m, also included in Zhang 1988).

Summarizing the increase in upper timberline elevations along this cross-section, an increase of 700-800 m at a horizontal distance of ca. 300 km can be ascertained. The increase is already quite steep (300-400 m) on the Himalayan south slope. Mass elevation effects of the Tibetan highlands result in a continuous rise north of the crestline towards the Nyainqentanglha Shan.

N Myanmar – Arunachal Pradesh – SE Tibet

Still a bit distant from major mass elevations, outer Himalayan ranges of northern Myanmar have upper timberlines between 3700 and 3900 m as observed in the mountain ranges along upper Mali Hka (27°50'N/97°25'E) and Nmai Hka (e.g. at Imaw Bum; 26°15'N/98°19'E) (Ward 1921; 1944-45; 1946a; b; 1957). However, timberline (*Rhododendron* spp.) may exceed 4000 m as in the range between Nu Jiang (Salween) and Irrawaddy at 28°N (figures 2 ands 12) (Stamp 1925; Handel-Mazzetti 1927a). Proceeding northward to the SE escarpment of the Tibetan Plateau, upper timberline elevations increase to 3900-4100 m at the slopes of the Hkakabo Razi massif (5881 m), e.g., in Seingkhu/Adung V. (28°08'N/97°22'E), above Tahawndam, or in Jité V. (28°27'N/97°57'E) on the Tibetan side (Ward 1932, 1933, 1939, 1944-45). Species-rich moist evergreen broadleaved forests with a great variety of mosses and epiphytes characterize these mountain ranges in lower altitudinal belts, followed by *Rhododendron*-coniferous forests higher up with a prominent undergrowth of *Arundinaria* spp. Several *Rhododendron* spp. (*Rh. selense*, *Rh. praestans*, *Rh. fulvoides*, *Rh. cerasinum* a.o.), *Abies delavayi*, *Betula utilis*,

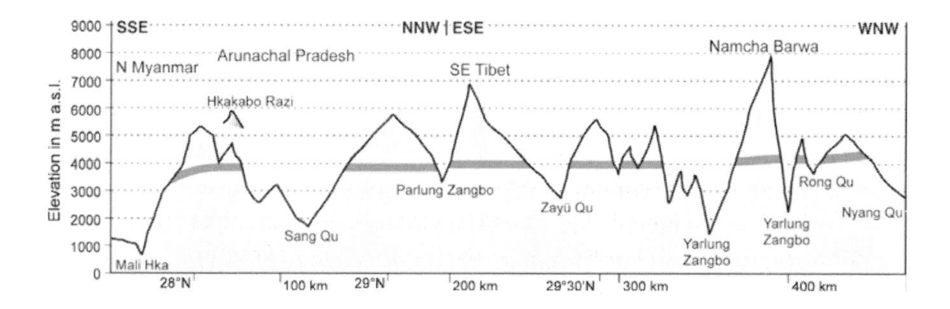

Figure 12 Altitudinal position of timberline ecotone on north-facing slopes along the cross-section N Myanmar – Arunachal Pradesh – SE Tibet

Juniperus spp., *Salix*, *Sorbus* and *Prunus* spp. are principal upper timberline species at the above localities. Comparable timberline elevations are developed in Lohit V. and Mishmi Hills on Arunachal Pradesh territory, adjacent to the west. Rao and Panigrahi (1961) reported a timberline of *Rhododendron* spp., *Juniperus* spp., and *Abies densa* at ca. 4000 m above Walong (28°03'N/ 97°05'E), Ward (1929; 1930) observed *Abies delavayi* and *Rhododendron* spp. at 3800-3900 m in the Mishmi Hills and in Delei V. (28°12'N/97°10'E). In the upper Lohit catchment in SE Tibet, upper timberline, mainly composed of *Rhododendron* spp., *Betula utilis*, *Larix griffithiana*, *Abies spectabilis*, *Abies georgii* var. *smithii*, *Abies delavayi* var. *motuoensis*, slightly increases to 3900-4250 m in upper Zayü Qu (29°30'N/96°20'E; Frenzel et al. 2003; see also Ward 1933-34), north of Lepa (28°57'N/97°05'E; Kaulback 1938), and in upper Sang Qu (Zayü Qu E; 29°10'N/97°20'E) (Ward 1934; Kaulback 1938). Similar altitudes are reached in upper Parlung Zangbo catchment (W of Shugden Gompa; 29°30'N/96°47'E) adjoining to the north (3900 m; *Betula utilis*, *Larix griffithiana*, *Picea balfouriana*, *Picea likiangensis*; Ward 1934; 1936) as well as in the Motuo region (Yarlung Zangbo) adjoining to the west (3900-4200 m; *Rhododendron campanulatum*, *Rh. barbatum*, *Betula utilis*, *Larix griffithiana*, *Abies delavayi* var. *motuoensis*; Chang 1981). In the latter region, which is also called Mêdog region, Xu (1992) describes the timberline ecotone as being situated between 4000 and 4300 m, formed in the upper part by several *Rhododendron* and *Salix* spp. as well as by scattered *Abies delavayi* var. *motuoensis*.

Approaching the Tibetan highlands to the W and NW, upper timberline elevations considerably increase and may reach more than 4700 m. The massifs of Namcha Barwa (7756 m) and Gyala Peri (7142 m) have maximum upper timberlines (*Juniperus*) of 4450 m and 4650 m (29°53'N/94°53'E) at southern exposures (Frenzel et al. 2003). Contrastingly, north-facing slopes show timberlines of only 3900 to 4100-4200 m (*Rhododendron* spp., *Abies delavayi*; Ward 1926; Schweinfurth 1957b; Peng et al. 1997; Bräuning 1999; see also Taylor 1947; Fang et al. 1996; Chapman and Wang 2002). At Sui La (29°50'N/ 95°24'E), NE of Namcha Barwa, Peng et al. (1997) recorded dark coniferous forests (*Abies*) up to 4300 m, Bräuning (1999) worked in *Abies* and *Larix* cf. *griffithiana* stands at 4000-4050 m above Bomi (29°47'N/95°42'E). Further west in Nyang Qu, maximum timberline elevations range between 4400 and 4600 m, e.g. at Nyingchi and Gongbogyamda, where Li, B (1993) listed *Picea linzhiensis*, *Larix griffithiana*, *Betula utilis*, *B. platyphylla*, *Juniperus tibetica* and *J. saltuaria* as primary timberline species. Bräuning (1999) observed *Abies* cf. *delavayi* (N-exp.) and *Juniperus* spec. (S-exp.) at 4300 m at Nyingchi (29°35'N/94°46'E). Miehe and Miehe (2000) recorded *Juni-*

perus tibetica even at 4720 m at Kongbo (30°03'N/93°59'E), in a northern tributary of Nyang Qu. At Atsa Lake further NW (30°40'N/93°15'E), junipers reach 4500 m (Ward 1936). Above Mainling (29°04'N/93°57'E), south of Yarlung Zangbo, Bräuning (1999) discovered junipers at 4200 m (S-exp.), and Li, B (1993) found *Abies georgii* var. *smithii*, *B. utilis*, *Juniperus indica*, and *Larix griffithiana* growing up to 4300 m. General figures for the upper timberline in this region as given by Chang (1981) and Zheng et al. (1981) are 4000-4400 m at northern exposures (*Abies delavayi*, *Picea balfouriana*, *Rhododendron* spp.) and 4300-4600 m on south-facing slopes (*Juniperus tibetica*).

Likewise, remarkable altitudes are reached north of the eastern Nyainqentanglha Shan, e.g., west of Qamdo (31°05'N/96°57'E) at the upper Lancang Jiang (Mekong), where Frenzel et al. (2003) observed *Picea balfouriana* at 4500 m (N-exp.) and *Juniperus tibetica* at 4700 m (S-exp.) (see also Teichman 1922). Further west at Riwoqe (31°14'N/96°29'E), Bräuning (1999) investigated *Picea balfouriana* and *Juniperus* spec. at 4400 m, and at Banbar (31°09'N/94°50'E), junipers at 4300 m (S-exp.). Proceeding southward towards the lower river gorge country in SE Tibet, i.e. along the mountain ranges between Nu Jiang (Salween), Lancang Jiang (Mekong), and Jinsha Jiang (Yangtsekiang), upper timberlines still range between 4300 and 4600 m. Summarizing early observations, Weigold (1935) determined 4400 m as uppermost tree limit. Frenzel et al. (2003) recorded *Picea balfouriana* at 4500 m west of Gartog (29°40'N/98°31'E). Ward (1934; 1935; 1936) travelled between the Zayu and Salween valleys, where he observed *Rhododendron* spp., *Betula cylindrostachya*, and *Picea likiangensis* at 4200-4300 m (N-exp.), and *Juniperus incurva* at 4600 m (S-exp.). Chang (1981) and Li, W (1993) give a general figure of 4300-4600 m (*Abies squamata*, *A. georgii*, *Picea balfouriana*, *Juniperus tibetica*, *J. convallium*) in the wider region.

Summarizing upper timberline elevations along this cross-section, it is evident that there is a gradual increase from the outer Himalayan ranges in northern Myanmar to the fringes of the Tibetan highlands (SE-NW) that amounts to ca. 300 m at a horizontal distance of ca. 200 km. The massifs of Namcha Barwa and Gyala Peri lead to a rather abrupt rise in elevation. Increasing timberline elevations continue in northern directions (Nyainqentanglha Shan and beyond) to a maximum of about 4700 m so that the increase sums up to 900-1000 m at a horizontal distance of 400-500 km.

NW Yunnan – Sichuan

The easternmost mountain ranges of the Himalaya system as defined in this review include the Hengduan Shan in NW Yunnan and the adjoining ranges in western Sichuan (figures 2 and 13). As stated above, upper timberline elevations in the Myanmar-Yunnan mountain ranges (Gaoligong Shan) at about 27-28°N vary from 3900 to ca. 4100 m (Ward 1913; 1923; Stamp 1925; Handel-Mazzetti 1921; 1927a, b). North of 28°N a progressive increase in upper timberline elevations can be assessed. Above Changputhang (28°00'N/ 98°29'E), Handel-Mazzetti (1927a) observed the upper timberline above 4100 m, primarily composed of *Abies forrestii*, *Rhododendron* spp., and *Arundinaria melanostachya*. Similar timberline elevations are reached in the mountain range between Nu Jiang (Salween) and Lancang Jiang (Mekong) at this latitude (southern Taniantaweng Shan) (cf. Yang and Zheng 1990). At Si La, between Tseku and Tschamutong (28°01'N/98°45'E), *Rhododendron* spp., *Salix* spp., and *Abies forrestii* were recorded at 4200-4300 m at northern exposures (Handel-Mazzetti 1927a). Comparable conditions prevail at Nyingser La (27°58'N/98°43'E), where Rock (1926; 1947) and Handel-Mazzetti (1927b) described *Rhododendron* scrub with *Arundinaria* undergrowth, all clad with *Usnea longissima* beardmosses, as well as *Juniperus* and *Cerasus* krummholz at the same altitude. 4100-4350 m is a general figure in this region according to Handel-Mazzetti (1931). Further north, upper timberline is somewhat depressed in the rain shadow of Moirigkawagarbo (6809 m) E slope (28°28'N/98°36'E), where Ward (1923) observed the timberline ecotone (*Rhododendron* spp., *Betula utilis*) between 3600 and 3800 m (see also Wilson 1913). Under already drier conditions (at least in the lower altitudinal belts) in the mountain ranges between Lancang Jiang and Jinsha Jiang (Yangtze), upper timberline around 28°N lies at similar altitudes as in the ranges to the west (cf. Diels 1913). In upper Yondze Kha (28°08'N/99°07'E), Rock (1947) found *Rhododendron* spp., *Picea* spec., and *Abies* spec. between 3900 and 4100 m.

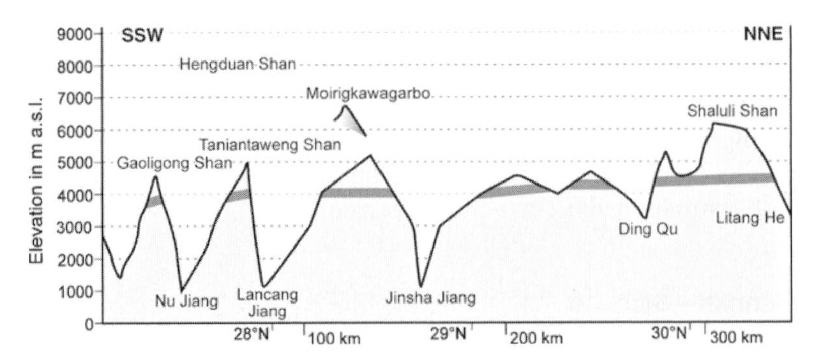

Figure 13 Altitudinal position of timberline ecotone on north-facing slopes along the cross-section NW Yunnan – Sichuan

Proceeding from the river gorge country northward to the eastern Tibetan highlands in Sichuan, upper timberline elevations considerably increase up to 4500-4700 m. Rock (1931) described *Larix potanini*, *Abies*, and *Picea* growing up to 4600 m superseded upslope by 7-10 m tall *Rhododendron* spp. attaining 4700 m (Konkaling Mts.; 28°30'N/100°E). Handel-Mazzetti (1921) found *Abies delavayi* and *Sorbus vilmorini* growing above 4400 m further north around 29°N. Kreitner (1893) already noted an elevation of 4700 m at a northern slope at Gambu Gongkar (29°55'N/99°40'E). Further timberline records were given by Ward (1918), who noted *Larix* and *Picea* above 4500 m, and by Schäfer (1938), who found *Rhododendron* spp., *Abies* spp., and *Picea* spp. at 4500-4600 m near Batang (c. 30°N/99°E). The latter reported an even higher locality (4700 m; N-exp.) with the same species at Zogqen (32°07'N/98°51'E). SE and S of Zogqen, Bräuning (1999) recorded *Picea balfouriana* at 4350 m (31°49'N/99°08'E; 31°58'N/98°51'E). The same author found *Juniperus* spec. attaining 4500 m at a southern slope NE of Batang (30°20'N/99°33'E). NW of Litang (30°30'N/100°E), *Picea retroflexa* and *Juniperus* spec. were noted at 4600-4700 m on a north-facing slope by Heim (1933). In a more recent regional overview, Li and Walker (1986) described *Abies* forests as the uppermost forests on shady slopes (above *Picea likiangensis*), ascending to 4450 m (see also Ku and Cheo 1941; Song 1983 who give the same figure), where *Abies georgei* forms a mosaic with *Larix potaninii* var. *macrocarpa* and the tree-like *Rhododendron fictolacteum* and *Rh. decorum*. The transition zone to the alpine belt is marked by *Rhododendron* shrubs (*Rh. adenogynum*, *Rh. polifolium*), achieving 1-3 m in height. Sunny slopes have *Juniperus saltuaria* stands up to 4500 m. Further E and NE, towards the eastern margin of the Tibetan Plateau, upper timberline elevations decrease again to 4200-4000 m and below as observed e.g., at Gongga Shan (7556 m; 29°42'N/102°06'E) by Imhof (1947), Messerli and Ives (1984) and Thomas (1999) or at Zitsa Degu (33°08'N/103°56'E; *Rhododendron* spp., *Abies faxoniana*, *Betula utilis*) by Winkler (1997) (see also Lehmkuhl and Liu 1994; Fang et al. 1996; Bräuning 1999; Frenzel et al. 2003).

Along this cross-section upper timberline increases by 700-800 m at a horizontal distance of about 300-400 km. However, the increase is moderate (200-300 m) within the river gorge country. The major increase (500-600 m) is to be observed ascending the Tibetan highlands.

Summary of Upper Timberline Observations

Upper timberline positions across the entire Himalayan system basically show two gradients of increasing timberline elevations: one is developed from NW to SE along the mountain arc, the other one is a peripheral-central gradient, which is, however, modified at huge high mountain massifs (e.g., Nanga Parbat, Dhaulagiri/Annapurna) with an elevated timberline compared to areas located immediately to the north. As a rule, south-facing slopes show higher upper timberlines than north-facing slopes as long as anthropogenic interferences have not blurred or even destroyed the natural setting. Unfortunately, this is the case almost everywhere. Massive human impacts on southern slopes in combination with the drought-related lack of easily recognizable altitudinal vegetation belts (at least in the NW) make timberline observations much more difficult on south-facing slopes. Thus, it is no surprise that only 25 % of more than 300 upper timberline observations in the author's database refer to southern exposures. The floristic change from deciduous *Betula*- to evergreen *Rhododendron*-dominated upper timberlines along the NW-SE gradient is coupled with decreasing winter cold and strongly increasing humidity levels in the SE, favouring competitive abilities of *Rhododendron* species. There are only minor physiognomic changes along that gradient. High coniferous forests giving way to medium-sized broadleaved tree stands and finally to a krummholz belt is the dominant timberline pattern on north-facing slopes throughout the mountain arc. Remnant vegetation at southern slopes consists mostly of open coniferous stands that dissolve into isolated patches or single crippled trees at higher altitudes. A third gradient relates to the time of observations: Generally, older observations from the first half of the twentieth century indicate higher timberline elevations compared with contemporary records or with those from recent decades. This must be attributed partially to measurement problems before modern altimeters had been developed. Older timberline records have to be interpreted with caution. But a more likely explanation is the considerable human impact that has led to a timberline depression in many localities in the past century.

3 Relationship of Upper Timberline to Ecological Conditions and Processes

3.1 Macroclimate and Heat Deficiency

The observed NW-SE gradient of increasing timberline elevations in the HKH mountain system corresponds to the macroclimatic pattern of increasing temperatures in the same direction. The mountain system extends across a considerable range of geographical latitudes (26-37°N) and thus reflects the general fact that upper timberline is highest in the Subtropics and drops from there to the temperate zones as well as to the equator. However, the latitude-related timberline gradient in the HKH mountain system is far from being linear, even if only the Himalayan south slope is considered (figure 14). The complex configuration of mountain ranges and valleys, their differentiated position against prevailing wind and precipitation regimes along the mountain arc, and the resulting pattern of local and regional climatic and related edaphic and biotic conditions modify the latitude-timberline relationship to such an extent that at 28°N, for instance, timberline elevations from 3650 up to 4500 m occur (cf. figure 14). Nevertheless, the NW-SE timberline gradient corroborates the underlying principle that the upper timberline is in general caused by heat deficiency. At 36°N, mean annual temperature at timberline is 1.1°C (station Damé high, 3780 m, 36°01'N/74°35'E; data kindly provided by Working Group Prof. M. Winiger, University of Bonn), whereas at similar altitudes at 29°N in the SE (station Daochen, 3728 m, 29°03'N/100°18'E) mean annual temperature already amounts to 4.0°C (data in Miehe et al. 2001). Applying the regional lapse rate for Yunnan of 0.58°C 100 m^{-1} (Li, W 1993), a mean annual temperature of 1.1°C would only be reached at ca. 4230 m. The latitude-related increase in mean temperatures accounts for much higher temperature sums at same elevations above sea level and thus causes the upper timberline to rise.

The peripheral-central gradient, which is to be observed more or less pronounced within each macroclimatic unit along the entire mountain arc, must also be attributed to higher temperature sums at same elevational levels.

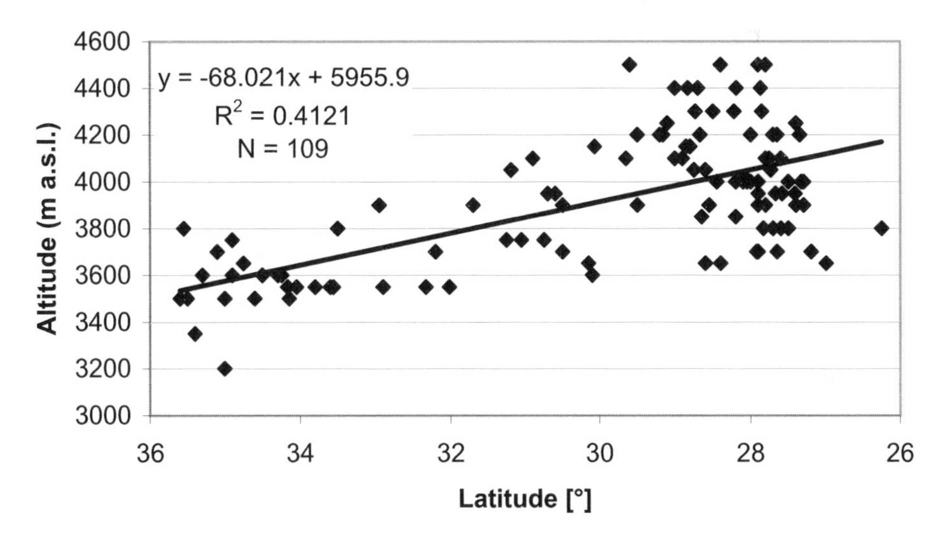

Figure 14 Relationship between latitude and altitudinal position of timberline on the Himalayan south slope (only data from north-facing slopes)

However, in contrast to latitude-related solar radiation and energy fluxes, higher mean elevations of the interior Himalayan and Transhimalayan mountain ranges cause higher temperatures along this gradient. This so-called 'mass-elevation effect' (De Quervain 1904; Haffner 1997), i.e. the effect of extensive mountain massifs as elevated heating surfaces leading to positive thermal anomalies compared with marginal ranges or free air, results in elevated altitudinal vegetation limits. The heating effect is most pronounced over the Tibetan Plateau (up to > 4°C at the ground) (Flohn 1968), primarily due to a considerable sensible heat transfer from the surface (average altitude 4500 m). Additionally, convective activity and orographic precipitation over the humid areas of SE Tibet and E Himalaya contributes to the energy transfer to the atmosphere by releasing latent heat of condensation. Yeh (1982) calculated that sensible and latent heat terms are almost equal east of 85°E.

The mass-elevation effect is always combined with the lee effect of increasingly continental climate character along the peripheral-central gradient. The combined effects of continentality and mass-elevation clearly overcompensate the effect of the latitudinal gradient of air temperature regarding the altitudinal position of upper timberline. Along all of the cross-sections, cloud cover and precipitation considerably decreases towards the inner mountain ranges and further towards the rain shadow areas in the Transhimalaya (table 1). Only in the arid NW and in the humid SE, the total decrease is more moderate. The sequence of high mountain chains acting as topographic barriers against moist air masses accounts for this decrease of annual precipitation that already amounts to more than 50 %, sometimes 80-90 % between the outer and inner ranges. Decreasing cloudiness and precipitation is coupled with higher sunshine duration and irradiation, earlier snow melt and higher soil temperatures, and finally leads to a more favourable topoclimate with higher summer temperatures and a longer growing season. This continentality effect, very pronounced in inner Himalayan valleys (Flohn 1970), is obviously even more important with regard to the ascent of isotherms over large Himalayan mountain masses and to rising altitudinal vegetation limits than the mass-elevation effect itself. The above finding of a much elevated timberline immediately north of huge high mountain massifs rising to more than 8000 m (e.g. Nanga Parbat, Dhaulagiri/Annapurna), distinctly modifying the otherwise gradually ascending peripheral-central timberline gradient, suggests that pronounced rain shadow and continentality effects of insurmountable ranges exert a stronger influence on timberline position than mass-elevation alone.

As stated above, heat deficiency is the key element within the explanation chain of the upper timberline phenomenon. However, the knowledge of how

Table 1 Decrease of annual precipitation along peripheral-central cross-sections from the Himalayan S slope to the Transhimalaya (after data in Miehe 1990; Miehe et al. 2001; Schickhoff 1993)

Cross-section	Himalayan S slope Station	mm a^{-1}	Inner mountain ranges Station	mm a^{-1}	Decrease in %	Transhimalaya Station	mm a^{-1}	Total decrease in %
I: Paktia - Chitral	Fort Lockhart (1998 m; 33°33' N/70°55' E)	862	Chitral (1499 m; 35°51' N/71°50' E)	528	-39	Eshkashem (2620 m; 36°42' N/71°34' E)	125	-78
II: Kaghan - Hunza	Balakot (991 m; 34°33' N/73°21' E)	1545	Gilgit (1460 m; 35°55' N/74°20' E)	141	-91	Taxkorgan (3090 m; 37°47' N/75°14' E)	68	-96
III: Jammu - Central Karakorum	Bhadarwah (1690 m; 32°58' N/75°43' E)	1134	Skardu (2181 m; 35°18' N/75°41' E)	208	-82	Xaidulla (3700 m; 36°21' N/78°00' E)	32	-97
IV: Himachal Pradesh - Zanskar	Simla (2202 m; 31°06' N/77°10' E)	1408	Kyelong (3500 m; 32°35' N/77°04' E)	545	-61	Leh (3500 m; 34°09' N/77°34' E)	83	-94
V: Uttaranchal	Mukteswar (2311 m; 29°28' N/79°39' E)	1318	Joshimath (1875 m; 30°33' N/79°34' E)	971	-26	Gar (4232 m; 32°07' N/80°04' E)	54	-96
VI: W-Central Nepal	Lumle (1642 m; 28°18' N/83°48' E)	5138	Manang (3420 m; 28°40' N/84°01' E)	465	-91	Mustang (3705 m; 29°11' N/83°58' E)	172	-97
VII: E-Central Nepal - S Tibet	Tarke Ghyang (2480 m; 28°00' N/85°33' E)	3373	Nyalam (3810 m; 28°11' N/85°58' E)	709	-79	Tingri (4300 m; 28°38' N/87°05' E)	271	-92
VIII: E Nepal (Khumbu Himal)	Aisealukharka (2143 m; 27°21' N/86°45' E)	2343	Namche Bazar (3490 m; 27°49' N/86°43' E)	976	-58	Lhazê (4000 m; 29°05' N/87°36' E)	263	-89
IX: Sikkim/Bhutan - S-Central Tibet	Darjeeling (2128 m; 27°03' N/88°16' E)	2601	Lingshi (4100 m; 27°52' N/89°26' E)	505	-81	Nagarzê (4432 m; 28°58' N/90°24' E)	336	-87
X: N Myanmar - SE Tibet	Putao (409 m; 27°20' N/97°25' E)	3993	Zayu (2328 m; 28°39' N/97°28' E)	754	-81	Baxoi (3260 m; 30°03' N/96°55' E)	223	-94
XI: NW Yunnan - Sichuan	Gongshan (1591 m; 27°42' N/98°40' E)	1668	Dêqên (3589 m; 28°27' N/98°53' E)	664	-60	Batang (2589 m; 30°00' N/99°06' E)	465	-72

low air temperatures and related factors such as length of growing season, frost, duration of snow cover, or soil temperatures affect tree growth and regeneration in the HKH mountain system is still very fragmentary. Miehe (1997) assumes a period of 100 days of mean daily temperatures higher than 5°C as the lower limit of the alpine belt. Establishing coincidences of timberlines with climatic parameters is hampered by the lack of climate stations at higher altitudes. Extrapolating available temperature data from nearby stations at lower altitudes using lapse rates can only provide rather inaccurate information since lapse rates may vary considerably from season to season (e.g. high rates during snow melt, low rates in full monsoon) and even at small spatial scales due to topography, exposure, snow cover, humidity, inversions etc. Nevertheless, as long as climate stations at timberline elevations are not installed in higher density, and as long as modelling approaches (e.g. Böhner 1996) have not been applied more extensively, available data have to be used to get at least a rough indication of thermal conditions at upper timberline. Recent climatological studies in the drier NW of the HKH mountain system (Weiers 1995; De Scally 1997; Jacobsen 1998; Cramer 2000; see also Flohn 1969) considerably improved the knowledge of vertical temperature gradients. Summer lapse rates average 0.7°C 100 m^{-1}, a rate which may be applied to the dry, but still partially forested areas of Tibet as well. The more humid areas in W, Central and E Himalaya as well as in S and SE Tibet, Yunnan, and Sichuan have mean summer lapse rates of c. 0.6°C 100 m^{-1} (cf. Flohn 1970; Kikuchi and Ohba 1988; Miehe 1990; Ohsawa 1990; Li, W 1993; Schickhoff 1993; Peng et al. 1997; Thomas 1999). These differentiated lapse rates were used to extrapolate mean temperatures of the warmest month and of the four warmest months (tetratherm) of high altitude climate stations to nearest timberline elevations (table 2).

The regionally differentiated results (table 2) reveal remarkable differences in mean temperatures at timberline. Due to the higher degree of continentality, the mean temperature of the warmest month, which often range around 10°C in northern hemisphere continental mountains (cf. Holtmeier 2003), is distinctly higher in the NW Himalaya and Karakorum (roughly between 10 and 13°C) compared to the more humid and monsoon-influenced regions. The same holds true for the tetratherm (between 8 and 11°C in the NW) which is considered a more adequate indicator of thermal conditions at upper timberline. It is interesting to note that the upper timberline in Tibet and E Himalaya develops at much lower mean temperatures. Apparently this must be mainly attributed to much lower winter temperatures in the NW (cf. table 2). Station pairs at similar altitudes show big differences in winter coldness. E.g., mean temperature of the coldest month is -12.9°C at Baldihel (3900 m;

Table 2 Climate data of high altitude stations and relationship between altitudinal position of upper timberline and temperature parameters (after data kindly provided by J. Böhner, University of Göttingen, and Working Group Prof. M. Winiger, University of Bonn, and data in Schickhoff 1993; Weiers 1995; Bräuning 1999)

Station (lat./long.)	Altitude [m a.s.l.]	Mean annual precipitation [mm]	Mean annual temperature [°C]	Mean temperature of coldest month [°C]	Mean temperature of warmest month [°C]	Elevation of nearest timber-line [m a.s.l.] / aspect	Extrapolated mean tempera-ture of warmest month [°C]	Extrapolated mean temperature of four warmest months [°C]
NW Himalaya/Karakorum								
Garmashbar (36°31' N/73°32' E)	3600	533	0.4	-12.8	14.2	4000 (S)	11.4	8.3
Baldihel (36°21' N/74°48' E)	3900	547	-0.5	-12.9	12.2	4000 (S)	11.5	9.1
Karimabad (36°18' N/74°40' E)	2300	137	11.2	-0.6	21.9	4000 (S)	10.0	8.1
Naltar (36°09' N/74°10' E)	2880	358	6.0	-6.3	17.9	3800 (N)	11.5	9.1
Dame high (36°01' N/74°35' E)	3780	546	1.1	-10.7	13.0	3750 (N)	13.2	11.0
Skardu (35°18' N/75°41' E)	2181	208	11.5	-2.9	24.1	4000 (S)	11.4	9.3
Astore (35°22' N/74°54' E)	2166	576	9.7	-2.3	21.1	3800 (N)	11.3	9.4
Dras (34°26' N/75°46' E)	3066	651	1.8	-15.7	17.0	3700 (N)	13.2	10.7
Battakundi (34°56' N/73°43' E)	2670	1010	8.0	-1.6	18.9	3750 (N)	12.4	10.0
Naran (34°54' N/73°39' E)	2362	1210	9.2	-1.2	19.0	3650 (N)	11.3	10.2
Kalam (35°22' N/72°35' E)	2290	900	10.8	-0.1	20.0	3700 (N)	11.5	10.3
W/Central/E Himalaya								
Lokpal (30°44' N/79°38' E)	4267	2179	-0.8	-9.6	7.1	4000 (N)	8.7	7.5
Jomosom (28°47' N/83°43' E)	2744	260	11.7	3.8	18.6	4400 (N)	9.5	8.9
Sermathang (27°57' N/85°36' E)	2625	3880	10.9	4.1	15.9	3700 (N)	10.0	9.4
Tengboche (27°50' N/86°46' E)	3857	1032	3.7	-3.8	9.4	4300 (N)	7.0	6.0
Darjeeling (27°03' N/88°16' E)	2128	2601	12.9	6.1	17.1	3700 (N)	11.1	10.8
Lachen (27°43' N/88°32' E)	2697	1741	9.3	2.7	15.6	4100 (N)	7.9	6.6
Yadong (27°44' N/89°05' E)	4300	408	0.0	-8.9	7.7	4000 (N)	9.5	8.7

Table 2 continued

Station (lat./long.)	Altitude [m a.s.l.]	Mean annual precipitation [mm]	Mean annual temperature [°C]	Mean temperature of coldest month [°C]	Mean temperature of warmest month [°C]	Elevation of nearest timber-line [m a.s.l.] / aspect	Extrapolated mean tempera-ture of warmest month [°C]	Extrapolated mean temperature of four warmest months [°C]
S-Central Tibet								
Damxung (30°29' N/91°06' E)	4200	502	1.0	-10.8	10.6	4300 (S)	9.9	8.8
Lhasa (29°42' N/91°08' E)	3658	443	7.7	-2.1	15.5	4600 (S)	8.8	7.9
Nagarzê (28°58' N/90°24' E)	4432	383	2.6	-5.4	10.9	4650 (S)	9.4	7.6
Nyalam (28°11' N/85°58' E)	3810	709	3.5	-3.9	10.5	4300 (N)	7.6	6.6
Nedong (29°18' N/91°48' E)	3552	394	8.3	-0.7	15.7	4600 (S)	8.4	7.4
Nyingchi (29°34' N/94°28' E)	3000	663	8.6	0.3	15.7	4400 (N)	7.3	6.4
Mainling (29°04' N/93°57' E)	3000	687	8.2	-0.3	15.6	4300 (N)	7.8	7.0
Gyitang (28°25' N/92°28' E)	3900	276	4.9	-4.6	13.0	4400 (S)	9.5	8.7
SE Tibet/Yunnan/Sichuan								
Qamdo (31°11' N/96°59' E)	3241	470	7.6	-2.4	16.2	4500 (N)	8.7	7.3
Dêgê (31°44' N/98°34' E)	3201	612	6.4	-2.8	14.5	4350 (N)	7.6	6.3
Garzê (31°38' N/99°59' E)	3394	636	5.6	-4.4	14.0	4350 (N)	8.3	7.0
Batang (30°00' N/99°06' E)	2589	465	12.7	3.9	19.7	4500 (S)	8.2	7.4
Litang (30°00' N/100°16' E)	3949	722	3.0	-6.0	10.5	4600 (N)	6.6	5.6
Daochen (29°03' N/100°18' E)	3728	636	4.0	-5.9	11.9	4400 (N)	7.9	7.1
Dêqên (28°27' N/98°53' E)	3589	664	4.8	-3.0	11.9	4200 (N)	8.2	7.4

36°21'N/74°48'E) versus -4.6°C at Gyitang (3900 m; 28°25'N/92°28'E), and -12.8°C at Garmashbar (3600 m; 36°31'N/73°32'E) versus -0.7°C at Nedong (3552 m; 29°18'N/91°48'E). Low winter temperatures, later snow melt and shorter growing seasons obviously overcompensate the advantage of higher mean summer temperatures so that upper timberline in the NW develops at lower elevations than expected when solely looking at mean temperatures. Much lower mean temperatures of the warmest month (between 7.6 and 8.7°C) were calculated for the upper timberline in SE Tibet, Yunnan, and Sichuan (cf. table 2), roughly corresponding to findings of Li, W (1993) and Frenzel et al. (2003), who give a figure of 8°C for the warmest month. Altogether, relationships between mean temperatures and upper timberline in the HKH mountain system support the general view of Holtmeier (2003) that mean temperatures differ too much to be considered appropriate indicators of thermal conditions and that any isotherms should not be considered the causal factor for upper timberlines.

3.2 Carbon Balance

Adverse climatic conditions at timberline impede photosynthesis and carbon assimilation. Research on how heat deficiency affects carbon balance of Himalayan trees growing at their upper limit has not been conducted so far. Some information is available, however, on patterns of biomass and net primary productivity of timberline vegetation. Whereas central Himalayan forests can potentially support high forest standing crop and net primary production up to the lower subalpine belt (cf. Singh and Singh 1992; Singh et al. 1994), declining temperatures towards higher altitudes have pronounced adverse effects on tree growth. Garkoti and Singh (1995) reported a decrease in forest biomass in the Nanda Devi region from 308.3 t ha^{-1} at 2750 m (*Acer cappadocicum*) to 177.5 t ha^{-1} at 3150 m (*Betula utilis*) and to 40.5 t ha^{-1} at 3300 m (*Rhododendron campanulatum*) and a decrease in net primary productivity from 19.5 to 15.2 and to 10.0 t ha^{-1} a^{-1} in these stands respectively. Net assimilation rate (production efficiency) increased along the gradient as did the root:shoot ratio. The authors concluded that enhanced carbon costs of root production for sustaining shoot growth may limit plant height and biomass in the severe timberline environment. These carbon costs differ between deciduous trees like the birch and evergreen rhododendrons (Garkoti and Singh 1994). *Betula* stands show lower nutrient use efficiency (net primary productivity per unit of foliage nitrogen), e.g., they require markedly larger root biomass for given unit of nutrient uptake and show lower retranslocation of nutrients from senescing leaves. The strategy to produce a large root biomass to support nutrient-rich leaves with high photosynthetic rates is to their disadvantage when nutrient availability in the soil is low. Since low soil temperatures at timberline limit nutrient availability, in particular in spring, higher carbon costs of deciduous trees lower their competitive abilities as against evergreen rhododendrons with lower leaf tissue nutrient status and lower nutrient uptake requirements from soil during leaf production. Nutrient use efficiency appears to be an important plant trait affecting competitive abilities of deciduous versus evergreen Himalayan timberline trees.

Long-term negative carbon balances at upper timberline are unlikely in the more humid Himalayan regions. On the other hand, evidence for a very low dry matter production that might be coupled with a temporarily insufficient carbon balance was found in junipers at S-exposed upper timberlines in the arid NW of the HKH mountain system. *Juniperus turkestanica* trees at 3600 m above the Batura Glacier (Hunza), reaching an age of more than 1000 years, show a mean growthring increment of only 0.37 mm a^{-1} and 15-20 rings per cm on average, whereas growthring increments of juniper trees from lower altitudes (2700-3000 m) are mostly higher than 1.50 mm a^{-1} (Schickhoff 2000b). Esper (2000) assessed mean growth rates of below 0.30 mm a^{-1} at upper timberline (3900 m) in Baltistan. According to Klötzli (1991), single stems at altitudes around 4000 m in Morkhun V., Hunza, may even have growth rates as low as c. 0.01 mm a^{-1} (approximately 100-120 rings per cm). Such extraordinarily low growth rates reflect particularly severe living conditions (e.g., very cold winter months, relatively low precipitation, high irradiation and temperature, high evaporation and transpiration, low soil moisture, wind, increased water and nutrient stress) that may lead to an exhaustion of carbohydrate reserves in extreme years.

3.3 Freezing and Frost Drought

Within the timberline-environment nexus, freezing and related frost drought are further important aspects reflecting effects of heat deficiency on tree growth. Resistance of trees and tree seedlings to damages by low temperatures and related phenomena (frost, desiccation, ice particle abrasion, parasitic snow fungi etc.) is critical for their growth and survival. The extent of damages during winter and late winter largely depends on the length and favourability of the preceding growing season (Holtmeier 2003). Despite the major importance of frost tolerance and winter hardiness, the state of knowledge regarding Himalayan timberline tree species is rather meagre. Only one study has been conducted using winter material from Langtang and Khumbu Himal (Sakai and Malla 1981). The study showed that deciduous timberline trees developed greater winter hardiness than evergreen rhododendrons and coniferous trees which both show comparable freezing resistance. *Salix* spp. (*S. denticulata*, *S. daltoniana*) and *Betula utilis* developed the greatest freezing resistance among the most prominent timberline species (up to -70 and -50°C in xylem), followed by *Rhododendron campanulatum*, *Juniperus recurva*, *Sorbus* spec., *Pinus wallichiana* (all between -25 and -35°C in leaves, leaf buds, cortex and xylem), and *Abies spectabilis* and *Larix potanini* (-20 to -25°C). The authors attribute the comparatively low level of winter hardiness in E Himalayan rhododendrons and conifers to the high humidity and moderate winter climate. It has to be stressed, however, that frost tolerance is by far sufficient to prevent the risk of frost damages in normal years. Differences in winter hardiness are reflected in the distribution pattern of *Betula utilis* and rhododendrons along the mountain arc (see above), corresponding to the general pattern of declining frost tolerance at increasing oceanity of the climate (Sakai 1983). Proceeding from the E Himalaya to E Tibet and Sichuan, where again lower winter temperatures prevail, rhododendrons again give way to temperate deciduous and coniferous trees at timberline (cf. Ohsawa 1990).

Among conifers, *Juniperus* spp. are obviously most cold-tolerant. This is not only indicated by their distribution at the highest timberlines in Tibet and in the NW of the HKH mountain system, but also by their capability forming timberlines at south-facing slopes without protecting snow cover for most of the winter season. Cramer (2000) assessed the difference in the date of snow melt between southern and northern aspects in Bagrot V./Karakorum to be 2-3 months at timberline elevation. Since higher incoming radiation, earlier snow melt and higher air and needle temperatures increase the risk of frost drought, junipers at south-facing upper timberlines must have developed not only a high level of frost tolerance, but also a sufficient frost drought tolerance. It is obvious from the small-scale distribution pattern at differing slope orientations throughout the Himalaya that junipers show a much more pronounced frost drought tolerance compared to rhododendrons on shady slopes (cf. Schickhoff 1993; Miehe 1990). Still, the knowledge of how susceptible Himalayan timberline trees and their seedlings are to frost drought is very scarce. Ecophysiological studies in this respect have not been conducted so far.

3.4 Soil Temperatures

Likewise, basic knowledge deficits still exist regarding the effects of soil temperatures at Himalayan timberlines. Analysing the role of soil temperatures, that have a profound influence on tree seedling growth and survival and are again influenced by numerous soil physical and chemical parameters, snow cover, plant cover, and topography, is already hampered by the fact that only very few measurements of soil temperatures exist at all in the HKH mountain system, let alone soil temperatures at timberline elevations. Recently, climate data (1991-2001) of the Pak-German research program 'Culture Area Karakorum' (CAK) in northern Pakistan shed some light on soil thermal conditions at upper timberline. The station Damé high (3780 m, N-exp., 36°01'N/74°35'E), situated 30 m above the upper timberline in Bagrot V./ Karakorum, recorded higher mean soil temperatures compared to mean air temperatures throughout the year (figure 15). The mean annual soil temperature at 5 cm depth is 5.0 °C, comparable to the figure of 4.8°C assessed for the *Pinus hartwegii* timberline under similar hygrothermal conditions on Mexican volcanoes (Körner 1998a). At 20 cm depth, 4.6°C was recorded compared to a mean air temperature of 1.1°C. The annual range of mean soil temperatures at 5 cm depth is from -2.7°C (January) to 15.6°C (July). 73 days with freezing and thawing of the uppermost topsoil were recorded on average between 1991 and 1999 (Weber 2000). Freeze-thaw frequency decreases to only 3-4 days at a depth of 30 cm so that damages by freeze-thaw effects are largely restricted to the upper rooting zone. Remarkable

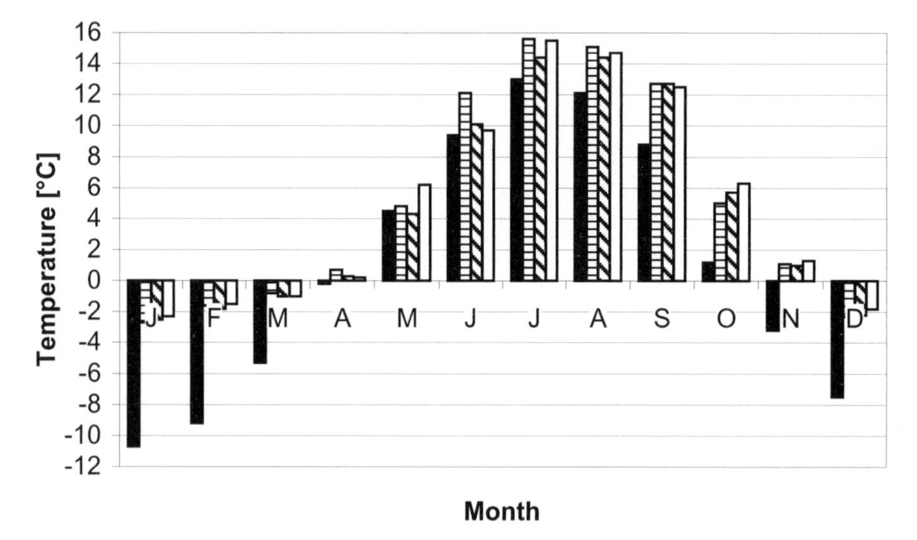

Figure 15 Mean monthly air and soil temperatures at different depths as recorded at upper timberline in Bagrot V./Karakorum (30 m above actual upper timberline), station Damé high (3780 m, N-exp., 36°01'N/74°35'E) (after data kindly provided by Working Group Prof. M. Winiger, University of Bonn)

differences exist between northern and southern aspects regarding freeze-thaw frequency. At 5 cm soil depth under mat vegetation at 3600 m, Weber (2000) recorded 42 freeze-thaw days on a south-facing slope in the winter of 1998/99, whereas the snow-protected north-facing slope had only 2 freeze-thaw days. This result underlines the statement that timberline trees occurring at southern aspects must have developed a higher frost drought tolerance.

Mean soil temperatures are above zero from April through November, but only from May through October temperatures are assumingly sufficient to guarantee a substantial soil biological activity and nutrient mineralisation (cf. figure 15). The mean soil temperature of the warmest month amounts to 15.6°C at 5 cm depth versus a mean air temperature of only 13.0°C. Moreover, soil temperatures at 5 cm depth show much higher diurnal amplitudes (mean up to 27.5°C) with lower minima and higher maxima because high irradiation is transferred to sensible heat at the soil surface whereas the air (mean diurnal amplitude 7-8°C) is affected by turbulent heat exchanges (cf. Cramer 2000). The mean maximum of soil temperature at 3780 m in August (34.0°C) is even higher than those of the valley station Dainyor at the foot of a south-facing slope (1520 m; 28.1°C).

However, since mean temperatures do not exist in nature (Holtmeier 2003), only a look at locally and temporally occurring extreme temperatures really illustrates the heat and water stress seedlings and juvenile trees are subjected to at upper timberline in the Karakorum. For instance, soil temperature at Damé high increased from 5°C in the morning to 37°C at midday on a cloudless day in August 1991, when solar radiation increased to more than 1000 W m^{-2}. Only two hours after the build-up of a convectional cloud cover, soil temperature dropped again to 23°C. The highest soil temperatures (at 5 cm depth) recorded at Damé high were 52.3°C in July 1992 and 46.4°C in August 1992 (Cramer 2000). It has been assumed and corroborated by regeneration studies (Schickhoff 2002) that conifer seedlings, more hygrophilous *Picea* and *Abies* seedlings in particular, can hardly tolerate such temperatures and subsequent effects. Soil temperatures of 50°C and more might exceed lethal values of seedlings (cf. Larcher 1994). At the same time, increased evaporation and transpiration further reinforce the soil moisture deficit in the rooting zone so that seedlings are at risk to dry up. Thus, altering the equable forest climate beneath a protective closed canopy by creating large forest gaps threatens a successful regeneration in high altitude forests in a subtropical high mountain environment like the Karakorum (Schickhoff 2000 a, b, c; 2002). The importance of an equable forest climate is best expressed by almost identical mean air and soil temperatures within the closed forest below upper timberline, whereas above timberline diurnal and annual temperature fluctuations are by far more pronounced in the soil compared to the air (cf. Cramer 2000).

Somewhat lower soil temperatures were recorded in upper subalpine fir-spruce forest blanks (3900 m) under more humid conditions in W Sichuan. Hou (1990) reported an annual range of soil temperature from -5°C to 13°C at 10 cm depth. After mid-November, soil temperature dropped below -5°C. The lapse rate of soil temperature between 3000 and 3900 m amounts to 0.77°C 100 m^{-1} for January and 0.82°C 100 m^{-1} for August. Regrettably, no substantial further information on soil temperatures at Himalayan timberline elevations exists so far. In this respect, Himalayan timberline research is still in its infancy.

3.5 Wind

Likewise, the knowledge of the influence of wind on tree growth and site conditions at Himalayan timberlines is very deficient. It can be concluded from wind measurements as well as from observations, however, that wind effects on timberline are less pronounced compared to other high mountain ranges at middle or high latitudes, at least in the NW of the mountain system. E.g., mean annual wind speed at high mountain stations in the European Alps or in the Rocky Mountains range between 7 and more than 10 m sec^{-1} (cf. Holtmeier 2003), whereas respective values at the stations Baldihel (3900 m; 36°21'N/74°48'E) and Damé high (3780 m, N-exp., 36°01'N/74°35'E) were only 2.8 and 1.9 m sec^{-1} respectively (after data kindly provided by Working Group Prof. M. Winiger, University of Bonn). Even gusts rarely exceed 16-17 m sec^{-1} (cf. Cramer 2000) compared to more than 70 m sec^{-1} in the Rocky Mountains. Although mean wind speed is slightly higher during winter months when westerlies prevail, physiological or mechanical stresses such as wind pressure and abrasion assumingly play only a minor role. The same is true for effects of heavy storms. Wind throw, wind-induced uprooting of trees, or crown/stem breakage were very rarely observed by the author, neither in the central Himalaya nor in the NW Himalaya/Karakorum/Hindu Kush. On the other hand, wind effects on site conditions like the enhancement of evaporation, in particular at south-facing slopes, or the relocation of snow and thus increased frost drought risk should not be underestimated.

Some observations are documented referring to the position of timberline and the distribution of timberline trees with respect to the mosaic of wind-exposed and wind-protected sites. Miehe (1990) reports from Langtang Himal that uppermost closed *Betula utilis* stands with undergrowth of *Rhododendron campanulatum* show a distinct contraction to leeward sides. Wind-exposed slopes at timberline ecotone are devoid of trees whereas closed uppermost birch groves, somewhat deformed by wind pressure, reach higher altitudes at more wind-protected sites. In Khumbu Himal, *Rhododendron* thickets near the upper limit of forests at wind-blown slopes are replaced by alpine dwarf shrubs which are obviously more drought- and frost drought-resistant (Miehe 1989). According to his observations, south-facing and wind-exposed slopes may even show a naturally lower timberline than shady lee ones. Depressed upper timberlines at wind-exposed slopes were also observed in the Thak Khola (Dhaulagiri/Annapurna Himal) (Miehe 1982, 1984), where a particularly pronounced diurnal valley wind system is developed. Here, the difference between the altitudinal timberline position at wind-blown ridges (*Pinus wallichiana*) and in the wind shadow of major peaks may be up to 500 m.

3.6 Exposure and Snow Cover

As obvious from the upper timberline descriptions along different cross-sections (see above), altitudinal position of timberline and the distribution pattern of tree species greatly differs between sunny exposures and shady slopes (photo 8). Generally, south-facing slopes show higher upper timberlines than north-facing slopes. The magnitude of differences in the altitudinal position of timberline between southern and northern aspects under natural conditions is hard to assess since southern aspects are often heavily influenced by anthropogenic activities throughout the mountain system. Human impact often blurred natural conditions to such an extent that previous authors tried to explain anthropogenic patterns, which were not recognized as such, by natural factors. E.g., south-facing slopes in the Central and E Himalaya were described as having a less-elevated timberline compared to northern aspects or even as being naturally devoid of tree growth due to high solar radiation, xeric conditions and exposure to frost and drought in winter (cf. Schäfer 1938; v. Wissmann 1960/61; Haffner 1979; Miehe 1982; 1984). In the meantime it has become obvious that human impact was strongly under-estimated and that many non-forested southern aspects are potentially wooded. Miehe et al. (1998) established the hypothesis for S and E Tibet that south-facing slopes are potential woodland, given a humid vegetation period and a minimum annual precipitation of 340 mm. As for the difference in the altitudinal position of timberline between southern and northern aspects under natural conditions it may be assumed from the numerous above observations that this difference is up to several hundred meters, also considering that exposure-dependent differences are much greater in subtropical high

Photo 8 Striking contrasts between north- (*Pinus wallichiana*, *Picea smithiana*) and south-facing slopes (*Juniperus excelsa*) near timberline in Sai V./Karakorum (Schickhoff, 1993-09-06)

mountains compared to temperate or boreal regions. In todays' cultural landscapes, actual differences of up to 250 m were observed on the Himalayan south slope, and up to 300-450 m in S Tibet (see above).

Exposure to solar radiation and related site conditions (e.g. thermal regime, humidity, duration of snow cover, soil moisture, etc.) cause the above described pattern of characteristic tree species distribution at north- and south-facing timberlines. Natural contrasts between northern and southern aspects are most pronounced under semi-arid to semi-humid conditions and in regions with long snow cover duration as in the NW of the HKH mountain system (see also v. Wissmann 1960/61). In Bagrot V./Karakorum, Cramer (1997) assessed considerable differences in radiation intensity in winter months. On a sunny day in January, the south-facing slope received a shortwave radiation of about

1000 W m^{-2}, the north-facing slope only a diffuse radiation of 77 W m^{-2}. At noon in mid-April, the respective figures were 1223 and only 544 W m^{-2}. Correspondingly, southern aspects may become free of snow above 3000 m already in January, whereas snow cover lasts until May at the same altitude on north slopes. Timberline vegetation at north-facing slopes (coniferous and birch forests) benefits from the more equable topoclimate, the long-lasting snow cover and the more favourable soil water budget. On the other hand, juniper trees at south-facing slopes have to cope with a combination of unfavourable climatic effects (high irradiation, high evaporation, drought stress). Anthropogenic interferences on sunny slopes reinforce these natural contrasts. With increasing humidity towards the E Himalaya, exposure differences are more and more blurred. In the Black Mountains of Central Bhutan, *Abies densa* forests even cover both northern and southern aspects in the timberline ecotone (Miehe and Miehe 2000).

3.7 Environmental Conditions and Growth Forms

At north-facing slopes from the central to the NW of the HKH mountain system, *Betula utilis* and tall shrub species of the genus *Salix* appear to be best adapted to a thick and long-lasting snow cover by developing growth forms adjusting to mechanical influences of snow (snow pressure, snow creep, sliding snow, avalanches). This adaptive ability combined with the ability to recover from damages by forming basal sprouts is one of the major factors for their higher competitiveness at upper timberline compared to conifers like firs and spruces. Birches usually show upright growth forms until closed stands, interspersed with some conifers, reach their upper limit and disintegrate into isolated tree patches. Further upslope, growth forms become stunted, often multi-stemmed and depressed and under most severe climatic conditions *Betula* individuals may develop only mat-like growth forms. On steeper slopes and along avalanche lanes, snow creep and sliding snow induce strong stem deformation leading to the characteristic curved stems of *Betula utilis* ('Umlegebirken'; photo 9), which have been described from many localities (e.g.,Troll 1939; Schweinfurth 1957a; Miehe 1982; 1990; Schickhoff 1993; 2002). Generally, thickness and duration of snow cover is critical for the distribution pattern of *Betula utilis* at local and regional scales in the HKH mountain system.

Likewise, *Rhododendron campanulatum*, already showing inherently bent and stunted growth forms, is able to further develop environmentally controlled growth forms in response to thick snow cover. Stem bases are often strongly deformed and bent down by snow pressure (Schmidt-Vogt

1990a). In contrast, other *Rhododendron* spp. at timberline like *Rh. barbatum* do not show this adaptive ability. The snow adaptation of *Rh. campanulatum* seems to be critical for its distribution far into the snow-rich W Himalaya (cf. Miehe 1991b). Only the often multi-stemmed dwarf *Sorbus* trees at the upper timberline ecotone more or less retain their physiognomy (Miehe 1990; Miehe and Miehe 2000), which might be comparatively more genetically fixed. Nothing has been reported so far about how natural regeneration of evergreen trees at snow-rich sites might be endangered by parasitic snow fungi, a common phenomenon in subalpine coniferous forests of temperate and boreal high mountains (cf. Holtmeier 2003).

South-facing slopes lack a long-lasting snow cover. Nevertheless, growth form changes of juniper species at timberline elevations reflect increasingly severe climatic conditions. Approaching the upper distribution limit, *Juniperus recurva* is able to modify its growth form from an upright growing tree

Photo 9 Curved stems of *Betula utilis* due to snow creep and sliding snow in Nilt V./Rakaposhi North Slope, Karakorum at 3600 m (Schickhoff, 1992-07-20)

to a krummholz-like appearance, head-high and multi-branched near the ground, and finally to small stunted mat-like shrubs of 0.5 m, taking advantage of the favourable microclimate near the ground (Miehe 1990). Likewise, *Juniperus indica* was observed to change its growth form from tree to low-growing shrub, and extending as such into the lower alpine belt (Miehe 1997). Juniper trees of the dry northwest of the HKH mountain system (*J. excelsa*, *J. turkestanica*) are comparatively less able to develop such far-reaching growth modifications. However, they reflect severe conditions at upper timberline by displaying crippled, gnarled, multi-branched scrub-like growth forms (1-2 m), contrasting to upright juniper trees at lower altitudes reaching 12-15 m (Schickhoff 2002).

3.8 Soils

Little is known about soils and its varying physical and chemical properties as a site factor at Himalayan timberlines. Like elsewhere in high mountains, different soils form a small-scale mosaic whereas the timberline vegetation might be quite homogeneous over varying soil types. Differences in edaphic conditions on north- and south-facing slopes and respective vegetation patterns suggest that soil physical parameters, to a large extent determined by mountain topography, exert a greater influence compared to soil chemical parameters. Colluvial soils are common at upper timberline, and the duration of morphodynamically stable periods at a site determines to a great extent the intensity of soil formation. In drier regions, the upper forest belt at north-facing slopes shows the most favourable ecological preconditions for soil formation and is the core zone of recent pedogenesis (cf. Reineke 2001; Schickhoff 2002 for the Karakorum). Similarly favourable conditions were assessed in the upper forest belt on the Himalayan south slope, e.g. in Kali Gandaki V., where Kunze (1982) found the highest soil biological activity between 3300 and 3400 m.

Under semi-humid and humid conditions on the Himalayan south slope, north-facing slopes at timberline elevations often have relatively deep, moist, medium to strongly acidic soils (mostly Cambisols or Podzols) that accumulate large amounts of only partly decomposed or largely undecomposed organic debris and show more or less advanced podzolization processes (cf. Taylor et al. 1936; Hoon 1938; Dhir 1969; Verma et al. 1990; Schickhoff 1993; 1996a; Bäumler and Zech 1994; Baillie et al. 2004). Comparable soils were described from Namcha Barwa in Tibet (Peng et al. 1997), and from Yunnan and Sichuan (Zhu 1990; Li 1993; Thomas 1999). Soils at south-facing slopes, in contrast, are subjected to rather deviating soil-forming conditions. Higher irradiation and evaporation and a less favourable soil water budget result in a retarded soil formation, in particular where the vegetation cover is sparse. Dry, shallow, more coarse-textured, slightly acidic or slightly alkaline soils (often Regosols) with high cation exchange capacity and base saturation are widely distributed at southern aspects at timberline. Soil physical and chemical parameters at north- and south facing slopes near timberline in the Karakorum are summarized in table 3. Here, soil moisture is obviously more important for seedling establishment and growth of junipers at south-facing slopes than other soil physical or chemical properties. Junipers are often found growing on rocky outcrops where their long roots can deeply penetrate into rock crevices to make use of soil moisture stored in fine soil accumulations and protected from evaporation (Schickhoff 1993). This distribution pattern is also favoured by ornithochorous seed dispersal: birds like crows (*Corvus* spec.) or jackdaws (*Pyrrhocorax* spec.) often rest on fissured rocks feeding on ripe juniper berries.

3.9 Regeneration

There is virtually no information on the effects of environmental conditions on natural regeneration of Himalayan timberline tree species. As stated above, forest degradation and altering site conditions especially on south-facing slopes may prevent regeneration at all in drier regions (Schickhoff 2000a; 2002). Likewise, only very few observations on regeneration strategies (seed-produced regeneration, seed banks, seed dispersal, vegetative reproduction) are documented. For instance, junipers are able to regenerate from tree stumps (Miehe and Miehe 2000). Recently, *Pinus wallichiana* seedlings have been observed to occur c. 250 m above the current upper timberline, and this was attributed to climate change (Dubey et al. 2003). More detailed studies on regeneration at timberline elevations are strongly needed.

4 Human Impact on Upper Timberline

Undoubtedly, the natural environment of the HKH mountain system has been reshaped by human impact in many regions since Neolithic times (Schickhoff 1995a). Archaeological and paleobotanical findings suggest that the foundation of permanent settlements and further anthropogenic influences date back up to 4500-5500 B.P. (e.g., Agrawal 1992; Jacobsen and Schickhoff 1995; Schlütz 1999; Miehe et al. 2002). Seasonal migration of mountain

Table 3 Soil physical and chemical parameters at north- and south facing slopes near timberline in the Karakorum (after data in Schickhoff 2002)

Plot	Plant community	Altitude (m a.s.l.)	Exp.	pH ($CaCl_2$)	$CaCO_3$ (wt %)	C_{org} (wt %)	N_t	C/N	CEC	K	Na (cmol$_c$/kg)	Ca (cmol$_c$/kg)	Mg	Mn	BS (%)	Gravel	Sand (wt %)	Silt (wt %)	Clay
North-facing slopes																			
Faqirkot II	*Pinus wallichiana*	3220	NNE	5.1	0.08	3.3	0.11	18	9.0	0.3	0.09	7.3	1.0	0.07	96	6	9	71	20
Raikot I	*Pinus wallichiana*	3230	NNW	5.5	0	6.4	0.18	21	17,4	0.4	0.09	14.9	2.0	0.06	99	1	33	55	12
Raikot II	*Pinus wallichiana*	3220	NNW	5.5	0	3.0	0.09	20	9.7	0.3	0.10	8.1	1.1	0.04	99	1	23	63	14
Thak I	*Pinus wallichiana*	3200	N	5.4	0	5.9	0.16	21	16.0	0.6	0.07	12.9	2.3	0.08	99	16	21	62	17
Talu Brok	*Pinus wallichiana*	3330	N	5.3	0	2.6	0.07	21	8.3	0.4	0.08	6.3	1.2	0.31	95	1	22	64	14
Bilamik	*Pinus wallichiana*	3550	NE	5.3	0	2.3	0.07	19	4.3	0.3	0.09	3.5	0.4	0.03	98	1	44	48	8
Kar Gah	*Picea smithiana*	3190	NNW	5.2	0.03	4.6	0.14	19	13.3	0.4	0.09	10.7	2.0	0.07	99	1	27	54	19
Kalapani	*Abies pindrow*	3260	NNE	5.2	0.08	6.7	0.25	16	23.7	0.2	0.16	20.7	2.3	0.18	99	11	31	45	24
Bayal Camp	*Betula utilis*	3560	NNE	4.9	0	6.1	0.17	21	17.1	0.4	0.10	10.8	5.5	0.07	98	0	20	62	18
South-facing slopes																			
Morkhun	*Juniperus turkestanica*	3500	SSW	7.9	0	7.4	0.21	21	19.4	0.5	0.04	16.5	2.4	0	100	60	69	24	7
Batura I	*Juniperus semiglobosa*	3190	SSW	7.2	0	9.1	0.35	15	29.5	0.5	0.01	26.7	2.3	0	100	1	40	47	13
Batura II	*Juniperus turkestanica*	3610	ESE	7.4	0	21,9	0.63	20	56.9	0.6	0.15	53.5	2.6	0.01	100	5	33	53	14
Ganji	*Pinus wallichiana*	3300	ESE	5.9	0	14,8	0.29	30	25.8	0.7	0.09	20.1	4.7	0.08	99	2	32	52	16
Gor	*Pinus wallichiana*	3520	SSE	5.6	0	13,8	0.45	18	25.7	0.5	0.11	23.1	2.0	0.03	100	4	16	65	19

nomads from Himalayan forelands to the alpine pastures as well as local mountain pastoralism is assumed to have occurred in the NW Himalaya since 3000 B.P. at the latest (Jettmar 1993), pointing to adverse effects on upper timberlines for millennia. Although human interferences may have been insignificant until colonial times due to very low population densities - the forests of India and the Himalaya were considered to be more or less untouched and inexhaustible in the 17th century and in the first decades of the 18th century (cf. Stebbing 1922) - the natural resources of upper forest belts and alpine grazing lands have been more or less integrated into local, regional and supra-regional economies for a long time. Considering the long period of time since animal husbandry, timber logging, fuelwood collection and the like are integral parts of village economies, and considering population growth and other far-reaching socioeconomic changes in the 19th and 20th centuries, it becomes obvious that present landscape patterns at timberline elevations are cultural landscape patterns and must be interpreted as transformed by local/regional

Photo 10 Present landscape patterns at timberline elevations are often largely transformed by anthropogenic utilization systems (Kaghan V./W Himalaya; 3400 m) (Schickhoff, 1990-08-13)

utilization systems (photo 10). It is very unlikely that totally undisturbed timberline ecotones still exist in High Asia (cf. Miehe and Miehe 2000). Timberline ecotones have been transformed to a locally and regionally varying extent and thus indicate anthropogenic utilization patterns.

At a local scale, timberlines at north- and south-facing slopes reflect contrasting utilization pressure. The pronounced effects of exposure to solar radiation (see above) result in a higher utilization pressure at south-facing slopes, in particular with regard to pastoral use (photo 11). Animals can graze at south-facing slopes for weeks or months in late winter and spring while north-facing slopes are still snow-covered. Depression of upper timberlines and downslope extension of alpine pastures is a widespread phenomenon at southern aspects throughout the HKH mountain system. Depression is mainly caused by long-term gradual degradation processes starting from peripheral degradation at forest margins. Grazing, wood-cutting, lopping, fires and other

Photo 11 High grazing pressure at the south-facing slope resulted in the destruction of timberline structures, whereas the shady slope is far less affected (Kaghan V./W Himalaya; 3300 m) (Schickhoff, 1990-08-26)

interferences lead to an increasing decline in canopy cover, to altered site conditions and regressive successions. Under drier site conditions and continuous browsing, trampling and cutting pressure grasses encroach on forest gaps and become highly competitive. Tree regeneration fails in the long term, in particular on dry sites where microclimate and edaphic conditions may irreversibly alter, and remaining trees die or are cut (Schickhoff 1995b; 2000b; 2002). Thus, the forest margin recedes downslope step by step, whereas pasture areas are enlarged. Finally, uppermost forest stands are completely replaced by alpine mats with a great variety of grazing weeds. In the Central Himalaya, Miehe (1997) found typical non-grazed alpine plant species descending 500-1000 m below the potential upper timberline. Thus, the depression of upper timberline at south-facing slopes may amount to more than 500 m, in particular in localities where extensive alpine grazing grounds are not available. It can be assumed that at least in the semi-humid to humid regions, perhaps also in some semi-arid regions, presently non-forested southern aspects are potentially wooded. Fire-clearing of entire slopes is rare nowadays, but was a frequent concomitant of previous settlement periods (Iwata and Miyamoto 1996; Beug and Miehe 1999). However, fires are frequent in the subalpine belt and lead to considerable damage in juniper forests which are highly susceptible to fire. Forests are burned to increase pasture areas, and herders set pastures on fire to stimulate grass growth and improve grazing conditions (cf. Schmidt-Vogt 1990b). Overgrazing and fire are assumed to be the main agents for lowering timberline in Nepal Himalaya (Schmidt-Vogt 1993; 1995).

Juniper stands at south-facing timberlines are not only seriously affected by grazing, but also by the excessive demand for fuelwood, at least in drier regions with low forest density. Fuelwood still makes up the lion's share of domestic energy requirements in most regions of the HKH mountain system. The use of substitute fuels is slowly increasing, but mostly people still rely on wood for cooking and heating. In areas where deadwood, fallen material or small branches have become scarce, even green trees are cut. Juniper trees are preferably cut as fuelwood because of their comparatively higher calorific value. In many locations near permanent settlements, only very few remnants of the once extensive juniper woodlands are left due to the need for fuelwood (e.g. Nüsser 2000; Schickhoff 2001). But also at upper timberline, accessible juniper groves have been degraded to a large extent to meet the fuelwood demand of pasture settlements and temporary shelters of herders. Additionally, some minor uses of junipers such as using branches as bedding material in shelters and stalls or for religious purposes (the leaves are commonly used as incense by Buddhists) are a drain on many subalpine stands. The comparatively low productivity hinders the regeneration of degraded juniper stands.

North-facing slopes are characterized by a lower utilization potential for pastoral use, in particular in spring. Nevertheless, a considerable depression of upper timberline can be observed at northern aspects as well (photo 12). The effects of exposure only modify resource utilization. As a net result gradual degradation processes also lead to a lowering of timberline. Although grazing pressure is lower on north-facing slopes, intensive pastoralism and the extension of alpine grazing grounds is the major cause for receding upper forest margins. Moreover, other forest uses cause forest degradation and destruction in the long term. The wood of *Betula utilis* is used for construction of shelters, small bridges and to a lesser extent as firewood. Most important minor uses include the multi-purpose utilization of the bark (writing and

Photo 12 Depression of upper timberline by c. 300 m in Saiful-Muluk V. (Kaghan/W Himalaya). Intensive pastoralism has completely destroyed the *Betula utilis* belt and has degraded the uppermost *Abies pindrow* forest (Schickhoff, 1990-07-17)

wrapping paper, pipe stem, umbrella, roofing material, etc.), and lopping of leaf fodder. *Rhododendron campanulatum* is mainly used as firewood. High coniferous stands of firs, spruces, and pines are primarily exploited for timber (photo 13). Selective timber logging is often uncontrolled and the cut wood inefficiently used (Schmidt-Vogt 1990a; Schickhoff 2002). Cutting of stemwood is also carried out in order to produce a great variety of smaller items (inside furnishings, roofing shingles, windows, doors, floors, boxes, plywood, etc.). Long-term anthropogenic interferences in the timberline ecotone on north-facing slopes may ultimately lead to a depression of timberline of several hundred meters. For instance, the *Betula utilis* belt in Kaghan V./West Himalaya has been largely decimated, uppermost coniferous forests depleted, and the difference between actual and potential timberline was assessed to be up to 300 m (Schickhoff 1995a).

At a regional scale, the intensity of interferences in the timberline ecotone often depends on the utilization potential of alpine pasture areas with regard to local, regional, and supraregional utilization systems. Differences in pastoral utilization intensity between the West Himalaya and the Karakorum and the contrasting state of upper timberlines well reflect these inter-

Photo 13 Timber logging in a *Picea smithiana* stand near timberline in Nilt V./ Rakaposhi North Slope, Karakorum at 3210 m (Schickhoff, 1992-07-19)

relationships. On the West Himalayan south slope alpine grazing lands are of comparatively limited extension, but on the other hand the valleys are densely populated and livestock numbers are extraordinarily high, resulting in an enormous grazing pressure by local mountain pastoralism. Additionally, grazing pressure is intensified every summer when mountain nomads (nomadic Gujars, Bakrwals, Gaddis, recently also Afghan nomads) with their large flocks migrate from the winter grazing areas in the Himalayan foothills to the high pastures in the upper valleys (Casimir and Rao 1985; Tucker 1986; Grötzbach 1989; Schickhoff 1993; Irfanullah 2002). This impact was traditionally very high, but has decreased in recent decades due to the increasing abandonment of nomadic lifestyle (Uhlig 1976; Schickhoff 1995a; Ehlers and Kreutzmann 2000; see also Farooquee and Rao 2001 for the Mon Pa in E Himalaya). As stated above, timberline vegetation has been adversely affected by the combined effects of mountain nomadism and local pastoralism, and the upper timberline has been depressed by several hundred meters. In the Karakorum, alpine grazing lands are much more extensive. At the same time, utilization pressure is much lower because of comparatively low population and livestock density, and because the high pastures are not frequented by mountain nomads. Besides, recent socio-economic transformation processes have led to a decline of local pastoralism in the frame of combined mountain agriculture (e.g. Kreutzmann 1989; Stöber and Herbers 2000; Fischer 2000). As a result, upper timberline has not been as much affected as on the West Himalayan south slope. Considerable depressions of upper timberline (north-facing slopes) were assessed to a locally limited extent only (Schickhoff 2002). The birch belt has been more negatively affected beyond the drought limit of high coniferous forests where local wood deficits result in a higher utilization pressure.

The accessibility of high valleys is another major factor determining the intensity of human impacts on upper timberline at a regional scale. Due to road construction and opening-up of formerly secluded high mountain regions, external influences could penetrate and affect high altitude forests even in peripheral valleys. Improved transport facilities can have significant effects on the intensity of forest utilization near timberline since market-oriented timber logging might become an economically feasible option. High altitude forest degradation in the Karakorum well exemplifies the intensified forest use after opening-up of the region (Schickhoff 1995; 1997; 2002). Accessibility is also a prerequisite for the development of mountain tourism. In recent decades, felling trees and collecting firewood for an increasing number of lodges, camping groups, trekking and mountaineering expeditions have reached appreciable dimensions. Since fuel is very scarce above timberline, considerable sections of timberline forests and scrub (chiefly junipers), also in the lower alpine belt, have been cleared in order to meet the fuelwood needs. Sometimes even roots are grubbed up and alpine cushion plants are used as firewood. Touristic impacts concentrate on main trekking and mountaineering routes, e.g., in northern Pakistan (Grötzbach 1993), Nanda Devi Biosphere Reserve (Khacher 1978), Annapurna and Langtang Himal (Soliva 2002), or in Mt. Everest region (Stevens 2003). Byers (1997) estimated on the basis of repeat photographs that more than half of the shrub juniper growth in the vicinity of Dingboche (Mt. Everest region) has been lost since the early 1960ies. Recently, rules and regulations have been established governing fuel use. In most of the touristic areas in the HKH mountain regions, expeditions and trekking groups are now obliged to carry in kerosene, gas or petrol for team members and porters, and collecting of fuelwood was restricted to fallen deadwood. However, rules and regulations are evaded in many cases (cf. Soliva 2002), and the high touristic fuelwood demand remains to be a heavy burden for high altitude forests.

5 Outlook: Timberline Ecotones in a Changing Environment

The present review reveals that basic research deficits concerning many aspects of Himalayan timberlines are still prevalent. We are still at the very beginning of the way to better understand timberline ecotones in the HKH mountain system. Whereas information on altitudinal position, physiognomy, and floristics is more or less sufficiently documented in the widely scattered literature, studies on ecological interrelationships are very rare. If carried out, such studies quite often focus on single aspects of the ecological complexity of upper timberlines rather than on ecological interactions of varied factors. More systematic, interdisciplinary research is strongly needed to get further insights into the many abiotic, biotic, and anthropogenic factors, to understand how these factors are interrelated and how complex ecological and socio-economic processes at various spatial and temporal scales are finally expressed in present spatial and physiognomic timberline structures. A complex landscape-ecological approach as advocated by Holtmeier (2003) seems to be most promising in order to encompass the ecological complexity and great heterogeneity of timberline ecotones. As for anthropogenic interferences, human impact on south-facing slopes is a major research deficit that has to be tackled, also from a paleoecological perspective, in order to better understand the evolvement of the present-day cultural landscape.

Timberlines have always been fluctuating due to either slow or abruptly changing environmental conditions. Recently, two agents of change have been emerging that might exert a comparatively strong influence on timberline dynamics in the HKH mountain system: i) accelerating socio-economic transformation processes in a globalizing world, and ii) climate change. As a rule, socio-economic change is connected with changing land use patterns and land use intensities. As exemplified above by the decline of high pasturing in the Karakorum, strategies of resource utilization that directly influence timberline ecotones are being adapted within the framework of overall changing livelihood strategies. Negative effects on timberline structures may be weakened by decreasing grazing pressure, but can be reinforced by other developments. Apart from pastoral use, forest use in the upper subalpine belt and tourism are present human impacts that may rapidly change in intensity in response to socio-economic change and thus influence timberline dynamics. Likewise, timberline structures are presently supposed to be modified under the influence of ongoing warming trends. Most of the regions within the HKH mountain system experienced an increase of mean annual temperatures in recent decades that amounts to up to 0.9°C (Böhner 1996). Glaciers have retreated on a large scale in the past 30 years (e.g. Qin et al. 2000; Karma et al. 2003), monsoon rainfall is supposed to decrease in vast regions (Duan et al. 2002), and an upward shift of timberline tree species (*Pinus wallichiana*) was already observed and attributed to climatic changes (Dubey et al. 2003). Timberline structures might be subjected to change more rapidly by these emerging agents than in previous decades. The scientific community should use recent developments as an opportunity to strongly intensifying Himalayan timberline research in order to tackle the many basic research deficits pertaining to timberline ecology and to analyse responses of timberline to contemporary environmental change.

Acknowledgements

I am most grateful to Prof. Dr. Friedrich-Karl Holtmeier (University of Münster) for encouraging me to do research in the Himalaya when I still was a student of geography and landscape ecology at his department and for supervising my diploma and doctoral theses. I am also highly indebted to Prof. Dr. Matthias Winiger (University of Bonn) for his support within the framework of the joint Pakistani-German research project 'Culture Area Karakorum (CAK)'. Studies in the Himalaya and Karakorum were financially supported by grants from the DFG (German Research Council), DAAD (German Academic Exchange Service), and the Office for Promotion of Graduates, University of Münster.

References

Agrawal DP (1992) Man and Environment in India Through Ages. New Delhi
Aitchison JET (1881) The vegetation of Kuram and Hariab valleys. Indian For 5: 179-188
Arno SF (1984) Timberline. Mountain and Arctic Forest Frontiers. Seattle
Bäumler R, Zech W (1994) Soils of the high mountain region of Eastern Nepal: classification, distribution and soil-forming processes. Catena 22: 85-103
Baillie IC, Tshering K, Dorji T, Tamang HB, Dorji T, Norbu C, Hutcheon AA, Bäumler R (2004) Regolith and soils in Bhutan, Eastern Himalayas. Europ J Soil Sci 55: 9-27
Beug HJ, Miehe G (1999) Vegetation History and Human Impact in the Eastern Central Himalaya (Langtang and Helambu, Nepal). Diss Bot 318. Berlin Stuttgart
Billiet F, Léonard J (1986) Voyage botanique au Cachemire et au Ladakh (Himalaya occidental). Jardin Bot Nat Belgique: 1-47
Bishop BC (1990) Karnali under Stress. Livelihood Strategies and Seasonal Rhythms in a Changing Nepal Himalaya. Univ of Chicago, Geogr Res Pap 228-229
Böhner J (1996) Säkulare Klimaschwankungen und rezente Klimatrends Zentral- und Hochasiens. Göttinger Geogr Abh 101
Bolliger J, Kienast F, Zimmermann NE (2000) Risks of global warming on montane and subalpine forests in Switzerland – a modeling study. Reg Environ Change 1: 99-111
Bräuning A (1999) Zur Dendroklimatologie Hochtibets während des letzten Jahrtausends. Diss Bot 312. Berlin Stuttgart
Braun G (1996) Vegetationsgeographische Untersuchungen im NW-Karakorum (Pakistan). Kartierung der aktuellen Vegetation und Rekonstruktion der potentiellen Waldverbreitung auf der Basis von Satellitendaten, Gelände- und Einstrahlungsmodellen. Bonner Geogr Abh 93
Breckle SW (1972) Alpenrosen im Hindukusch? Jahrb Ver Schutz Alpenpfl u -tiere 37: 140-146
Breckle SW (1973) Mikroklimatische Messungen und ökologische Beobachtungen in der alpinen Stufe des afghanischen Hindukusch. Bot Jahrb Syst 93: 25-55
Breckle SW (1975) Ökologische Beobachtungen oberhalb der Waldgrenze des Safed Koh (Ost-Afghanistan). Vegetatio 30: 89-97
Breckle SW, Frey W (1974) Die Vegetationsstufen im Zentralen Hindukusch. Afghanistan Journal 1: 75-80
Brockmann-Jerosch H (1919) Baumgrenze und Klimacharakter. Beitr Geobot Landesaufnahme Schweiz 6. Zürich
Bruce CG (1914) Kulu and Lahoul. London
Buchanan W (1919) A recent trip into the Chumbi Valley, Tibet. Geogr J 53: 403-410
Byers A (1996) Historical and contemporary human disturbance in the upper Barun Valley, Makalu-Barun National Park and Conservation Area, East Nepal. Mountain Res Developm 16: 235-247
Byers A (1997) Landscape change in Sagarmatha National Park, Khumbu, Nepal. Himal Res Bull 17: 31-41

Calciati C (1914) Esplorazione della valli Kondus e Hushee nel Karakoram Sud-Orientale. Boll della Reale Soc Geogr, Ser V, vol III, pt II: 995-1014 e 1076-1085

Carpenter C, Zomer R (1996) Forest ecology of the Makalu-Barun National Park and Conservation Area, Nepal. Mountain Res Developm 16: 135-148

Casimir MJ, Rao A (1985) Vertical control in the western Himalaya: some notes on the pastoral ecology of the nomadic Bakrwal of Jammu and Kashmir. Mountain Res Developm 5: 221-232

Champion HG (1936) A Preliminary Survey of the Forest Types of India and Burma. Indian For Rec (NS), Silviculture, vol 1, no 1. Delhi

Chandra R (1949) A trip to Bara Banghal in Kangra District. Indian For 75: 501-504

Chang DHS (1981) The vegetation zonation of the Tibetan Plateau. Mountain Res Developm 1: 29-48

Chapman GP, Wang YZ (2002) The Plant Life of China. Berlin Heidelberg New York

Chaudhary RP (1998) Biodiversity in Nepal (Status and Conservation). Bangkok

Chaudhary RP, Kunwar RM (2002) Vegetation composition of Arun Valley, East Nepal. In: Chaudhary RP, Subedi BP, Vetaas OR, Aase TH (eds) Vegetation and Society. Their Interaction in the Himalayas, Kathmandu-Bergen, pp 38-55

Chaudhuri, AB (1992) Himalayan Ecology and Environment. New Delhi

Cooper RE (1933) Botanical tours in Bhutan. Notes Roy Bot Gard Edinburgh 18: 67-118

Cramer T (1997) Climatic gradients in the Karakorum and their effects on the natural vegetation. In: Stellrecht I, Winiger M (eds) Perspectives on History and Change in the Karakorum, Hindukush and Himalaya. Culture Area Karakorum Scientific Studies 3: 265-276

Cramer T (2000) Geländeklimatologische Studien im Bagrottal, Karakorumgebirge, Pakistan. Geo Aktuell Forschungsarbeiten 3. Göttingen

Däniker A (1923) Biologische Studien über Wald- und Baumgrenze, insbesondere über die klimatischen Ursachen und deren Zusammenhänge. Vierteljahresschr Naturforsch Ges Zürich 63

De Quervain A (1904) Die Hebung der atmosphärischen Isothermen in den Schweizer Alpen und ihre Beziehung zu den Höhengrenzen. Gerlands Beitr Geophys 6: 481-533

De Scally FA (1997) Deriving lapse rates of slope air temperature for meltwater runoff modeling in subtropical mountains: an example from the Punjab Himalaya, Pakistan. Mountain Res Developm 17: 353-362

Dhar U, Kachroo P (1983) Alpine Flora of Kashmir Himalaya. Jodhpur

Dhir RP (1969) Pedological characteristics of some soils of the NW-Himalaya. J Ind Soc Soil Sci 15: 61-69

Dickoré WB (1991) Zonation of flora and vegetation of the northern declivity of the Karakoram/Kunlun Mountains (SW Xinjiang China). GeoJournal 25: 265-284

Dickoré WB, Miehe G (2002) Cold spots in the highest mountains of the world – diversity patterns and gradients in the flora of the Karakorum. In: Körner C, Spehn E (eds) Mountain Biodiversity. A Global Assessment, London-New York, pp 129-147

Dickoré WB, Nüsser M (2000) Flora of Nanga Parbat (NW Himalaya, Pakistan). An Annotated Inventory of Vascular Plants with Remarks on Vegetation Dynamics. Englera 19. Berlin-Dahlem

Diels L (1913) Untersuchungen zur Pflanzengeographie von West-China. Beibl Bot Jahrb 109: 55-88

Dobremez JF (1976) Le Népal. Ècologie et Biogéographie. Paris

Dobremez JF, Jest C (1971) Carte écologique du Népal I. Région Annapurna-Dhaulagiri. Doc Carte Végétation Alpes 9: 147-190

Dobremez JF, Shrestha TB (1980) Carte écologique du Népal. Région Jumla-Saipal 1:250 000. Cahiers Népalais 9. Paris

Donner W (1972) Nepal. Raum, Mensch und Wirtschaft. Wiesbaden

Drew F (1875) The Jummoo and Kashmir Territories. A Geographical Account, London

Duan K, Yao T, Pu J, Sun W (2002) Response of monsoon variability in Himalayas to global warming. Chinese Sci Bull 47: 1842-1845

Dubey B, Yadav RR, Singh J, Chaturvedi R (2003) Upward shift of Himalayan pine in Western Himalaya, India. Current Science 85: 1135-1136

Dudgeon W, Kenoyer LA (1925) The ecology of Tehri Garhwal: a contribution to the ecology of the western Himalaya. J Ind Bot Soc 4: 233-285

Duthie JF (1893-94) Report on a botanical tour in Kashmir. Rec Bot Surv India 1(1): 1-18, 1(3): 25-47

Eberhardt, E (2004) Plant Life of the Karakorum. The Vegetation of the Upper Hunza Catchment (Northern Areas, Pakistan). Diversity, Syntaxonomy, Distribution. Diss. Bot. 387. Berlin-Stuttgart

Ehlers E, Kreutzmann H (2000) High mountain ecology and economy: potential and constraints. In: Ehlers E, Kreutzmann H (eds) High Mountain Pastoralism in Northern Pakistan. Erdkundl Wissen 132: 9-36

Ellenberg H (1966) Leben und Kampf an den Baumgrenzen der Erde. Naturwiss Rdsch 19: 133-139

Esper J (2000) Paläoklimatische Untersuchungen an Jahrringen im Karakorum und Thien Shan Gebirge (Zentralasien). Bonner Geogr Abh 103

Fang JY, Ohsawa M, Kira T (1996) Vertical vegetation zones along 30° N latitude in humid East Asia. Vegetatio 126: 135-149

Farjon A, Miehe G, Miehe S (2000) The taxonomy, distribution and ecology of *Juniperus* in High Asia. In: Forest and Walnut Research Institute of National Academy of Sciences / Kyrgyz-Swiss Forestry Support Programme (eds) Proceedings – International Symposium on Problems of Juniper Forests, 6th-11th August 2000, Osh, Kyrgyzstan, Bishkek, pp 70-90

Farooquee NA, Rao KS (2001) Changing values among Mon Pa pastoralists and their ecological implications for rangelands in the eastern Himalayas. Nomadic Peoples NS 5: 168-174

Finsterwalder R (1938) Die geodätischen, gletscherkundlichen und geographischen Ergebnisse der deutschen Himalaya-Expedition 1934 zum Nanga Parbat. Deutsche Forschung, vol 2. Berlin

Fischer D (1970) Waldverbreitung, bäuerliche Waldwirtschaft und kommerzielle Waldnutzung im östlichen Afghanistan. Afghanische Studien 2. Meisenheim

Fischer R (2000) Coming down from the mountain pastures: decline of high pasturing and changing patterns of pastoralism in Punial. In: Ehlers E, Kreutzmann H (eds) High Mountain Pastoralism in Northern Pakistan. Erdkundl Wissen 132: 59-72

Flohn H (1968) Contributions to a Meteorology of the Tibetan Highlands. Atmospheric Sciences Paper 130. Fort Collins

Flohn H (1969) Zum Klima und Wasserhaushalt des Hindukush und der benachbarten Hochgebirge. Erdkunde 23: 205-215

Flohn H (1970) Beiträge zur Meteorologie des Himalaya. In: Hellmich W (ed) Khumbu Himal. Ergebnisse des Forschungsunternehmens Nepal-Himalaya, vol 7/2, Innsbruck, pp 25-45

Franklin JF Swanson FJ Harmon ME (1992) Effects of global climatic change on forests in northwestern North America. In: Peters RL, Lovejoy TE (eds) Global Warming and Biological Diversity, New Haven, pp 244-257

Freitag H (1971) Die natürliche Vegetation Afghanistans. Beiträge zur Flora und Vegetation Afghanistans I. Vegetatio 22: 285-344

Frenzel B, Bräuning A, Adamczyk S (2003) On the problem of possible last-glacial forest-refuge-areas within the deep valleys of eastern Tibet. Erdkunde 57: 182-198

Frey W, Probst W (1982) Zwischen Tagab und Alisang – eine vegetationskundliche Exkursion im westlichen Grenzgebiet des monsunbedingten Waldgürtels auf der SE-Abdachung des zentralen Hindukusch. Afghanistan Journal 9: 62-72

Frey W, Probst W, Shaw A (1976) Die Vegetation des Jokham-Tals im Zentralen Afghanischen Hindukusch. Afghanistan Journal 3: 16-21

Gammie GA (1894) Report on a botanical tour in Sikkim, 1892. Rec Bot Surv India I(2): 1-24

Garkoti SC, Singh SP (1994) Nutrient cycling in the three central Himalayan forests ranging from close canopied to open canopied treeline forests, India. Arct Alp Res 26: 339-348

Garkoti SC, Singh SP (1995) Variation in net primary productivity and biomass of forests in the high mountains of Central Himalaya. J Veg Sci 6: 23-28

Goodfellow BR (1954) North of Pokhara. Himal J 18: 81-86

Gorrie RM (1933) The Sutlej Deodar, its ecology and timber production. Indian For Rec Silvicult 17 (4): 1-140

Grace J, Berninger F, Nagy L (2002) Impacts of climate change on the tree line. Ann Bot 90: 537-544

Gratzer G, Rai PB, Glatzel G (1999) The influence of the bamboo *Yushania microphylla* on regeneration of *Abies densa* in central Bhutan. Can J For Res 29: 1518-1527

Grey-Wilson C (1974) Plant-hunting on the frontier of Tibet – a quest in north-west Nepal. Quart Bull Alp Gard Soc 42: 143-164, 202-212, 308-318

Grierson AJC, Long DG (1983) Flora of Bhutan. Vol 1, part 1. Edinburgh

Griffith W (1847) Journals of Travels in Assam, Burma, Bootan, Afghanistan and the Neighbouring Countries. Calcutta

Grötzbach E (1965) Kulturgeographische Beobachtungen im Farkhar-Tal (Afghanischer Hindukusch). Die Erde 96: 279-300

Grötzbach E (1989) Kaghan – Zur Entwicklung einer peripheren Talschaft im Westhimalaya (Pakistan). In: Haserodt K (ed) Hochgebirgsräume Nordpakistans im Hindukusch, Karakorum und Westhimalaya. TU Berlin, Beitr u Mat z Reg Geogr 2: 1-18

Grötzbach E (1993) Tourismus und Umwelt in den Gebirgen Nordpakistans. In: Schweinfurth U (ed) Neue Forschungen im Himalaya. Erdkundl Wiss 112: 99-112

Gupta RK (1983) The Living Himalayas. Vol 1: Aspects of Environment and Resource Ecology of Garhwal. New Delhi

Gupta RK (1994) Resource ecology for soil conservation and sustainable development in the cold desert regions of West Himalaya. In: Pangtey YPS, Rawal RS (eds) High Altitudes of the Himalaya, Nainital, Biogeography, Ecology & Conservation, pp 138-178

Gupta RK, Singh JS (1962) Succession of vegetation types in Tons Valley of the Garhwal Himalaya. Indian For 88: 289-296

Gupta VC, Kachroo P (1983) Life-form classification and biological spectrum of the flora of Yusmarg, Kashmir. Trop Ecol 24: 22-28

Haffner W (1967) Ostnepal – Grundzüge des vertikalen Landschaftsaufbaus. In: Hellmich W (ed) Khumbu Himal. Ergebnisse des Forschungsunternehmens Nepal Himalaya, vol 1, Berlin Heidelberg New York, pp 389-426

Haffner W (1972) Khumbu Himalaya. Landschaftsökologische Untersuchungen in den Hochtälern des Mt. Everest-Gebietes. Erdwiss Forsch 4: 244-262

Haffner W (1979) Nepal Himalaya. Untersuchungen zum vertikalen Landschaftsaufbau Zentral- und Ostnepals. Erdwiss Forsch 12, Wiesbaden

Haffner W (1997) Hochasien: Der Effekt großer Massenerhebungen. Geogr Rdsch 49: 307-314

Halpin PN (1994) Latitudinal variation in montane ecosystem response to potential climatic change. In: Beniston M (ed) Mountain Environment in Changing Climates. Routledge, London, pp 180-203

Handel-Mazzetti H (1921) Übersicht über die wichtigsten Vegetationsstufen und -formationen von Yunnan und SW-Setschuan. Bot Jahrb Syst 56: 578-597

Handel-Mazzetti H (1927a) Das nordost-birmanisch-west-yünnanesische Hochgebirgsgebiet. In: Karsten G, Schenck H (eds) Vegetationsbilder, 17. Reihe, vol 7/8, Jena, pp 37-48

Handel-Mazzetti H (1927b) Naturbilder aus Südwest-China. Wien Leipzig

Handel-Mazzetti H (1931) Die pflanzengeographische Gliederung und Stellung Chinas. Bot Jahrb Syst, Pflzgesch, Pflzgeogr 64: 309-323

Hartmann H (1966) Beiträge zur Kenntnis der Flora des Karakorum. Engl Bot Jahrb 85: 259-328, 329-409

Hartmann H (1968) Über die Vegetation des Karakorum. 1.Teil: Gesteinsfluren, subalpine Strauchbestände und Steppengesellschaften im Zentral-Karakorum. Vegetatio 15: 297-387

Hartmann H (1990) Pflanzengesellschaften aus der alpinen Stufe des westlichen, südlichen und östlichen Ladakh mit besonderer Berücksichtigung der rasenbildenden Gesellschaften. Candollea 45: 525-574

Haserodt K (1980) Zur Variation der horizontalen und vertikalen Landschaftsgliederung in Chitral. In: Jentsch C, Liedtke H (eds) Höhengrenzen in Hochgebirgen. Arb Geogr Inst Univ Saarl 29, Saarbrücken, pp 233-250

Haserodt K (1989) Chitral (pakistanischer Hindukusch). Strukturen, Wandel und Probleme eines Lebensraumes zwischen Gletschern und Wüste. In: Haserodt K (ed) Hochgebirgsräume Nordpakistans im Hindukusch, Karakorum und Westhimalaya. TU Berlin, Beitr u Mat z Reg Geogr 2: 43-180

Hättenschwiler S, Körner C (1995) Responses to recent climate warming of *Pinus sylvestris* and *Pinus cembra* within their montane transition zone in the Swiss Alps. J Veg Sci 6: 357-368

Heim A (1933) Minya Gongkar. Bern

Heim A, Gansser A (1939) Thron der Götter. Erlebnisse der 1. Schweizer Himalaya-Expedition. Zürich Leipzig

Hermes K (1955) Die Lage der oberen Waldgrenze in den Gebirgen der Erde und ihr Abstand zur Schneegrenze. Kölner Geogr Arb 5

Heske F (1932) Die Wälder in den Quellgebieten des Ganges und der Plan zu ihrer geregelten Bewirtschaftung. Forstl Jahrb 83: 473-504, 535-631, 647-707

Heske F (1937) Im heiligen Land der Gangesquellen. Neudamm

Herzhoff B, Schnitzler H (1981) Die subalpinen Birken-*Rhododendron*-Wälder im Dagwan-Tal/West-Himalaya. Mitt Dtsch Dendrolog Ges 72: 171-186

Heuberger H (1956) Der Weg zum Tscho Oyu. Kulturgeographische Beobachtungen in Ost-Nepal. Mitt Geogr Ges Wien 98(1): 3-28

Hinckley TM, Goldstein GH, Meinzer F, Teskey RO (1985) Environmental constraints at arctic, temperate-maritime and tropical treelines. In: Turner H, Tranquillini W (eds) Establishment and Trending of Subalpine Forest: Research and management. Ber Eidg Anst forstl Versuchswes 270: 21-30

Holtmeier FK (1965) Die Waldgrenze im Oberengadin in ihrer physiognomischen und ökologischen Differenzierung. Ph D-Thesis, University of Bonn

Holtmeier FK (1966) Die ökologische Funktion des Tannenhähers im Zirben-Lärchenwald und an der Waldgrenze im Oberengadin. J Ornithol 4: 337-345

Holtmeier FK (1973) Geoecological aspects of timberline in northern and central Europe. Arct Alp Res 5: 45-54

Holtmeier FK (1974) Geoökologische Beobachtungen und Studien an der subarktischen und alpinen Waldgrenze in vergleichender Sicht (Nördliches Fennoskandien/Zentralalpen). Erdwiss Forsch 8. Wiesbaden

Holtmeier FK (1985) Die klimatische Waldgrenze – Linie oder Übergangssaum (Ökoton)? – Ein Diskussionsbeitrag unter besonderer Berücksichtigung der Waldgrenzen in den mittleren und hohen Breiten der Nordhalbkugel. Erdkunde 39: 271-285

Holtmeier FK (1987) Human impacts on high altitude forests and upper timberline with special reference to middle latitudes. In: Fujimori T, Kimura M (eds) Hu-

man impacts and management of mountain forests. Proceedings of an international workshop, Japan, 5-13 Sept 1987, pp 2-20
Holtmeier FK (1989) Ökologie und Geographie der oberen Waldgrenze. Ber Reinhold-Tüxen-Ges 1: 15-45
Holtmeier FK (1993a) The upper timberline: ecological and geographical aspects. In: Anfodillo T, Urbinati C (eds) Ecologia delle foreste di alta quota, pp 1-26
Holtmeier FK (1993b) Der Einfluß der generativen und vegetativen Verjüngung auf das Verbreitungsmuster der Bäume und die ökologische Dynamik im Waldgrenzbereich. Geoökodynamik 14: 153-182
Holtmeier FK (1994) Ecological aspects of climatically-caused timberline fluctuations – review and outlook. In: Beniston M (ed) Mountain Environment in Changing Climates, London, pp 220-232
Holtmeier FK (1995) Waldgrenze und Klimaschwankungen – Ökologische Aspekte eines vieldiskutierten Phänomens. Geoökodynamik 16: 1-24
Holtmeier FK (2000) Die Höhengrenze der Gebirgswälder. Arb Inst Landschaftsökol, Univ Münster, vol 8. Münster
Holtmeier F-K (2003) Mountain Timberlines – Ecology, patchiness, and dynamics. Advances in Global Change Research, 14. Kluwer Academic, Dordrecht
Holtmeier FK, Broll G (1992) The influence of tree islands on microtopography and pedoecological conditions in the forest-alpine tundra ecotone on Niwot Ridge, Colorado Front Range, USA. Arct Alp Res 24: 216-228
Hooker JD (1855) Himalayan Journals. Notes of a Naturalist in Bengal, The Sikkim and Nepal Himalayas, The Khasia Mountains, &c. London
Hoon RC (1938) Study of the soils in the hilly areas of Kashmir – an investigation of soil profiles under deodar, blue pine, silver fir and chir. Indian For Rec (NS) 3(6): 195-261
Hou G (1990) An evaluation of soil fertility of fir-spruce forest blanks in subalpine areas of West Sichuan. In: Yang Y, Zhang J (eds) Protection and Management of Mountain Forests. Proceedings of the 5th IUFRO-Workshop on Ecology of Subalpine Zones, Beijing, pp 82-87
Howard-Bury CK (1922a) Mount Everest: The Reconnaissance, 1921. London
Howard-Bury CK (1922b) The Mount Everest expedition. Geogr J 59: 81-99
Imanishi K (1954) Annapurna and Manaslu, 1952. Himal J 18: 176-177
Imhof E (1900) Die Waldgrenze in der Schweiz. Gerlands Beitr Geophys 4: 241-330
Imhof E (1947) Der Minya Konka. Eine geographische Skizze. Geogr Helv 2: 243-255
Irfanullah S (2002) Gujars in the Pakistani Hindu Kush-Himalayas: conflicts and dilemmas about lifestyles and forest use. Nomadic Peoples (NS) 6: 99-109
Iwata S, Miyamoto S (1996) History of deforestation in the Himalayas. Tropics 5: 243-262
Jacobsen JP (1998) Investigations into the vertical temperature and precipitation gradients in two test areas in northern Pakistan. In: Stellrecht I (ed) Karakorum-Hindukush-Himalaya: Dynamics of Change, Köln, pp 145-162
Jacobsen JP, Schickhoff U (1995) Untersuchungen zur Besiedlung und gegenwärtigen Waldnutzung im Hindukusch/Karakorum. Erdkunde 49: 49-59

Jettmar K (1993) Voraussetzungen, Verlauf und Erfolg menschlicher Anpassung im nordwestlichen Himalaya mit Karakorum. In: Schweinfurth U (ed) Neue Forschungen im Himalaya. Erdkundliches Wissen 112: 31-47

Jobbágy EG, Jackson RB (2000) Global controls of forest line elevation in the northern and southern hemispheres. Global Ecol & Biogeogr 9: 253-268

Kala CP, Singh SK, Rawat GS (2002) Effects of sheep and goat grazing on the species diversity in the alpine meadows of western Himalaya. The Environmentalist 22: 183-189

Kalakoti BS, Pangtey YPS, Saxena AK (1986) Quantitative analysis of high altitude vegetation of Kumaun Himalaya. J Indian Bot Soc 65: 384-396

Kanai H (1963) Phytogeographical observations on the Japano-Himalayan elements. J Fac Sci Univ Tokyo, Sect 3, Bot 8(8): 305-339

Karma Ageta Y, Naito N, Iwata S, Yabuki H (2003) Glacier distribution in the Himalayas and glacier shrinkage from 1963 to 1993 in the Bhutan Himalayas. Bull Glaciol Res 20: 29-40

Kaul V, Sarin YK (1974) Studies on the vegetation of the Bhadarwah Hills. I Altitudinal zonation Bot Notiser 127: 500-507

Kaulback R (1938) A journey in the Salween and Tsangpo basins, south-eastern Tibet. Geogr J 91: 97-122

Kawakita J (1956) Vegetation. In: Kihara H (ed) Fauna and Flora of Nepal Himalaya. Scientific Results of the Japanese Expeditions to Nepal Himalaya, 1952-1953, vol 2, Kyoto, pp 1-66

Kerstan G (1937) Die Waldverteilung und Verbreitung der Baumarten in Ost-Afghanistan und in Chitral. In: Scheibe A (ed) Deutsche im Hindukusch, Berlin, pp 141-167

Khacher L (1978) The Nanda Devi Sanctuary – 1977. J Bombay Nat Hist Soc 75: 868-887

Kikuchi T (1991) Micro-scale vegetation patterns on talus in the alpine region of the Himalaya. In: Ohba H, Malla SB (eds) The Himalayan Plants, vol 2, Tokyo, pp 1-9

Kikuchi T, Ohba H (1988) Daytime air temperature and its lapse rates in the monsoon season in a Himalayan high mountain region. In: Ohba H, Malla SB (eds) The Himalayan Plants, vol 1, Tokyo, pp 11-18

Kitamura S (1960) Flora of Afghanistan. Results of the Kyoto Univ Scientific Expedition to the Karakoram and Hindukush, 1955, vol II, Kyoto University, Kyoto

Klasner FL, Fagre DB (2002) A half century of change in alpine treeline patterns at Glacier National Park, Montana, USA. Arct Antarct Alp Res 34: 49-56

Kleinert C (1974) Dolpo – das höchste Siedlungsgebiet im Nepal-Himalaya. Geogr Rdsch 26: 359-363

Klötzli F (1991) Niches of longevity and stress. In: Esser G, Overdieck D (eds) Modern Ecology. Basic and Applied Aspects, Amsterdam, pp 97-110

Klötzli F (2004) Karakorum (Hunza-Tal) – ein Schuttgebirge? In: Burga, CA, Klötzli, F, Grabherr, G (eds) Gebirge der Erde. Landschaft, Klima, Pflanzenwelt, Stuttgart, pp 315-324

Klötzli F, Schaffner R, Bosshard A (1990) Pasture development and its implications in the Hunza Valley. Prepared for AKRSP/IUCN. IUCN, Gland, Switzerland
Körner C (1998a) A re-assessment of high elevation treeline position and their explanation. Oecologia 115: 445-459
Körner C (1998b) Worldwide positions of alpine treelines and their causes. In: Beniston M (ed) The Impacts of Climate Variability on Forests, Heidelberg, pp 221-229
Körner C (1999) Alpine Plant Life. Functional Plant Ecology of High Mountain Ecosystems. Berlin Heidelberg
Körner C (2003) Ein morphologiebedingter Wärmemangel bestimmt die Waldgrenze. In: Winiger M (ed) Carl Troll: Zeitumstände und Forschungsperspektiven. Kolloquium im Gedenken an den 100. Geburtstag von Carl Troll. St. Augustin, Colloquium Geographicum 26: 114-119
Kreitner G (1893) Geographie. In: Die wissenschaftlichen Ergebnisse der Reise des Grafen Bela Szechenyi in Ost-Asien 1877-1880, vol 1, Wien, pp 1-304
Kreutzmann H (1989) Hunza. Ländliche Entwicklung im Karakorum. Abh Anthropogeogr 44. Berlin.
Ku CC, Cheo YC (1941) A preliminary survey of the forests in western China. Sinensia 12: 81-133
Kullman L (2004) The changing face of the alpine world. Global Change Newsletter 57: 12-14
Kunze C (1982) Geoökologische Untersuchungen im Kali-Gandaki-Tal. Untersuchungen zur bodenbiologischen Aktivität. Giessener Beitr z Entwicklungsforsch I/8: 187-194
Larcher W (1994) Ökophysiologie der Pflanzen. Stuttgart
Lehmkuhl F, Liu S (1994) An outline of physical geography including Pleistocene glacial landforms of eastern Tibet (Provinces Sichuan and Qinghai). GeoJournal 34(1): 7-30
Leuschner C (1996) Timberline and alpine vegetation on the tropical and warm-temperate oceanic islands of the world: Elevation, structure, floristics. Vegetatio 123: 193-206
Li B (1993) The alpine timberline of Tibet. In: Alden J, Mastrantonio JL, Odum S (eds) Forest Development in Cold Climates, New York London, pp 511-527
Li W (1993) Forests of the Himalayan-Hengduan Mountains of China and Strategies for Their Sustainable Development. ICIMOD, Kathmandu
Li X, Walker D (1986) The plant geography of Yunnan Province, southwest China. J Biogeogr 13: 367-397
Linchevsky A, Prozorovsky AV (1949) The basic principles of the distribution of the vegetation of Afghanistan. Kew Bull 1949(2): 179-214
Lischke H, Guisan A, Fischlin A, Williams J, Bugmann H (1998) Vegetation responses to climate change in the Alps: Modeling studies. In: Cebon P, Dahinden U, Davies H, Imboden DM, Jaeger CC (eds) Views from the Alps. Regional Perspectives on Climate Change, Cambridge London, pp 309-350
Lloyd P (1950) New British exploration in Nepal. Geogr J 116: 172-182

Longstaff TG (1907) Notes on a journey through the western Himalaya. Geogr J 29: 201-211

Longstaff TG (1908) A mountaineering expedition to the Himalaya of Grahwal. Geogr J 31: 361-395

Longstaff TG (1928) The Nanda Devi group and the sources of the Nandagkini. Geogr J 71: 417-430

Ludlow F (1937) The birds of Bhutan and adjacent territories of Sikkim and Tibet. The Ibis 1/1937: 1-46

Ludlow F (1944) The birds of south-eastern Tibet. The Ibis 1944: 43-86, 176-208, 348-389

Maikhuri RK, Nautiyal S, Rao KS, Saxena KG, RL Semwal (1998) Traditional community conservation in the Indian Himalaya: Nanda Devi Biosphere Reserve. In: Kothari A, Pathak N, Anuradha RV, Taneja B (eds) Communities and Conservation. Natural Resource Management in South and Central Asia, New Delhi London, pp 403-423

Malla SB, Shrestha AB, Rajbhandari SB, Shrestha TB, Adhikari PM, Adhikari SR (1997) Flora of Langtang and Cross Section Vegetation Survey (Central Zone). Bull Dept Med Plants Nepal 6 (2nd Ed). Kathmandu

Malyshev L (1993) Levels of the upper forest boundary in northern Asia. Vegetatio 109: 175-186

Mani MS (1978) Ecology and Phytogeography of High Altitude Plants of the Northwest Himalaya. London

Meebold A (1909): Eine botanische Reise durch Kashmir. Engl Bot Jahrb 43, Beibl 99: 63-90

Messerli B, Ives JD (1984) Gongga Shan (7556 m) and Yulongxue Shan (5596 m). Geoecological observations in the Hengduan Mountains of southwestern China. In: Lauer W (ed) Natural Environment and Man in Tropical Mountain Ecosystems. Erdwiss Forsch 18: 55-77

Messerli B, Ives JD (eds) (1997) Mountains of the World – A Global Priority. New York London

Metz JJ (1998) The ecology, use, and conservation of temperate and subalpine forest landscapes of West Central Nepal. In: Zimmerer KS, Young KR (eds) Nature's Geography. New Lessons for Conservation in Developing Countries, Madison, Wisconsin, pp 287-306

Meusel H, Schubert R (1971) Beiträge zur Pflanzengeographie des West-Himalayas. Teil 1-3. Flora 160: 137-194, 373-432 u 573-606

Miehe G (1982) Vegetationsgeographische Untersuchungen im Dhaulagiri- und Annapurna-Himalaya. Diss Bot 66. Vaduz

Miehe G (1984) Vegetationsgrenzen im extremen und multizonalen Hochgebirge (Zentraler Himalaya). Erdkunde 38: 268-277

Miehe G (1989) Vegetation patterns of Mt. Everest as influenced by monsoon and föhn. Vegetatio 79: 21-32

Miehe G (1990) Langtang Himal. Flora und Vegetation als Klimazeiger und -zeugen im Himalaya. A prodromus of the vegetation ecology of the Himalayas. Diss Bot 158. Berlin-Stuttgart

Miehe G (1991a) Die Vegetationskarte des Khumbu Himal (Mt. Everest-Südabdachung) 1:50.000. Gefügemuster der Vegetation und Probleme der Kartierung. Erdkunde 45: 81-94

Miehe G (1991b) Der Himalaya, eine multizonale Gebirgsregion. In: Walter H, Breckle SW (eds) Ökologie der Erde, vol 4, Stuttgart New York, pp 181-230

Miehe G (1997) Alpine vegetation types of the central Himalaya. In: Wielgolaski FE (ed) Polar and Alpine Tundra. Ecosystems of the World 3, Amsterdam, pp 161-184

Miehe G (2004) Himalaya. In: Burga, CA, Klötzli, F, Grabherr, G (eds) Gebirge der Erde. Landschaft, Klima, Pflanzenwelt, Stuttgart, pp 325-348

Miehe G, Miehe S (1994) Zur Waldgrenze in tropischen Gebirgen. Phytocoenologia 24: 53-110

Miehe G, Miehe S (1996) Die obere Waldgrenze in tropischen Gebirgen. Geogr Rdsch 48: 670-676

Miehe G, Miehe S, Huang J, Otsu T (1998) Forschungsdefizite und -perspektiven zur Frage der potentiellen natürlichen Bewaldung in Tibet. Peterm Geogr Mitt 142: 155-164

Miehe G, Miehe S (2000) Comparative high mountain research on the treeline ecotone under human impact. Erdkunde 54: 34-50

Miehe G, Winiger M, Böhner J, Zhang Y (2001) The climatic diagram map of High Asia. Purpose and concepts. Erdkunde 55: 94-97

Miehe G, Miehe S, Schlütz F (2002) Vegetationskundliche und palynologische Befunde aus dem Muktinath-Tal (Tibetischer Himalaya, Nepal). Erdkunde 56: 268-285

Miehe S, Cramer T, Jacobsen JP, Winiger M (1996) Humidity conditions in the Western Karakorum as indicated by climatic data and corresponding distribution patterns of the montane and alpine vegetation. Erdkunde 50: 190-204

Miehe S, Miehe G, Huang J, Otsu T, Tuntsu T, Tu Y (2000) Sacred forests of South-Central Xizang and their importance for the restauration of forest resources. In: Miehe G, Zhang Y (eds) Environmental Changes in High Asia. Marburger Geogr Schr 135: 228-249

Nakao S (1955) Ecological Notes. In: Kihara H (ed) Fauna and Flora of Nepal Himalaya. Scientific Results of the Japanese Expeditions to Nepal Himalaya, 1952-1953, vol 1, Kyoto, pp 278-290

Neubauer HF (1954/55) Versuch einer Kennzeichnung der Vegetationsverhältnisse Afghanistans. Ann Naturhist Hofmus Wien 60: 77-113

Nüsser M (1998) Nanga Parbat (NW-Himalaya) Naturräumliche Ressourcenausstattung und humanökologische Gefügemuster der Landnutzung. Bonner Geogr Abh 97

Nüsser M (2000) Change and persistence: Contemporary landscape transformation in the Nanga Parbat Region, Northern Pakistan. Mountain Res Developm 20: 348-355

Nüsser M, Dickoré WB (2002) A tangle in the triangle: Vegetation map of the eastern Hindukush (Chitral, northern Pakistan). Erdkunde 56: 37-59

Numata M (1967) Notes on botanical trip in eastern Nepal, I. Chiba Univ, J Coll Arts Sci 5(1): 57-74

Ogino K, Honda K, Iwatsubo G (1964) Vegetation of the upper Swat and the East Hindukush. In: Kitamura S (ed) Plants of West Pakistan and Afghanistan. Results of the Kyoto University Scientific Expedition to the Karakoram and Hindukush 1955, vol III, Kyoto University, Kyoto, pp 247-268

Ohsawa M (1987) Life Zone Ecology of the Bhutan Himalaya. Chiba

Ohsawa M (1990) An interpretation of latitudinal patterns of forest limits in South and East Asian mountains. J Ecol 78: 326-339

Ohsawa M, Shakya PR, Numata M (1973) On the occurrence of deciduous broad-leaved forests in the cool-temperate zone of the humid Himalayas in eastern Nepal. Japan. J Ecol 23: 218-228

Osmaston AE (1922) Notes on the forest communities of the Garhwal Himalaya. J Ecol 10: 129-167

Ozenda P, Borel JL (1991) Mögliche ökologische Auswirkungen von Klimaveränderungen in den Alpen. CIPRA Kleine Schriften 8/91. Vaduz

Paffen KH, Pillewizer W, Schneider H-J (1956) Forschungen im Hunza-Karakorum. Erdkunde 10: 1-33

Paulsen J, Weber UM, Körner C (2000) Tree growth near treeline: Abrupt or gradual reduction with altitude? Arct Antarct Alp Res 32: 14-20

Peng B, Pu L, Bao H, Higgitt DL (1997) Vertical zonation of landscape characteristics in the Namjagbarwa Massif of Tibet, China. Mountain Res Developm 17: 43-48

Petermann A (1861) Der Kintschindjunga und der Sikkim-Himalaya überhaupt. Peterm Mitt 1861: 3-11

Peters RL (1990) Effects of global warming on forests. For Ecol Managem 35: 13-33

Plesnik P (1991) System of timberlines on the earth. Geografický Casopis 43: 134-149

Polunin O (1950a) An expedition to Nepal. J Roy Hort Soc 75(8): 302-315

Polunin O (1950b) Plant hunting in the Nepal Himalayas. Geogr Mag 23: 132-147

Polunin O (1952) The natural history of the Langtang Valley. In: Tilman HW (ed) Nepal Himalaya, Cambridge, pp 242-265

Polunin O (1956-57) A Kashmir journey. Gard Chronicle 140: 546-547, 628-629; 141: 66-67

Polunin O, Stainton A (1984) Flowers of the Himalaya. Delhi

Pradhan S, Saha GK, Khan JA (2001) Ecology of the red panda *Ailurus fulgens* in the Singhalila National Park, Darjeeling, India. Biol Conserv 98: 11-18

Price MF, Messerli B (2002) Fostering sustainable mountain development: from Rio to the International Year of Mountains, and beyond. Unasylva 208: 6-17

Puri GS, Gupta RK, Meher-Homji VM, Puri S (1989) Forest Ecology. Vol II: Plant Form, Diversity, Communities and Succession. 2nd Ed. New Delhi-Bombay-Calcutta

Qin D, Mayewski PA, Wake CP, Kang S, Ren J, Hou S, Yao T, Yang Q, Jin Z, Mi D (2000) Evidence for recent climate change from ice cores in the central Himalaya. Ann Glaciol 31: 153-158

Rafiq RA (1996) Taxonomical, Chorological and Phytosociological Studies on the Vegetation of Palas Valley. Ph D-thesis, University of Vienna

Rao RS (1963) A botanical tour in the Sikkim State, the eastern Himalaya. Bull Bot Surv India 5(2): 165-205

Rao RS, Panigrahi G (1961) Distribution of vegetational types and their dominant species in eastern India. J Indian Bot Soc 40: 274-285

Rao TA (1961) A botanical tour in Kashmir State. Rec Bot Surv India 18(2): 1-67

Rathjens C (1969) Verbreitung, Nutzung und Zerstörung der Wälder und Gehölzfluren in Afghanistan. Jahrb d Südasien-Inst d Univ Heidelberg 3: 7-18

Rathjens C (1974) Die Wälder von Nuristan und Paktia. Standortbedingungen und Nutzung der ostafghanischen Waldgebiete. Geogr Ztschr 62: 295-311

Rau MA (1974) Vegetation and phytogeography of the Himalaya. In: Mani MS (ed) Ecology and Biogeography in India, The Hague, pp 247-280

Rawal RS, Dhar U (1997) Sensitivity of timberline flora in Kumaun Himalaya, India: conservation implications. Arct Alp Res 29: 112-121

Rawal RS, Pangtey YPS (1993) Composition and structure of timber line vegetation in a part of Central Himalayas. In: Parkash R (ed) Advances in Forestry Research in India, vol VIII, Dehra Dun, pp 68-90

Rawal RS, Pangtey YPS (1994) Distribution and structural-functional attributes of trees in the high altitude zone of Central Himalaya, India. Vegetatio 112: 29-34

Reineke T (2001) Bodengeomorphologie des oberen Bagrot-Tales (Karakorum/Nordpakistan). Ph D-thesis, Univ of Bonn

Repp G (1963) Waldökologische Studien im westlichen Himalaya. Mitt Flor-Soz Arb-gem NF 10: 209-222

Richter M, Pfeifer H, Fickert T (1999) Differences in exposure and altitudinal limits as climatic indicators in a profile from western Himalaya to Tian Shan. Erdkunde 53: 89-107

Rock JR (1926) Through the great river trenches of Asia. Nat Geogr Mag 50: 133-186

Rock JR (1931) Konka Risumgongba, holy mountain of the outlaws. Nat Geographic 60: 1-65

Rock JR (1947) The Ancient Na-Khi Kingdom of Southwest China, 2 vols, Cambridge, Mass

Sakai A (1983) Comparative study on freezing resistance of conifers with special reference to cold adaptation and its evolutive aspects. Can J Bot 61: 2323-2332

Sakai A, Malla SB (1981) Winter hardiness of tree species at high altitudes in the East Himalaya, Nepal. Ecology 62: 1288-1298

Sargent C, Sargent O, Parsell R (1985) The forests of Bhutan: a vital resource for the Himalayas? J Trop Ecol 1(4): 265-286

Schäfer E (1938) Ornithologische Ergebnisse zweier Forschungsreisen nach Tibet. Journal für Ornithologie 86 (spec. issue). Berlin

Schickhoff U (1993) Das Kaghan-Tal im Westhimalaya (Pakistan). Studien zur landschaftsökologischen Differenzierung und zum Landschaftswandel mit vegetationskundlichem Ansatz. Bonner Geogr Abh 87

Schickhoff U (1994) Die Verbreitung der Vegetation im Kaghan-Tal (Westhimalaya, Pakistan) und ihre kartographische Darstellung im Maßstab 1:150.000. Erdkunde 48: 92-110

Schickhoff U (1995a) Himalayan forest-cover changes in historical perspective. A case study in the Kaghan Valley, Northern Pakistan. Mountain Res Developm 15: 3-18

Schickhoff U (1995b) Verbreitung, Nutzung und Zerstörung der Höhenwälder im Karakorum und angrenzenden Hochgebirgsräumen Nordpakistans. Peterm Geogr Mitt 139: 67-85

Schickhoff U (1996a) Contributions to the synecology and syntaxonomy of West Himalayan coniferous forest communities. Phytocoenologia 26: 537-581

Schickhoff U (1996b) Die Wälder der Nanga-Parbat-Region. Standortsbedingungen, Nutzung, Degradation. In: Kick W (ed) Forschung am Nanga Parbat. Geschichte und Ergebnisse. Beitr u Mat z Reg Geogr 8: 177-189

Schickhoff U (1997) Ecological change as a consequence of recent road building: the case of the high altitude forests of the Karakorum. In: Stellrecht I, Winiger M (eds) Perspectives on History and Change in the Karakorum, Hindukush and Himalaya. Culture Area Karakorum Scientific Studies 3, Köln, pp 277-286

Schickhoff U (2000a) The impact of Asian summer monsoon on forest distribution patterns, ecology and regeneration north of the main Himalayan range (E-Hindukush, Karakorum). Phytocoenologia 30: 633-654

Schickhoff U (2000b) Persistence and dynamics of long-lived forest stands in the Karakorum under the influence of climate and man. In: Miehe G, Zhang Y (eds) Environmental Changes in High Asia. Marburger Geogr Schr 135: 250-264

Schickhoff U (2000c) Umweltdegradierung und Biodiversitätsverluste in Hochgebirgsräumen Südasiens. Ber Reinhold-Tüxen-Ges 12: 153-172

Schickhoff U (2001) Juniper forests in Pakistan: Human impact on structure and dynamics. In: Forest and Walnut Research Institute of National Academy of Sciences / Kyrgyz-Swiss Forestry Support Programme (eds) Proceedings International Symposium on Problems of Juniper Forests, 6th-11th August 2000, Osh, Kyrgyzstan, Bishkek, pp 215-223

Schickhoff U (2002) Die Degradierung der Gebirgswälder Nordpakistans. Faktoren, Prozesse und Wirkungszusammenhänge in einem regionalen Mensch-Umwelt-System. Erdwiss Forsch 41. Stuttgart

Schickhoff U, Klaucke B (1988) Die Auswirkungen der Waldnutzung auf die Struktur und die Dynamik der Vegetation der Bergwälder im Helambu-Gebiet (Zentral-Himalaya). Diploma Thesis, Univ of Münster (unpubl)

Schlagintweit, A, H & R (1854-57) Beobachtungsmanuskripte während der Reisen in Indien und Hochasien. 43 Vols. (= Schlagintweitiana II, 1 der Bayerischen Staatsbibliothek, München)

Schlütz F (1999) Palynologische Untersuchungen über die holozäne Vegetations-, Klima- und Siedlungsgeschichte in Hochasien (Nanga Parbat, Karakorum, Nianbaoyeze, Lhasa) und das Pleistozän in China Diss Bot 315. Berlin Stuttgart

Schmid E (1938) Contributions to the knowledge of flora and vegetation in the Central Himalayas. J Ind Bot Soc 17: 269-278

Schmidt-Vogt D (1990a) High altitude forests in the Jugal Himal (Eastern Central Nepal). Forest types and human impact. Geoecol Res 6. Stuttgart

Schmidt-Vogt D (1990b) Fire in high altitude forests of the Nepal Himalayas. In: Goldammer JG, Jenkins MJ (eds) Fire in Ecosystem Dynamics, The Hague, pp 191-199

Schmidt-Vogt D (1993) Die Gebirgsweidewirtschaft in den Vorbergen des Jugal Himal. In: Schweinfurth U (ed) Neue Forschungen im Himalaya. Erdkundl Wiss 112, Stuttgart, pp 191-230

Schmidt-Vogt D (1995) Die Entwaldung im Nepal-Himalaya. Ursachen, Ausmaß, Folgen. In: Gaenszle M, Schmidt-Vogt D (eds) Nepal und die Himalaya-Region. Beitr z Südasienforsch 166: 89-100

Schweinfurth U (1957a) Die horizontale und vertikale Verbreitung der Vegetation im Himalaya. Bonner Geogr Abh 20

Schweinfurth U (1957b) The distribution of vegetation in the Tsangpo Gorge. Oriental Geographer 1: 59-73

Seybold S, Kull U (1985) A contribution to the floristics and vegetation of Zanskar (Kashmir). Bot Jahrb Syst 105: 263-277

Shebbeare EO (1934) The conifers of the Sikkim Himalaya and adjoining country. Indian For 60: 710-713

Shipton E (1952) The Everest 'Tigers'. The Sherpas and their country. Geogr Mag 25: 172-183

Shrestha TB (1985) Vegetation and people of western Nepal with special reference to Karnali zone. In: Majupuria TC (ed) Nepal – Nature's Paradise, Bangkok, pp 360-368

Singh G, Kachroo P (1976) Forest Flora of Srinagar and Plants of Neighbourhood. Delhi

Singh HB, Sundriyal RC, Sharma E (2003) Livestock grazing in the Khangchendzonga Biosphere Reserve of Sikkim Himalaya, India: Implications for management. Indian For 129: 611-623

Singh JB, Kachroo P (1983) Plant community characteristics in Pir Panjal forest range (Kashmir). J Econ Tax Bot 4: 911-937

Singh JS, Singh SP (1992) Forests of Himalaya. Structure, Functioning and Impact of Man. Nainital

Singh S (1929) The effect of climate on the conifers of Kashmir. Indian For 55: 189-203

Singh SP, Adhikari BS, Zobel DB (1994) Biomass, productivity, leaf longevity, and forest structure in the Central Himalaya. Ecol Monogr 64: 401-421

Smith WW (1913) The alpine and subalpine vegetation of South-East Sikkim. Rec Bot Surv India 4: 323-431

Smith WW, Cave GH (1911) The vegetation of the Zemu and Lhonak Valleys of Sikkim. Rec Bot Surv India 4(5): 141-260

Soliva R (2002) Der Naturschutz in Nepal. Eine akteurorientierte Untersuchung aus der Sicht der Politischen Ökologie. Kultur, Gesellschaft, Umwelt 5. Münster

Song Y (1983) Die räumliche Ordnung der Vegetation Chinas. Tuexenia 3: 131-157

Stainton JDA (1972) Forests of Nepal. New York

Stamp LD (1925) The Vegetation of Burma from an Ecological Standpoint. Calcutta

Stebbing EP (1922) The Forests of India. Vol I. London

Stevens GC, Fox JF (1991) The causes of treeline. Ann Rev Ecol Syst 22: 177-191

Stevens S (2003) Tourism and deforestation in the Mt. Everest region of Nepal. Geogr J 169: 255-277

Stöber G, Herbers H (2000) Animal husbandry in domestic economies: organization, legal aspects and present changes of combined mountain agriculture in Yasin. In: Ehlers E, Kreutzmann H (eds) High Mountain Pastoralism in Northern Pakistan. Erdkundl Wissen 132: 37-58

Sundriyal RC (1994) Vegetation dynamics and animal behaviour in an alpine pasture of the Garhwal Himalaya. In: Pangtey YPS, Rawal RS (eds) High Altitudes of the Himalaya. Biogeography, Ecology & Conservation, Nainital, pp 179-192

Szeicz JM, MacDonald GM (1995) Recent white spruce dynamics at the subarctic alpine treeline of north-western Canada. J Ecol 83: 873-885

Taylor EM, Mehta ML, Hoon RC (1936) A study of the soils in the hill areas of the Kulu Forest Division, Punjab. Indian For Rec (NS), 1(2): 289-346

Taylor G (1947) Plant collecting in south-eastern Tibet. J Roy Hort Soc 72: 130-144, 166-177

Teichman E (1922) Journeys through Kam (eastern Tibet). Geogr J 59: 1-19

Thomas A (1999) Overview of the geoecology of the Gongga Shan Range, Sichuan Province, China. Mountain Res Developm 19: 17-30

Thomson T (1852) Western Himalaya and Tibet. London

Tichy H (1954) Land der namenlosen Berge. Vienna

Tilman HW (1952) Nepal Himalaya. Cambridge

Tranquillini W (1979) Physiological Ecology of the Alpine Timberline – Tree Existence at High Altitudes With Special Reference to the European Alps. Ecol Stud 31. Berlin Heidelberg

Troll C (1937) Auszüge aus Tagebüchern. Rawalpindi-Srinagar-Nanga Parbat und zurück. Mai-August 1937. Geogr Inst, Univ Bonn (unpubl)

Troll C (1939) Das Pflanzenkleid des Nanga Parbat. Begleitworte zur Vegetationskarte der Nanga-Parbat-Gruppe (Nordwest-Himalaja) 1:50.000. Wiss Veröff d Dt Mus f Länderkde zu Leipzig (NF) 7: 151-193

Troll C (1948) Der asymmetrische Aufbau der Vegetationszonen und Vegetationsstufen auf der Nord- und Südhalbkugel. Jahresber Geobot Inst Rübel Zürich f 1947: 46-83

Troll C (1967) Die klimatische und vegetationsgeographische Gliederung des Himalaya-Systems. In: Hellmich W (ed) Khumbu Himal. Ergebnisse des

Forschungsunternehmens Nepal Himalayam, vol 1, Berlin Heidelberg New York, pp 353-388
Troll C (1972) The three-dimensional zonation of the Himalayan system. In: Troll C (ed) Landschaftsökologie der Hochgebirge Eurasiens. Erdwiss Forsch 4: 264-275
Troll C (1973): The upper timberlines in different climatic zones. Arct Alp Res 5(3, Pt 2): A 3-A18
Troll C, Schweinfurth U (1968) Die Karte des Khumbu-Himalaya (Ost-Nepal) 1: 50 000. Erdkunde 22: 29-33
Tucker RP (1986) The evolution of transhumant grazing in the Punjab Himalaya. Mountain Res Developm 6: 17-28
Tyson JB (1954) Oxford University Expedition to Tehri-Garhwal, 1952. Himal J 18: 87-92
Uhlig H (1976) Bergbauern und Hirten im Himalaya. Tagungsber u Abh d 40 Dt Geogr-Tages – Innsbruck, Wiesbaden, pp 549-586
Verma KS, Shyampura RL, Jain SP (1990) Characterization of soils under forest of Kashmir valley. J Ind Soc Soil Sci 38: 107-115
Visser PC (1928) Zwischen Karakorum und Hindukusch. Eine Reise nach dem unbekannten Herzen Asiens. Leipzig
Voigt M (1933) Kafiristan. Versuch einer Landeskunde auf Grund einer Reise im Jahre 1928. Breslau
Volk OH (1954) Klima und Pflanzenverbreitung in Afghanistan. Vegetatio 5-6: 422-433
Wagner G (1962) Diamirtal und Diamirgletscher. Geographische und glaziologische Beobachtungen am Nanga Parbat (Deutsche Diamir-Expedition 1961). Mitt Geogr Ges München 47: 157-192
Ward FK (1913) Wanderings of a naturalist in Tibet and western China. Scott Geogr Mag 29: 341-350
Ward FK (1918) The hydrography of the Yunnan-Tibet frontier. Geogr J 52: 288-299
Ward FK (1921) The distribution of floras in SE Asia as affected by the Burma-Yunnan ranges. J Indian Bot 2: 20-26
Ward FK (1923) From the Yangtse to the Irrawaddy. Geogr J 62: 6-20
Ward FK (1926) The Riddle of the Tsangpo-Gorges. London
Ward FK (1929) Botanical explorations in the Mishmi Hills. Himal J 1: 51-59
Ward FK (1930) The forests of the north-east frontier of India. Empire For J 9: 11-31
Ward FK (1932) Explorations on the Burma-Tibet frontier. Geogr J 80: 465-483
Ward FK (1933) Plant collecting at the source of the Irrawaddy. J Roy Hort Soc 58: 103-114
Ward FK (1933-34) Explorations in Tibet, 1933. Proc Linn Soc, London, 146th Sess: 110-113
Ward FK (1934) The Himalaya east of the Tsangpo. Geogr J 84: 369-397
Ward FK (1935) A sketch of the geography and botany of Tibet, being materials for a flora of that country. J Linn Soc London 50: 239-265
Ward FK (1936) A sketch of the vegetation and geography of Tibet (The Hooker Lecture). Proc Linn Soc, London, 148th Sess, pt 3: 133-160
Ward FK (1939) Ka Karpo Razi: Burma's highest peak. Himal J 9: 74-88

Ward FK (1940a) Botanical and geographical exploration in the Assam Himalaya. Geogr J 96: 1-13
Ward FK (1940b) Exploration in eastern Himalaya. J Roy Centr As Soc 27: 211-220
Ward FK (1944-45) A sketch of the botany and geography of North Burma I-III. J Bombay Nat Hist Soc 44: 550-574, 45: 16-30, 45: 133-148
Ward FK (1946a) Additional notes on the botany of North Burma. J Bombay Nat Hist Soc 46: 381-390
Ward FK (1946b) Botanical explorations in North Burma. J Roy Hort Soc 71: 318-325
Ward FK (1957) The great forest belt of North Burma. Proc Linn Soc London 136: 87-96
Wardle P (1971) An explanation for alpine timberline. New Zealand J Bot 9: 549-554
Wardle P (1974) Alpine timberlines. In: Ives JD, Barry RG (eds) Arctic and Alpine Environments, London, pp 371-402
Wardle P (1981) The upper limits of tree growth in oceanic and continental environments. In: Liu Dong-Sheng (ed) Geological and Ecological Studies of Qinghai-Xizang Plateau, vol 2, Beijing New York, pp 1953-1970
Wardle P (1993) Causes of alpine timberline: a review of the hypotheses. In: Alden J, Mastrantonio JL, Odum S (eds) Forest Development in Cold Climates, New York London, pp 89-103
Weber B (2000) Die Bodentemperatur als landschaftsökologischer Indikator am Beispiel des Karakorum-Gebirges (Pakistan). Diploma Thesis, Univ of Bonn (unpubl)
Webster GL, Nasir E (1965) The vegetation and flora of the Hushe Valley (Karakoram Range, Pakistan). Pak J For 15: 201-234
Weiers S (1995) Zur Klimatologie des NW-Karakorum und angrenzender Gebiete. Statistische Analysen unter Einbeziehung von Wettersatellitenbildern und eines Geographischen Informationssystems (GIS). Bonner Geogr Abh 92
Weigold H (1935) Südost-Tibet als Lebensraum. Jahrb Geogr Ges Hannover 1935: 203-247
Wendelbo P (1952) Plants from Tirich Mir. A contribution to the flora of the Hindukush. Nytt Mag f Bot 1: 1-70
Wien K (1937) Sikkim, die deutsche Himalaya-Expedition 1936. Ztschr f Erdkunde 1937: 586-600
Williams LHJ (1953) The 1952 expedition to western Nepal. J Roy Hort Soc 78: 323-337
Wilson EH (1913) A Naturalist in Western China With Vasculum, Camera, and Gun. London
Winkler D (1997) Waldvegetation in der Ostabdachung des Tibetischen Hochlands und die historische und gegenwärtige Entwaldung. Das Beispiel Jiuzhaigous (Zitsa Degu; NNW-Sichuan). Erdkunde 51: 143-163
Wissmann H v (1960/61) Stufen und Gürtel der Vegetation und des Klimas in Hochasien und seinen Randgebieten. Erdkunde 14: 249-272, 15: 19-44
Workman WH (1910) The Call of Snowy Hispar. London
Wright HL (1931) The forests of Kashmir. Emp For J 10: 182-189

Wynter-Blyth MA (1951) A naturalist in the north-west Himalaya. J Bombay Nat Hist Soc 50: 344-354 a 559-572

Xu F (1992) The vertical distribution of Mêdog's main type forest, Xizang. In: Yang Y, Zhang J (eds) Protection and Management of Mountain Forests. Proceedings of the 5th IUFRO Workshop of Ecology of Subalpine Zones, Chengdu, 5-14 Sept 1990, Beijing New York, pp 114-117

Yang Q, Zheng D (1990) On altitudinal land use zonation of the Hengduan Mountain region in southwestern China. GeoJournal 20: 369-374

Yeh DZ (1982) Some aspects of the thermal influences of Qinghai-Tibetan plateau on the atmospheric circulation. Arch Met Geophys Bioclim A 31: 205-220

Yoda K (1967) A preliminary survey of the forest vegetation of Eastern Nepal. II. General description, structure and floristic composition of the sample plots chosen from different vegetation zones. J Coll Arts Sci Chiba Univ 5(1): 99-140

Zhang B (1995) Geoecology and sustainable development in the Kunlun Mountains, China. Mountain Res Developm 15: 283-292

Zhang J (1988) The Vegetation of Xizang. Beijing

Zheng D (1990) A study of the altitudinal belts of vegetation in western Kunlun Mountains. Chinese J Arid Land Res 1: 227-237

Zheng D, Zhang Y, Yang Q (1981) Physico-geographical differentiation of the Qinghai-Xizang Plateau. In: Liu Dong-Sheng (ed) Geological and Ecological Studies of Qinghai-Xizang Plateau, vol 2, Beijing New York, pp 1851-1860

Zhu P (1990) Soil under alpine dark coniferous forests in western Sichuan. In: Yang Y, Zhang J (eds) Protection and Management of Mountain Forests. Proceedings of the 5th IUFRO-Workshop on Ecology of Subalpine Zones, Beijing, pp 71-75

Zimina RP (1973) Upper forest boundary and the subalpine belt in the mountains of the southern USSR and adjacent countries. Arct. Alp. Res. 5(3. Pt. 2): A29-A32.